Ecological Studies, Vol. 116

Analysis and Synthesis

Ecological Studies

Volumes published since 1989 are listed at the end of this book.

Springer
New York
Berlin
Heidelberg
Barcelona
Budapest
Hong Kong
London
Milan
Paris
Santa Clara
Singapore
Tokyo

Richard G. Lawford Paul B. Alaback
Eduardo Fuentes
Editors

High-Latitude Rainforests and Associated Ecosystems of the West Coast of the Americas

Climate, Hydrology, Ecology, and Conservation

With 83 Illustrations

Springer

Richard G. Lawford
Office of Global Programs
National Oceanic and Atmospheric Administration
1100 Wayne Avenue
Silver Spring, MD 20920-5603, USA

Paul B. Alaback
School of Forestry
Montana Forest and Conservation Experiment Station
University of Montana
Missoula, MT 59812-1063, USA

Eduardo Fuentes
Catholic University
Laboratory of Ecology
Castilla 114-D
Santiago
Chile

Cover Illustration: Schematic of the integrated effects of climatic, hydrological, and biological processes on ecosystems along the west coast of the Americas. Artwork prepared by Mr. Philip Gregory of the National Hydrology Research Centre in Saskatoon, Saskatchewan, Canada.

Library of Congress Cataloging in Publication Data
High-latitude rainforests and associated ecosystems of the West Coast of the Americas : climate, hydrology, ecology, and conservation / Richard G. Lawford, Paul Alaback, Eduardo Fuentes, editors.
p. cm.—(Ecological studies : vol. 116)
ISBN 0-387-94487-7 (hardcover)
1. Rain forest ecology—America. 2. Rain forests—America. 3. Rain forest conservation—America. I. Lawford, Richard G. II. Alaback, Paul B. III. Fuentes, Eduardo. IV. Series: Ecological studies : v. 116.
QH101.H55 1995
574.5′2642′097—dc20 95-5547

Printed on acid-free paper.

Acquiring editor: Robert C. Garber
Production coordinated by Chernow Editorial Services, Inc., and managed by Terry Kornak; manufacturing supervised by Joe Quatela.
Typeset by Best-set Typesetter Ltd., Hong Kong.
Printed and bound by Braun-Brumfield, Inc., Ann Arbor, MI.
Printed in the United States of America.

9 8 7 6 5 4 3 2 1

ISBN 0-387-94487-7 Springer-Verlag New York Berlin Heidelberg

Foreword

The high-latitude rainforests of the earth represent one of the last biomes to be impacted substantially by human efforts to "mine," in the worst cases, or sustainably harvest, in the best cases, biotic resources. Many of these areas are remote and have inhospitable climates, difficult terrain, and poorly developed transportation infrastructures. As a result, these regions, like the tropical wet forests, are among the last to experience large-scale forest clearing and wholesale losses of large mammals. These last frontiers are now under assault, and their character is changing so rapidly that we have not even had an opportunity to learn how these ecosystems were formed and how they are organized and function. This knowledge is fundamental for sustainable development and for minimizing the potentially detrimental impacts of the mechanized tools and practices now utilized for large-scale forest harvesting.

In this book, the nature of these forests and related systems, as well as the environment that produces them, is explored. It is done in a comparative context, with the goal of predicting the impacts of land use change in general for all of these systems, wherever they may occur in the world. The principal comparison is between the high-latitude rainforests of the west coasts of the Americas. These regions share comparable climates, physiography, and biome types. They have been separated through geological time and, thus, the biotic systems have had different origins even though they appear to function in a similar manner. Thus comparisons between these regions have provided a powerful tool for examining the generality of responses of systems, to perturbation for example, that share common physical evolutionary constraints.

Even though these disjunct biotic systems may have comparable responses to disruptions, the human-driven forces of change are not always similar. Differing societal and economic frameworks can lead to quite different land patterns of essentially identical biotic resources, providing many different scenarios for comparison. However, the rate and extent of change need be of a pace that is slow enough for us to learn from these inadvertent experiments. Unfortunately, this is not proving to be the case in all situations.

The assessment of the origins, status, function, and responses of high-latitude rainforests contained in this volume stems from the activities of a research consortium of scientists from both North and South America (AMIGO – America's Interhemisphere Geo-Biosphere Organization), who are examining the responses of the components of the earth system of the Americas to global change. Many of the analyses included in this book were first presented at a scientific meeting held in Victoria, Canada. The meeting was supported financially by the Canadian Royal Society's Global Change Program; the Canadian International Development Agency; the Inter-America Institute; the United Nations Education, Scientific and Cultural Organization; Environment Canada; and Forestry Canada.

Harold A. Mooney

Contents

Contributors

Paul B. Alaback	School of Forestry, Montana Forest and Conservation Experiment Station, University of Montana, Missoula, MT 59812-1063, USA
Juan J. Armesto	Laboratorio de Sistemática and Ecología Vegetal, Universidad de Chile, Casilla 653, Santiago, Chile
Mary T. Kalin Arroyo	Departamento de Biología, Facultad de Ciencias, Universidad de Chile, Casilla 653, Santiago, Chile
S. Mary P. Benbow	Department of Geography, University of Manitoba, Winnipeg, Manitoba R3T 2N2, Canada
Timothy J.B. Boyle	Canadian Forest Service, Hull, Quebec K1A 1G5, Canada

G.A. Bradshaw — USDA Forest Service, Pacific Northwest Research Station, Forest Sciences Laboratory, Corvallis, OR 97331, USA

Michael J. Brown — Forestry Tazmania, Hobart, B 7001, Australia

Lohengrin Cavieres — Departamento de Biología, Facultad de Ciencias, Universidad de Chile, Casilla 653, Santiago, Chile

Rosanne D. D'Arrigo — Tree-Ring Laboratory, Lamont-Doherty Geological Observatory, Palisades, NY 10964, USA

Toni L. De Santo — Forestry Sciences Laboratory, 2770 Sherwood Lane, Juneau, AK 99801, USA

Ana Maria Faggi — CEFYBO, CONICET, Serrano 665, 1414 Buenos Aires, Argentina

Maria R. Fiorella — USDA Forest Service, Pacific Northwest Research Station, Forest Sciences Laboratory, Corvallis, OR 97331, USA

Lars O. Hedin — Section of Ecology and Systematics, Cornell University, Ithaca, NY 14853, USA

Eugene D. Hetherington — Forestry Canada, Pacific and Yukon Region, Victoria, British Columbia V8Z 1M5, Canada

Gordon C. Jacoby — Tree-Ring Laboratory, Lamont-Doherty Geological Observatory, Palisades, NY 10964, USA

Richard G. Lawford — Office of Global Programs, National Oceanic and Atmospheric Administration, 1100 Wayne Avenue, Silver Spring, MD 20920-5603. Formerly of: Atmospheric Environment Service, 11 Innovation Boulevard, Saskatoon, Saskatchewan S7N 3H5, Canada

G.A. McBean — Atmospheric Environment Service, Environment Canada, Toronto, Ontario M3H 5T4 Canada

Peter L. Meserve — Department of Biological Sciences, Northern Illinois University, De Kalb, IL 60115-2861, USA

Alejandro Peñaloza — Departamento de Biología, Facultad de Ciencias, Universidad de Chile, Casilla 653, Santiago, Chile

Jennifer Read — Department of Ecology and Evolutionary Biology, Monash University, Clayton, Victoria, 3168, Australia

Magaly Riveros — Instituto de Botánica, Universidad Austral de Valdivia, Casilla 567, Valdivia, Chile

Carlos Sabag — Laboratorio de Sistemática and Ecología Vegetal, Universidad de Chile, Casilla 653, Santiago, Chile

C. Smith-Ramírez — Laboratorio de Sistemática Ecología Vegetal, Universidad de Chile, Casilla 653, Santiago, Chile

Doris Soto — Facultad de Pesquerias y Oceanografía, Instituto de Ciencias Naturales y Exactas, Universidad Austral de Chile, Campus Pelluco, Casilla 1327, Puerto Montt, Chile

David L. Spittlehouse — Research Branch, British Columbia Ministry of Forests, Victoria, British Columbia V8W 3E7, Canada

John G. Stockner — Department of Fisheries and Oceans, West Vancouver Laboratory, 4160 Marine Drive, Vancouver, British Columbia V7V 1N6, Canada

Frederick J. Swanson — USDA Forest Service, Pacific Northwest Research Station, Forest Sciences Laboratory, Corvallis, OR 97331, USA

Thomas T. Veblen — Department of Geography, University of Colorado, Boulder, CO 80309, USA

R.H. Waring — Department of Forest Science, College of Forestry, Oregon State University, Corvallis, OR 97331, USA

Mary F. Willson — Forestry Sciences Laboratory, 2770 Sherwood Lane, Juneau, AK 99801, USA

W.E. Winner — Department of Botany, College of Science, Oregon State University, Corvallis, OR 97331, USA

Introduction

Regional intercomparisons between ecosystems on different continents can be a powerful tool to improve understanding of the ways in which ecosystems respond to global change. Large areas are often needed to characterize the causal mechanisms governing interactions between ecosystems and their environments. Factors such as weather and climate patterns and land–ocean and land–atmosphere interactions all play important roles. Regional studies that hold as many factors constant over as large an area as possible are likely to yield the most useful insights into the functioning of ecosystems and their responses to outside stresses.

As a result of the strong physical north–south symmetry between the western coasts of North and South America, many researchers have used this area as a focus for comparative marine and terrestrial studies. The similarities in climate, coastal oceanography, and physiography between these two regions have been extensively documented. Examples of this symmetry are the mountains and oceans that introduce similar west-to-east climatic gradients and the latitudinal variations in radiation inputs that result in similar north–south changes in day length, diurnal cycles of temperature, and ecological adaptations to climate. However, climatic effects arising from increasing concentrations of atmospheric carbon dioxide are expected to occur more rapidly in the Northern Hemisphere than in the Southern Hemisphere, due to the buffering effect of the oceans in the south. Over time these asymmetries in the controls on ecosystems are

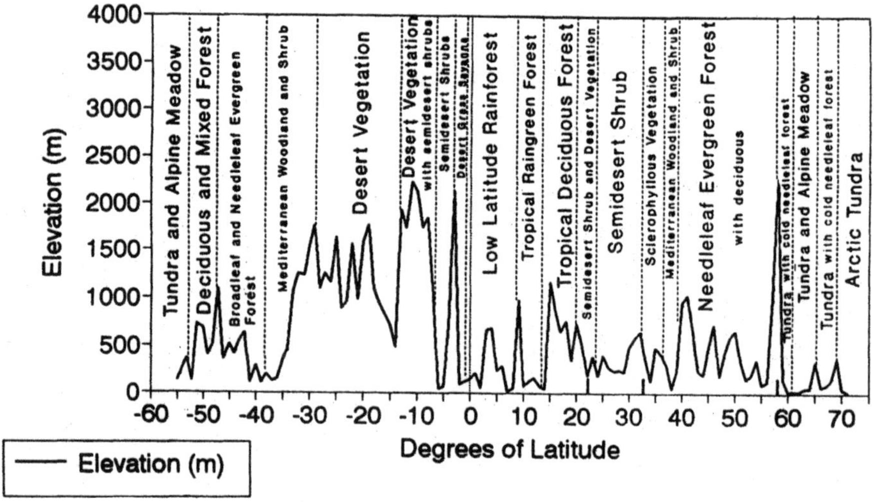

Figure 1. North-South transect from 70°N to 55°S showing the dominant coastal vegetation cover.

likely to enhance the differences and minimize the similarities between the hemispheres.

An understanding of parallel ecosystem processes and history in these two regions is needed to serve as a baseline to assess the differential effects of climatic change and economic development. Figure 1, which shows a schematic of the latitudinal distribution of vegetation, highlights the similarities and differences that exist in the distribution of biota along the two coasts.

Forests dominate the landscape in both hemispheres and play a key role in cycling and processing of nutrients between land, air, and sea, and in providing habitats for wildlife. As it is used in the title of this book, the term *high-latitude rainforests* refers to all rainforests outside the tropical regions. These forests contain some of the most extensive remaining examples of pristine landscapes in the temperate zone. Wide-ranging wildlife species, such as predatory birds and mammals, which have increasingly become rare or extinct in other temperate forest types are common in rainforests. Rainforests also include the vast majority of old-growth and ancient forests in each hemisphere and are intimately tied to some of the world's richest freshwater fish habitats. Coastal rainforests are a key focus for debate on conservation and management, yet little baseline information exists on their overall history, ecology, and response to management interventions.

High-latitude rainforests generally have persistent wet, cool, equitable climates where fire is a relatively infrequent factor in natural disturbance regimes. These rainforests are also highly structured and densely vegetated as compared to other temperate-latitude forest types, reflecting both a complex disturbance history and a moist, moderate climate.

Temperate and high-latitude rainforests are generally divided into four distinctive subtypes that can be defined by temperature, seasonality of rainfall, and natural disturbance regimes. The highest-latitude rainforest type, the *subpolar rainforest*, has high rainfall equitably distributed throughout the year but also a significant winter snowpack (especially in the Northern Hemisphere). Subpolar rainforests also have low species diversity and small, highly fragmented forest patches. *Perhumid rainforests* are warmer, more extensive, and much more diverse but still have abundant rainfall year-round. The *seasonal rainforest* type is transitional to mesophytic temperate forests, in that summer rainfall is low and catastrophic fires, although infrequent, play a key role in natural disturbance regimes. The *warm temperate rainforest* is the lowest-latitude extratropical rainforest type and has even less summer rainfall and a more frequent occurrence of fire than the seasonal rainforest type. This forest type is sometimes considered a wet subtype within the Mediterranean climatic zone.

Timber harvesting associated with development activities is a common environmental stress on forests in both hemispheres. Temperate rainforests on the west coasts of North and South America are being exploited as quickly as tropical forests, but because they occupy a much smaller fraction of the global land base than tropical regions they are potentially at greater risk. Not only do these rainforests play an essential role in the global carbon balance, because they serve as a carbon sink, but they also are an important source of revenue to the resource-dependent economies of American countries with coastal boundaries. In South America, the introduction of new tree species to improve forest productivity is further altering the biogeochemical balance, the regional hydrological cycle, and habitats for plants and animals.

In this book, we examine both physical processes that drive ecosystem functions in temperate and high-latitude rainforests and the historical factors that led to the development of this unique biome. Important local-scale effects arise from the complex biological interrelations that occur in these rainforests and from the effects of human uses on their associated ecosystems and landscapes.

Physical Processes and Patterns

The first section of this book deals with abiotic factors and processes that influence the distribution of biotic patterns in both space and time and the complex interactions between these factors. The north–south variations in forest structure are driven by climate as well as genetic pool structure and local geomorphology.

In order to establish the importance of latitudinal variations in climatic factors for the structure of these rainforests, Lawford analyzes transects of coastal patterns of temperature, precipitation, and other hydroclimatic variables along the western coasts of North and South America. Although the gross features of these patterns are similar between continents, there are many significant regional and local differences that control ecosystems at the landscape and smaller scales. For example, the displacement of the climatic equator northward from the geographical equator and the dramatic influence of the Andes mountains on local atmospheric circulation patterns led to broader bands of arid and semiarid conditions in Chile and Peru than in North America. Upper-air circulation patterns in the Southern Hemisphere are more consistent from season to season, leading to more stable zones of high precipitation along the west coast of South America. Although the estimates of hydroclimatic fields presented by Lawford have some inherent limitations, the general character of the fields reflects the large-scale controls on climate. However, it must be recognized that better models and more strategic data collection activities are needed to address the issues of land–atmosphere–biosphere links in these coastal areas.

McBean provides a more detailed examination of the dynamic controls on climate over the area of the western coast stretching from 30° N to 60° N. He notes that variability occurs on different space and time scales, with response times being dependent on the spatial scale of the phenomena being considered and the medium (e.g., atmosphere or ocean) in which the variation is occurring. The large-scale linkages between the atmospheric circulation pattern and the topography are reflected in the complex precipitation patterns observed along the coast. Differences in the dominant circulation regimes with latitude also influence the coastal oceanography. Upwelling of cold subsurface water is common between 30° N and 45° N due to the dominant offshore pressure pattern. Father north, upwelling is much more irregular.

Benbow continues the theme of atmosphere–ocean interactions, associating upwelling areas with certain climate influences, including those arising from the production of dimethylsulfide. This volatile sulfur compound, which is produced by marine phytoplankton, leads to the production of cloud condensation nuclei with high sulfide content. Benbow assesses the importance of phytoplankton–zooplankton interactions for DMS (dimethylsulfide) production. The major effects of atmospheric releases of DMS on climate include enhanced production of cloud and increases in the concentration of sulfur compounds in precipitation leading to differential chemical inputs to coastal forests.

Hedin and Hetherington document the differences between the atmospheric and geological inputs to coastal forests. In the Chilean rainforests, the inputs of sea-salt aerosols coupled with the low rates of weathering constitute a major biogeochemical constraint on forests. In contrast, inputs in North America forests represent a balance between atmospheric and geologic sources. Hedin and Hetherington suggest that temperate rainforests in the Southern Hemisphere may be sensitive to ENSO (El-Niño-Southern-Oscillation) events and especially sensitive to human activities.

The effects of climate on trees allow for the derivation of historical climate records from the analysis of tree-ring data. Records from the Northern Hemisphere are much more extensive than those in the Southern Hemisphere. As noted by D'Arrigo and Jacoby, these records allow for the analysis of indices of ocean–atmosphere interactions in the Pacific Ocean. These dendrochronological records are best for inferring temeratures from years prior to the instrumental record. Other issues that can be addressed through dendrochronological studies include forest responses to climatic change and CO_2-induced growth enhancement.

Waring and Winner review the need for real-time monitoring and call for a careful examination of forests in relation to current climate using meteorological data bases and remote sensing. Using their model, they are able to quantify the effects of frost, drought, and humidity on forests across the steep coastal climatic gradient experienced in Oregon. Based on the physiological adaptability of trees in North and South America and GCM (Global Climate Model) outputs, they conclude that climatic change will result in higher productivity in South American forests and lower productivity in North American forests.

Biotic Patterns

The second section of the book deals with the distribution of plants and animals and their interactions. Alaback examines the distinctive climate, history, and biota for the four temperate rainforest types in North America. Although the current forest composition is relatively new (2000–5000 years old), most of the dominant species are exceptionally well adapted to current climatic conditions. Dominant conifer species are widespread and genetically diverse throughout the coastal region. Few unique species occur in the region, except near the transition between rainforests and drier forests occurring in Mediterranean climates near the southern limit of forest. In North America, riparian rainforests are a key concentration of species richness and plant–animal interactions, although they have been heavily affected by human uses. Species richness in North American rainforests is less than in equivalent forest types in Chile, even though the regional species pool is greater in North America. Contrasts in evolutionary history, forest structure, species specialization, and genetic architecture suggest that their responses to climate change may diverge over time.

Arroyo and her associates examine the biodiversity patterns of forests in Chile and their historic and evolutionary causes as well as their relationship to current climatic conditions. Temperate rainforests of South

America are extremely isolated from other moist forest types yet have developed and maintained a high level of species diversity. These forests evolved under very different conditions than in North America, since they were once a part of a large, cool rainforest type that was continuous with tropical rainforests and they were subject to much less climatic variation than North American rainforests. Chilean rainforests have a much greater richness of unique specialized species than in analogous forests in North America. Woody genera also dominate southern rainforests to a much greater degree than in North America. Chilean rainforests are likely to be more sensitive to environmental stresses caused by climatic change and human activities than North American rainforests.

Veblen and Alaback study the role of natural disturbance in determining forest structure and composition in temperate forests in both hemispheres. Infrequent large-scale catastrophic disturbances as well as small-scale localized disturbances play key roles in structuring rainforests in both hemispheres. Windstorms are pervasive as chaotic disturbance agents in both hemispheres, but stem breakage is the dominant mechanism for tree mortality in North America, whereas rootthrow dominates in South America. Some disturbances, such as earthquakes and landslides, are of greater significance in South America, but others, such as wildfires and snow avalanches, are more important in North America. While a large degree of similarity exists between the material disturbance regimes of rainforests in North America and South America, many human-caused disturbances, such as the introduction of exotic plants and animals and the introduction of fire, have had a much greater effect in South America than in North America. Grazing, subsistence agriculture, and anthropogenic fire have had the greatest impact in South America, whereas logging for pulpwood export has had the greatest impact in North America.

Meserve describes the mammals and reptiles of the seasonal rainforests of North America and the perhumid rainforests of Chile. Different families dominate in each hemisphere. Lizards dominate in the south, while snakes dominate in the north. Fewer than half the number of mammal species in northern rainforests occur in Chilean rainforests. As with rainforest plant species, Chilean mammals have a much higher degree of uniqueness and have evolved in much greater isolation than their counterparts in North America. Logging, poaching, and resource development activities have resulted in much greater declines in mammal species diversity in Chile than in North America.

Willson and her collaborators examine the community structure of birds inhabiting rainforests in the perhumid zone of both southern Chile and coastal Alaska. They find that the rainforests of sourthern Chile support similar numbers of regularly breeding bird species as their counterparts in coastal Alaska, but relative abundances of birds tend to be higher in Chile. The structures of bird communities are significantly different in both regions as well, with a greater representation of upper-canopy insectivores in Chile and a greater diversity of leaf gleaners and understory birds in Alaska. Much of the contrast in community structure between Chile and Alaska is hypothesized to be due to the architectural differences between broadleaf and coniferous forests, but none of the measures tested are able to explain the differences in community structure that are observed.

Bird pollination and seed dispersal are important to the maintenance of forest species diversity in coastal temperate rainforests on Chiloé Island. Armesto and his coworkers found that one species of migrant frugiver bird accounts for nearly half the fruit consumption in these forests. Bird pollination is much more widespread and occurs throughout the year in Chile. The effects of logging and land-clearing activities on the maintenance of species richness are particularly prominent in this rainforest, since fragmentation tends to be associated with a decline in these key fruit-eating and pollinator bird species. Biotic interactions are more important to the maintenance of biodiversity in Chile than in North America.

Freshwater ecosystems show parallel trends with rainforests in terms of evolutionary history and unique community structures. Soto and Stockner compare the physical and biotic features of lakes in Chile and corresponding rainforest regions in southern British Columbia. The lakes in both regions are similar in chemistry, ontogeny, and geomorphology but contrast in temperature, productivity, and food web structure. These ecosystem features suggest that they are similar in chemistry, ontogeny, and geomorphology but contrast in temperature, productivity, and food web structure. These ecosystem features suggest that Chilean lakes may be more sensitive to introductions of exotic species and to the effects of logging on water chemistry than lakes in British Columbia.

Forest System Responses to Human Activities

The third and final section of the book summarizes techniques available for assessing the influence of human activities on the high-latitude ecosystems and discusses some of the management strategies and policies that are being undertaken or proposed to address the management of high-latitude forests. A new understanding of the importance of spatial gradients is leading to the development of new tools for analyzing and managing terrestrial ecosystems at the landscape level.

Bradshaw and her associates report on a technique for analyzing scale interactions in forest environments, using the concepts of patch dynamics and hierarchy theory. Landscape composition and rates of landscape change are determined by local patch–matrix interactions and patterns. When the rates of landscape change due to local processes are slow relative to large-scale changes, successional patterns will be more severely displaced. The results of these modeling studies raise additional questions about the role of small-scale processes such as stand level land conversions that may depend on latitude, geomorphology, or regional climate controls.

Spittlehouse notes the need to look at forest ecosystems and species in a broader sense, thereby eliminating the requirement to predict variability at the site level. By identifying and analyzing the sensitivity of key limiting factors, he assesses the impacts of future climatic change on forests. In the seasonal rainforest zone of southern British Columbia, for example, the key limiting factor to Douglas fir is available moisture. Consequently, an increase in temperature due to global change may or may not affect survival and distribution of Douglas fir, depending on the changes in precipitation that occur in conjunction with the warming.

Tasmania has a significant representation of both seasonal and perhumid temperate rainforest but contrasts with Chile and North America in history of human occupation and environmental management. These rainforests have many structural features in common with the rainforests of Chile, but they differ in disturbance history and degree of isolation. Brown and Read show that in contrast with North America and Chile, catastrophic disturbances play only a limited role in maintaining stand structure in Tasmania. Anthropogenic fire has played a key role in Tasmania for a much longer period than it has in rainforests in the Americas. Tasmania has one of the most comprehensive protected areas strategies of any rainforest region, with nearly 60% of the existing rainforest in established or proposed reserves and another 15% to 20% in deferred status. However, management of human-caused fire remains one of the most important challenges for the conservation of rainforests in Tasmania, as it does in Chile.

The range of human-use effects on forests involves the effects of logging on the wood, water, and air resources in the watershed. Hetherington assesses the influence of logging in a small stream ecosystem in a coastal temperate rainforest area over a 20-year period. In the logged areas, there was a reduction in evapotranspiration and increases in water yield and ground-water levels. However, the magnitude and direction of changes in peak flows, sediment transport, water quality, and fish populations were variable over time. This finding suggests that the rate and areal extent of logging is an important consideration in determining the overall impacts of logging.

Boyle notes that the coastal rainforests of western Canada represent some of that nation's most diverse and productive forests and account for as much as 40% of Canada's wood exports. They also have great ecological importance because of their high diversity of vertebrates in comparison to other Canadian forests. In addition, lichen and moss diversity is exceptionally high in coastal forests. A higher proportion of species are habitat or climate specialists in British Columbia rainforests than in forests of other provinces, and fewer threatened species occur in this area. Boyle observes that little protection exists for the most productive forests in

Canadian rainforests, even though they represent significant elements of regional biodiversity. Logging is generally thought to have the potential of constraining species diversity in the high-latitude rainforests. New governmental initiatives are being implemented to develop better guidelines to more effectively conserve biodiversity in this region.

In concluding the book, we draw a number of lessons from these comparative studies. While the biotic and abiotic patterns along the coasts of North and South America are generally understood, more detailed field studies and analyses are needed to understand the complex interactions between and within physical domains and biological communities. Natural systems are complex, and a small change can bring about large changes in ecosystem functioning and structure, depending on the state of the ecosystem. One of the best ways (and possibly the only way) to study such problems in their natural setting is to undertake regional comparative studies such as the ones described here. These studies must be interdisciplinary in nature, because expertise and paradigms from many disciplines must be brought to bear on the scientific problems arising from the linkages between ecosystem processes. Furthermore, this work must be done within a framework that allows for the propagation of perturbations from one scale to another, with movement from and to smaller scales. Finally, the human dimension must be incorporated into this research. Humans are the principal disturbers of ecosystems. However, through their institutional policies and personal actions, they can also address and correct the imbalances and disturbances that they have introduced. Consequently, our conclusions call for a stronger socioeconomic component in future regional comparative global-change studies so that more positive forces of change can be identified and studied.

Richard G. Lawford
Paul B. Alaback

1. Climate, Hydrology, and History

1. North–South Variations in West Coast Hydrometeorological Parameters and Their Significance for Earth Systems

Richard G. Lawford

Terrestrial processes coupled with atmospheric processes control the hydrological and biological characteristics of the western coasts of North and South America. Asymmetries exist between these two coasts in the distribution, richness, and diversity of biota. To some extent, these differences are a residual of the history of the two continents. They are also a function of the geomorphology and climate that control energy flows through ecosystems in these areas. An important question for the peoples of these regions, whose economies are derived from the local natural resources, relates to both the current rate of natural change and the sensitivity of these systems to anthropogenically driven changes that may result from local and global resource development. An understanding of the present is a key to predicting the future. Accordingly, we focus in this chapter on hydrometeorological controls, on existing ecosystems, with a view to identifying those aspects that may be most subject to change in the future.

Coastal mountains and the Pacific Ocean exert strong influences on atmospheric circulation patterns and on the climate and hydrology of the western coasts of the two continents. The similarities between the two coasts are unique and are not duplicated in either the central areas or the eastern coasts of the two continents, nor on any other pair of continents. The similarities in the gross features along the two coasts of the wind, ocean-current, and radiation patterns make the climate and hydrology somewhat symmetrical about the equator. However, asymmetries also exist. The nature of these symmetries and asymmetries and their consequences for the distribution of water resources and ecosystems along these coasts are examined here. Some aspects of this work are refinements of the ideas and preliminary data reported by Lawford (1993).

Overview of Hydrometeorological Patterns and Processes

Solar radiation, soil moisture, and nutrient supplies are all factors that influence ecosystem productivity and function. From an ecosystem perspective, precipitation is the most critical hydrometeorological parameter, because it determines the moisture availability and, together with surface geomorphology, controls runoff, river discharge, sediment transport, and nutrient cycling. Soil moisture and subsurface water movement are important related parameters and can influence biome characteristics. Waring and Schlesinger (1985) suggest that the ability of soils to provide moisture to plants during long drought periods is one of the major natural factors promoting forest growth over grasslands.

Temperature and insolation determine the rate of plant growth and, together with moisture availability, control biomass productivity. Plant functionality and biodiversity are dependent upon these parameters and their seasonal and interannual variations. Clouds play an important role in regulating insolation and generating precipitation.

Symmetry and Asymmetry

There are a number of reasons to anticipate significant symmetries between North and South America. First, the regular oscillation in the tilt of the earth's axis with a summer solstice in June when the sun reaches its zenith at 22.5° N, and a winter solstice in December when the same occurs at 22.5°S, should be conducive to symmetrical seasonal cycles in both hemispheres. Furthermore, the coastal ocean currents are similar, with the Antarctic drift current bifurcating as it reaches the shores of Chile near 51° S and the North Pacific current bifurcating west of Vancouver Island at approximately 49° N. A nearly continuous chain of mountains extends from Alaska in the north to the southern tip of Chile. In addition, the hemispheres have similar wind regimes, with strong westerly winds at mid-latitudes (30° to 55° in winter and 35° to 60° in summer) interacting with a mountain chain to produce higher precipitation amounts on the windward side of the mountains along the coast. However, it should be noted that winds in the Northern Hemisphere tend to undergo larger seasonal variations, with lower wind speeds in the summer.

It would be very surprising to find complete symmetry in these comparative analyses, in spite of the fact that the two regions show arguably

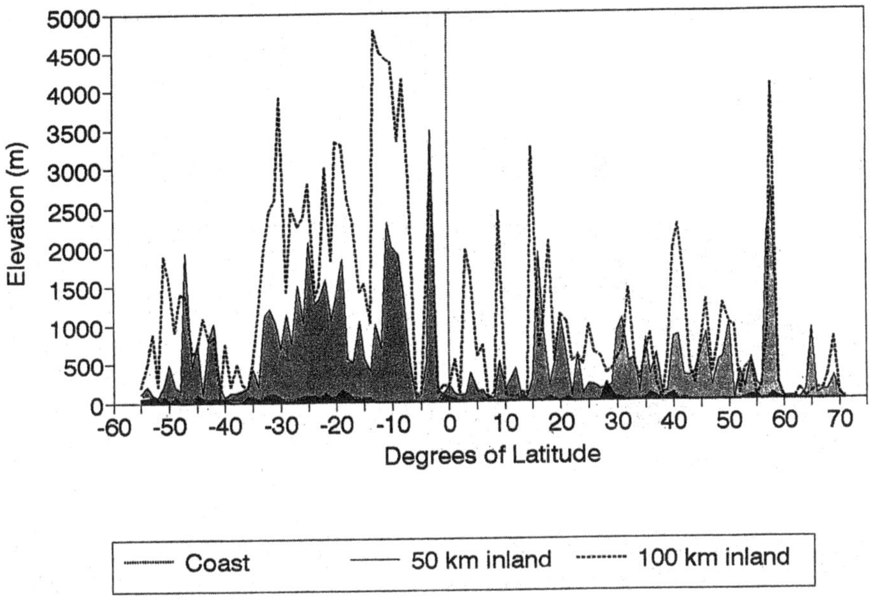

Figure 1.1. North–south transect on the west coasts of North and South America showing the average heights of terrain between 70° N and 55° S.

much greater similarity than any other two regions of equal area in each hemisphere. Sources of possible differences include the orientations and shapes of the mountains. The Andes are much higher and concentrated in a narrower band than are the western cordillera in the Northern Hemisphere. As shown in Figure 1.1, the Andes rise very rapidly with distance from the coast in the latitude band from 10° S to 35° S. The much larger land mass in the Northern Hemisphere at latitudes north of 30° N may also contribute to the asymmetry between the coasts of the two continents, because the majority of the Northern Hemisphere north of 30° N is covered by land, while the majority of the Southern Hemisphere south of 30° S is covered by water.

Changes in long-term energy input could also disrupt this symmetry. Over the millennia, the shape of the earth's orbit changes. At present, the orbit is elliptical and the earth is closest to the sun during December; consequently, the Southern Hemisphere receives more incoming solar radiation. The maximum amount of incoming shortwave radiation varies by as much as 7% between December and July, a variation that may be large enough to produce some differences in surface parameters.

Energy and Water Balances

The relationships between energy and water at a particular location are well known. The energy balance can be expressed as follows:

$$Q_{net} = SH + LH + GF \tag{1.1}$$

where Q_{net} = net radiation received at the surface
SH = sensible heat flux
LH = latent heat flux
GF = ground heat flux

In general, the sensible and latent heat fluxes are much larger than the ground heat flux, with the sensible heat flux being the dominant term in arid and semiarid regions.

The water balance can be expressed as follows:

$$R = P - E - S \tag{1.2}$$

where R = runoff
P = precipitation
E = evapotranspiration
S = storage

When this equation is averaged over a year or more, the changes in the storage term generally become insignificant. On a seasonal basis, however, storage of water in winter snow packs modulates the thermal and runoff regimes at higher latitudes. Precipitation sets an upper limit on the amount of evapotranspiration that can take place. In areas where the precipitation is less than the global average, the P and E terms are large and R is a residual term. However, in many "rainforest" areas along the west coasts of both North and South America, P is greater than E and runoff is a significant component of the water balance.

Study Methodology

Data Sets

In order to explore the north–south variations in the areas of interest, a 100-km wide strip of land along the western coasts of North and South America was selected and parameters were extracted from the atlases listed in Table 1.1. While more accurate data sources exist for some countries, it was felt preferable to use one data source for each data type for the entire region, provided that the source was reliable, because it increased confidence that differences were true differences in climate patterns rather than differences between data sources. However, where good national records exist, they have been used to assess the reliability of findings from this analysis. Energy and water budget parameters were based on maps prepared from global heat budget and water balance computations by Henning (1989) and others. In the case of the water

Table 1.1. Data Used in this Study

Parameter	Data Source	Types of Averages
Temperature	Climatic Atlas of North and Central America; Climatic Atlas of South America	Coastal transect, averages more than 100 km
Elevation	U.S. Military Series of World Maps	Coastal transect, averages more than 100 km
Precipitation	Climatic Atlas of North and Central America; Climatic Atlas of South America	Coastal transect, averages more than 100 km
Runoff	World Water Resources Atlas (L'vovich, 1979); Atlas of World Water Balance (Korzun et al., 1978); Surface Energy Balance (Henning, 1989)	Averages for the coast
Sensible Heat Flux	Henning (1989)	Averages for each latitude in the study zone
Latent Heat Flux	Henning (1989)	Averages for each latitude in the study zone
Net Radiation	Henning (1989)	Averages for each latitude in the study zone

balance computations, internal consistency between the parameters was not guaranteed for the study zone, since several data sources were used.

Energy fluxes and precipitation patterns were obtained from Henning (1989), who used surface climatological measurements to derive a set of global and continental maps showing all components of the surface energy and water cycles. Maps of the global water cycle produced by L'vovich (1979) and Korzun et al. (1978) were also used to derive the values of individual parameters along the coast.

Methodology

Data were abstracted from the maps listed in Table 1.1 for every 0.5° of latitude along the coast as well as at distances of 50 km and 100 km from the coast for a number of fields. On the basis of these values, averages were computed for each degree of latitude. Any averages estimated in this way along the coast are subject to error because of the large variability that exists in temperature and precipitation fields in this region. Furthermore, these averages are not always representative of specific locations along an east–west transect. The data derived in this manner extended for a continuous stretch from 71° N to 55° S. The 100-km wide strip, hereafter referred to as the *study zone*, provided a basis for comparing hydroclimatic conditions along the coasts of North and South America.

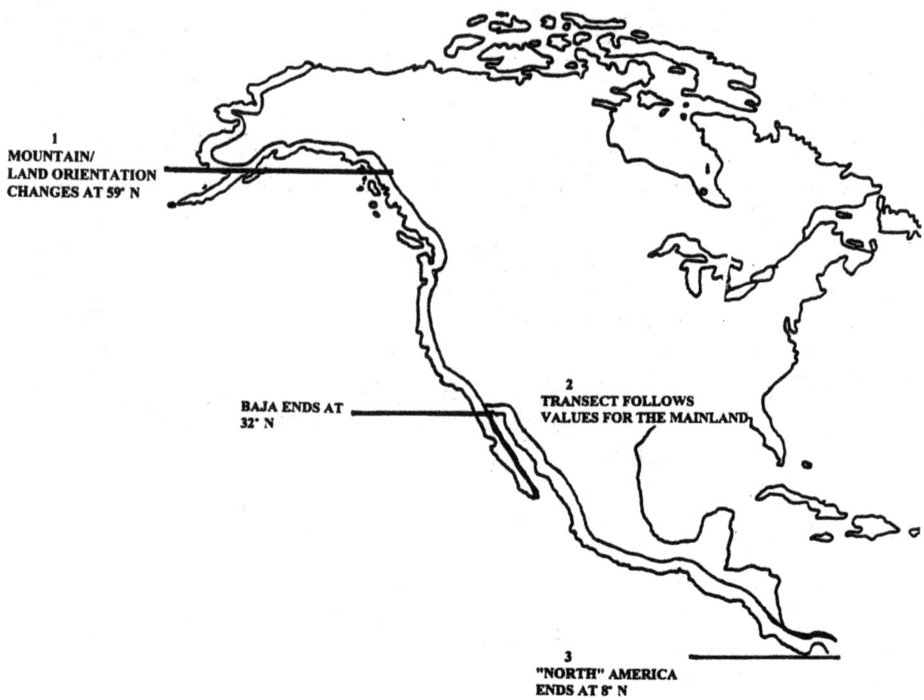

Figure 1.2. Latitudes where geographical discontinuities occurred in data abstraction along the study zone.

The geographical structure of this zone necessitated several decisions regarding the areas that were chosen in computing these averages. Figure 1.2 shows three areas where discontinuities occurred. The discontinuity at 59° N occurred because the coastal mountains changed direction, from north–south to east–west. South of 32° N, values from the west coast of mainland Mexico were included in the analysis instead of values from Baja, California. On some graphs, Baja is shown as a dashed line. The third discontinuity occurred at 8° N, where Central America joins South America. In this area, data from the west coast of South America were chosen for this comparative analysis.

Basic Topographic and Climatic Parameters

Elevation

While there are significant mountain ranges along the western coasts of North and South America, the Andes of South America are generally more imposing than the western cordillera of North America. Between latitudes 7° and 35°, the average elevation within 100 km of the coast is more than 1000 m higher in South America than in North America. At latitudes of 10° S to 12° S, elevations at distances of 100 km from the coast exceed 4500 m. The east–west topographical gradient in South America is particularly steep between 0 km and 50 km from the coast for latitudes from 7° S to 33° S, suggesting that orography may exert a significant control on the atmospheric circulation at these latitudes.

Vegetation Cover

Climate and water cycling in the coastal areas of these two continents affect the vegetation cover and the birds and animals that use this cover as habitat. Figure 1.3 shows the dominant vegetation type as a function of latitude. In the latitudes of primary interest for this chapter (35° to 60°), the most extensive vegetation cover is forest. The mountains in Chile south of 39° S are covered by broadleaf evergreen, deciduous, and needleleaf forests south to 54° S. In North America, needleleaf evergreen forests with some deciduous trees dominate north of 40° N. Between 40° N and 38° N, the coastal forests undergo a transition from broadleaf deciduous and needleleaf forests in the north and sclerophyllous or broadleaf evergreen shrubs in the south. The density of these shrubs decreases to a minimum between 32° N to 20° N latitude, where the vegetation becomes characteristically semidesert shrub. South of this band to 14° N, the density of trees increases where "rainforests" are dominant. South of the equator between 2° S and 30° S, the coastal areas

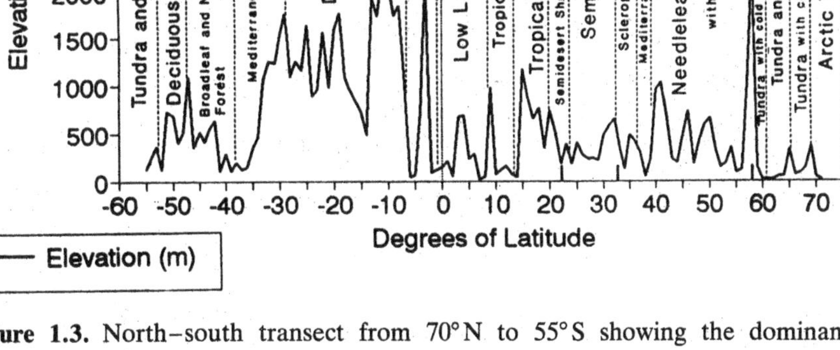

Figure 1.3. North–south transect from 70° N to 55° S showing the dominant coastal vegetative cover.

are barren deserts, with some semidesert shrubs and grasses on the higher terrain. South of 30° S, broadleaf evergreen shrubs increase, and south of 36° S, needleleaf evergreen shrubs are present.

At least two generalizations can be drawn from Figure 1.3. First, the boundaries between different vegetation covers are well defined, especially in the extratropics, where vegetation cover changes from rainforest to desert over distances of 5° of latitude. Second, distinct north–south asymmetries exist in the vegetative covers of the coasts of the two continents. Needleleaf forests (with some deciduous trees interspersed) are much more common at mid-latitudes in North America than in South America. On the other hand, desert and semidesert vegetation types are much more dominant in the 0°-to-30° latitude band in South America. Asymmetries between the coasts of the two hemispheres are most dramatic at the tropical and the subtropical latitudes.

Temperature

The average temperature values averaged for transects 0 km, 50 km, and 100 km from the coast are shown in Figure 1.4. No attempt was made to adjust the temperatures to a common altitude. Consequently, it must be recognized that a temperature for a given latitude can be an average of the temperatures at several different altitudes. In addition to mean annual temperatures, mean monthly temperatures in June and December are shown in Figure 1.4.

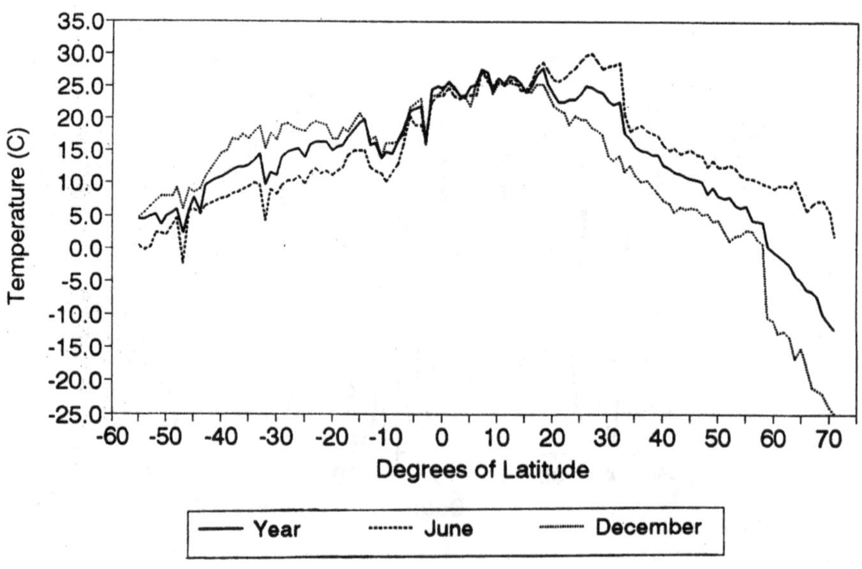

Figure 1.4. Variations in average temperature with latitude along the 70° N to 55° S transect for the entire year and for the months of June and December.

Temperatures are not symmetric on both sides of the equator. During their respective summers and winters, temperatures at low and mid-latitudes are warmer in the Northern Hemisphere than in the Southern Hemisphere. The difference is most marked in the latitudes from 55° N to 32° N. However, along the coast of Baja, California, temperatures are cooler and closer to the values on the coast of Chile than to those observed along the western coast of mainland Mexico. The mean annual temperature profile for South America shows more variability than the corresponding profile for North America. Comparisons with actual data along this coast as reported by Peña and Vidal (1993) suggest that the mean temperatures of individual stations at a given latitude varied by 1°C to 3°C from the average values. The decreases in temperature at 30° S and 45° S are associated with high mountains near the coasts, which affect the mean annual temperatures. Cooler temperatures in South America can be attributed to higher elevations along the coast and to cooler sea-surface temperatures off the coast. Coastal sea-surface temperatures exceed 25°C between 3° N and 22° N, and for a given latitude are generally 2°C to 4°C warmer in the North Pacific than in the South Pacific. The warmer temperatures in the Northern Hemisphere in this latitudinal band may contribute to the displacement of the mean annual position of the intertropical convergence zone (ITCZ) north of the equator.

Seasonal temperature changes influence vegetation types. Figure 1.4 shows the variations that occur between December and June along the coasts of North and South America. According to this graph, the largest winter-to-summer temperature changes (in excess of 15°C in the mean monthly temperatures) occur north of 56° N. Changes of 7°C to 10°C are observed between 23° N and 56° N and between 15° S and 44° S.

Cloud Cover

Cloud cover exerts an important control on the radiation balance and precipitation patterns along parts of the coast. An analysis of average annual cloud cover along the coast is shown in Figure 1.5. According to these large-scale averages, the coast of North America generally has less cloud than the coast of South America for corresponding latitudes between 20° and 35°. Just as a large east–west gradient exists in topography in South America, gradients also exist for clouds and precipitation at latitudes south of 40° S. Comparisons between data from Chilean climate stations and the averages in Figure 1.5 indicate that the averages may vary by 10 to 20% coverage from averages at specific stations. While both hemispheres show a minimum of coastal clouds at latitudes from 20° to 35°, the reduction is more marked in the Northern Hemisphere. In spite of the higher occurrence of cloud along the Chilean coast, it is noteworthy that the region is more arid in terms of rainfall and vegetation. More cloudiness is also found over high-precipitation regions near the

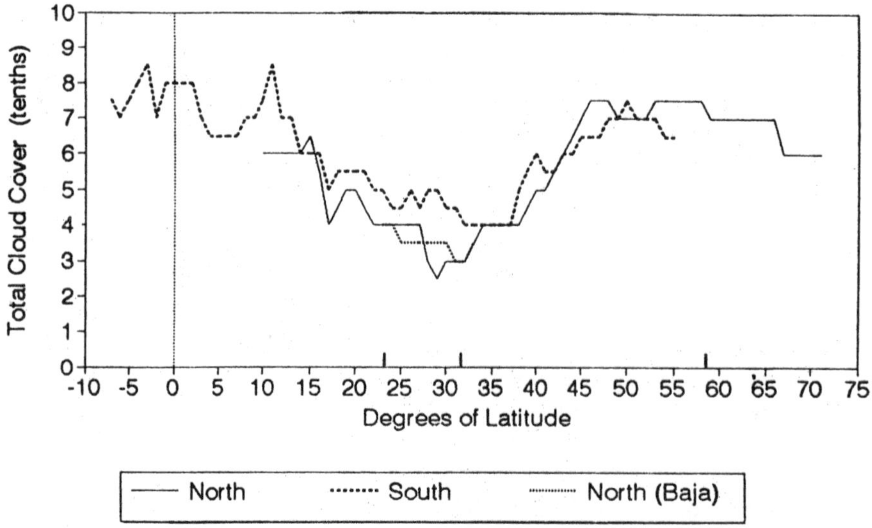

Figure 1.5. Variation with latitude of mean annual cloud cover for the study area.

equator and in the 40° to 55° latitude bands on the coasts of both continents. Nonprecipitating clouds occur frequently along certain parts of the coasts. Cereceda and Schemenauer (1991) have shown that there is a high incidence of coastal fog at latitudes from 30° S to 45° S (with a maximum frequency of occurrence between 35° S and 40° S), suggesting that cloud water plays an important role in the water balance of the region. Additional observations of cloud and fog water are needed at these latitudes to quantify the contribution of cloud water to the regional hydrology of higher elevations.

Cereceda and Schemenauer (1991) indicate that the potential for harvesting water from the clouds is very high along the coast between 20° S and 40° S. However, precipitation amounts are very low in this area. Abreu (1991) found that dynamic subsidence arises from the interactions of the upper air flow with the Andes mountains and from the thermal contrast between land and the ocean, particularly in the winter months. These findings suggest that while clouds may be more common at these latitudes, subsidence limits their development, so they do not produce significant amounts of precipitation. Other factors, such as the influence of dimethylsulfide (DMS) (Benbow, this volume), may also influence the formation of clouds along the coast of South America.

Precipitation

A transect of annual precipitation for the study zone is shown in Figure 1.6. The results indicate that there are three latitude bands where precipitation is well above the global average. These include the equatorial region, where the intertropical convergence zone (ITCZ) is responsible for the precipitation maximum. The ITCZ is a band of ascending air that migrates with season and is characterized by deep convection, which circles the globe. According to Figure 1.6, the highest values of mean annual precipitation in the study area are located at approximately 5°N, indicating that the climatological equator lies to the north of the geographical equator. This conclusion is corroborated by other authors who have mapped the distribution of precipitation for entire latitude circles.

The secondary precipitation maxima at mid-latitudes in both North and South America (35°N to 58°N and 35°S to 55°S) occur where strong westerly winds bring moisture from the ocean over the mountains. Again, reference to national data from Chile (Peña & Vidal, 1993) indicates that station-to-station precipitation averages in this area are much more variable than is implied by Figure 1.6. Closer to the equator, the most arid areas occur in the latitude bands from 15°S to 30°S and from 27°N to 32°N. In addition to being much broader, the arid region in South America is much drier than its counterpart in North America.

Figure 1.7, which shows the average monthly precipitation for December and June for the study zone, indicates that the precipitation maxima migrate north and south with the seasons. In the winter, the largest monthly averages of precipitation occur at latitudes between 36°S and 53°S, between 36°N and 38°N, and between 42°N and 49°N. In summer, the band of higher values shifts poleward and maximum amount decreases. This is particularly true for the west coast of North America, where the temperate latitude maximum decreases from 415 mm/month in winter (December) to 110 mm/month in summer (June). The dominant storm type also changes from synoptic storms in winter to a more convective regime in summer. In South America, relatively higher summer (December) precipitation values are found between 43°S and 53°S, with

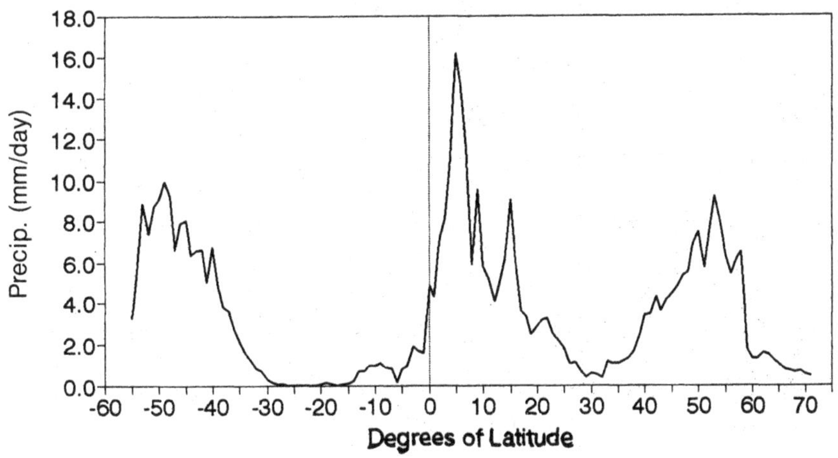

Figure 1.6. Variation with latitude of average precipitation per day for the study area.

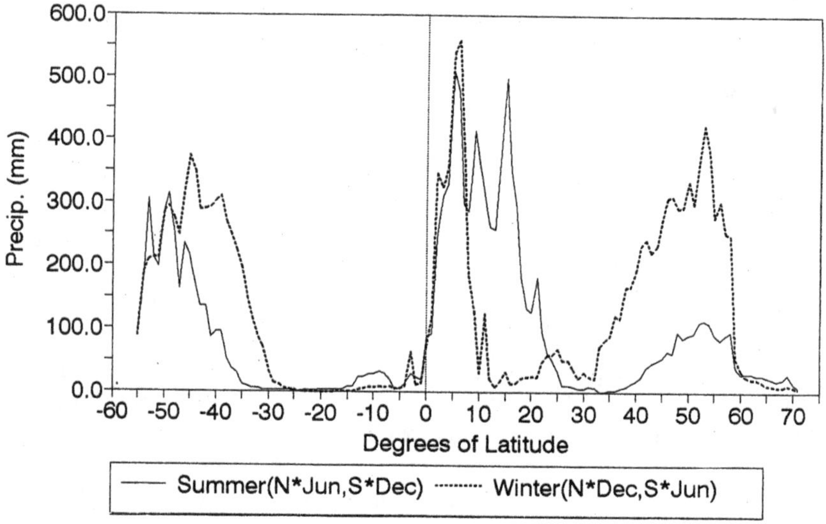

Figure 1.7. Variations with latitude of monthly precipitation for December and June for the study zone.

summer (December) maxima as large as 290 mm/month and winter (June) maxima of 400 mm/month.

Air flow over the mountains affects the areas of high precipitation for the mid-latitude regions. The rate at which air is forced upward as it crosses the mountains is determined by the velocity of the air flow, the angle of incidence between the air and the mountains, and the instability of the air mass. This rate governs precipitation production. McBean, (1994) has shown that the precipitation along the west coast of British Columbia is enhanced when troughing in the 500 mb geopotential height pattern occurs to the west of the coastal mountains, a factor which enhances the flow perpendicular to the mountains. According to Peixoto and Oort (1992), the maximum wind speed at 50 kPa over the Northern Hemisphere is 39 m/s^{-1} in winter and decreases to 20 m/s^{-1} in summer, while speeds in the Southern Hemisphere vary from 35 m/s^{-1} in winter to 24 m/s^{-1} in summer. This suggests that the wind-flow regime is more stable from winter to summer in the Southern Hemisphere. These data reflect the tendency of the average mid-tropospheric pressure gradient in the Northern Hemisphere summer to slacken, thereby reducing the velocity of the air flowing over the mountains. Consequently, the quantity of moisture extracted from the air is reduced. In the Southern Hemisphere, the average gradient wind remains stronger throughout the year, leading to higher precipitation rates in the summer. The more stable circulation regime in the Southern Hemisphere results from the ice-covered Antarctica and the sea ice surrounding it. Although the ice cover recedes somewhat during the Southern Hemisphere summer, it does not undergo the large winter-to-summer changes in area that are experienced by the snow pack and, to a lesser extent, by the sea ice in the Northern Hemisphere.

Temperature–Precipitation Trajectories

The variations of temperature and precipitation with latitude were analyzed to determine how these meteorological parameters could govern the dominant vegetation types. Figure 1.8 shows the distribution of each degree of latitude for the western coasts of North and South America as a function of mean annual temperature and annual precipitation. As noted earlier, the highest precipitation amounts occur in South America just north of the equator. According to the climate classification of biomes developed by MacMahon (1981), also shown in Figure 1.8, these areas should be dominated by tropical forests. Farther to the south, at latitudes between 5° S and 30° S, mean annual temperatures (MATs) range from 20.0°C to 15.0°C, while annual precipitation amounts (APs) range from 0 mm to 500 mm. This would suggest that the desert vegetation types should dominate over this latitude range. According to Figure 1.3, this zone is in fact dominated by desert vegetation.

At higher latitudes (45° N to 58° N in North America and 40° S to 55° S in South America), annual precipitation values range from 2000 mm to

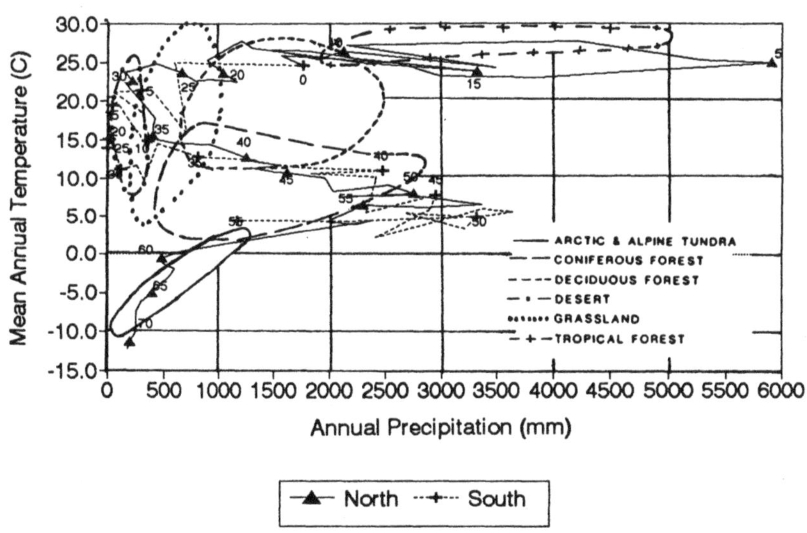

Figure 1.8. Plot of latitudes in a phase space of mean annual temperature and mean annual precipitation for the study area. A zonation of biome boundaries based on MacMahon (1981) is added.

3000 mm, and MATs range from 10°C to 5°C. At these latitudes, there seems to be little difference between North and South America. However, these latitudes are dominated by spruce and boreal forest in North America and by sandradeciduous trees in South America. This difference indicates that the history of forests in the two hemispheres is different and that the tree types that come into equilibrium with the regional climate are dependent to some extent on their history and on the gene types available regionally. It may also reflect the effects of different climatic histories in the past. It is important to recognize that tree growth is also dependent on other sources of moisture besides precipitation, such as fog and cloud water, and the availability of photosynthetic active radiation. Cloud-water measurements are not routinely available except for a few research sites on specific mountains along the coasts of Chile; consequently, it is difficult to determine the role of cloud water in regional water cycles and ecosystems.

The Energy Balance

As indicated by Equation 1.1, there should be a balance between the net radiation and the latent and sensible fluxes. Figure 1.9 shows the profile of the average annual values of the heat-balance components as computed and reported by Henning (1989). Note that while Henning's work is the most recently published atlas of its kind, the results in data-sparse

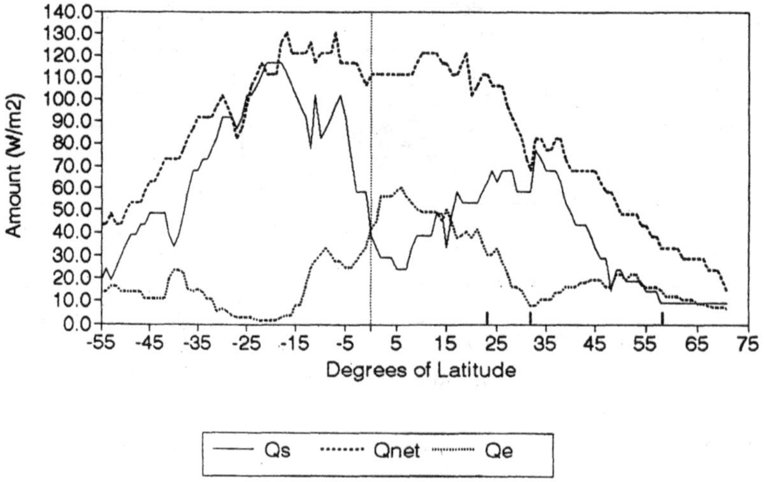

Figure 1.9. The latitudinal variation in average annual values of 1) net radiation, (Q_{net}); 2) sensible heat flux (Q_s); and 3) latent heat flux (Q_e) along the zone of study.

areas have not been extensively validated. According to Henning's computations, the west coast of South America has higher net radiation than North America in the latitude bands from 5° S to 20° S and 30° S to 40° S. These higher values for a particular latitude in South America result from the fact that the earth is closer to the sun during the South American summer.

Mean annual albedoes along the coast range from 15% to 25%, with the highest albedoes in South America occurring in arid regions between 15° S to 35° S and the highest albedoes in North America occurring north of 58° N, probably resulting from the winter snow cover at these latitudes. A secondary albedo maximum exists at 31°N in southern California.

A comparison of Figures 1.4 and 1.9 indicates that in spite of the higher net radiation in South America, air temperatures are lower. With higher surface temperatures in North America, there will be more long-wave radiation radiated to space, which will reduce the net radiation absorbed at the surface. This effect, combined with the lower insolation in the Northern Hemisphere, accounts for the lower net radiation values. It should be noted that the values for winter net radiation are very similar at equivalent latitudes in both hemispheres, with net radiation becoming negative north of 36° N in December and south of 33° S in June.

The distributions of annual sensible and latent heat fluxes show that the latent heat flux reaches its peak of 75 W/m^2 at 7° N, near the climatological equator. A latent heat flux of 75 W/m^2 corresponds to an evaporative loss of 108 cm per year. To the north of this maximum, latent heat fluxes decrease to 12 W/m^2 at 32° N. In the Southern Hemisphere north of 30° S, latent heat fluxes are at a maximum in January, while farther south, the maximum occurs in the spring between October and December. In the more arid regions of South America, the latent heat flux is less than 10 W/m^2 and decreases to 3 W/m^2 for the latitudes from 13° S to 31° S, with the minimum occurring between 13° S and 18° S. The sensible heat transfer is correspondingly very large at these latitudes, exceeding 115 W/m^2 at latitudes around 20° S.

As noted earlier, Abreu (1991) indicated that the cause of the aridity at these latitudes is the intense dynamic subsidence induced by the Andes mountains, particularly during the winter months. In addition, the occurrence of high sensible fluxes in an area where air temperatures are relatively low suggests that the low-level winds may be advecting the heated air to other regions and replacing it with cooler air from the ocean. Another possible explanation for these high values may lie in uncertainties in the values of sensible heat fluxes from Henning (1989).

Net radiation values in South American summers are higher than for corresponding latitudes in North America. In the higher latitudes, the average seasonal intensities increase to maximum values of 145 W/m^2 at 30° S during the November-to-January period and decrease to 70 to 75 W/m^2 in the winter (June to August). In North America, the highest values of net radiation occur in Baja California, with values reaching 120 to 130 W/m^2 in latitudes from 25° N to 30° N.

The energy budgets at most latitudes are quite well balanced. However, at latitudes of 19° S to 22° S and 24° S to 29° S and several other 1° latitude bands, the sums of the sensible and latent heat fluxes exceed the net radiation, indicating that either ground fluxes are significant or that uncertainties exist in the technique for generating the values, possibly because the sensible heat fluxes are overestimated.

The Water Balance on the West Coasts of South and North America

In Equation 1.1, the precipitation and evapotranspiration terms are generally large and control the water balance in a particular area. Precipitation and stream flow are measured regularly. As noted earlier, the storage-term parameters, when averaged over a year or longer, can be neglected at most latitudes, because changes in net water storage tend toward zero, except for the small component that infiltrates the ground-water system. Over longer intervals (>10 years), one would anticipate that the flux of water into the ground-water system would balance the flux of water out of the ground-water system to the surface.

If the storage term is neglected, runoff represents the difference between the precipitation and evaporation. When precipitation exceeds evaporation by a large amount, runoff will be large. For this investigation, runoff estimates from three independent sources were compared, namely, Henning's (1989) surface energy budget estimates (SEB), Korzun et al.'s (1978) World Water Balance (WWB) estimates and L'vovich's (1974) World Water Resources and their Future (WWRA) estimates. Figure 1.10 shows the runoff computed using each of these techniques. Although SEB and WWRA runoff estimates are of the same order of magnitude, some differences were found. As shown in Figure 1.10, SEB values reached a maximum of 1600 mm at 56° N, with secondary maxima occurring at 48° N, 40° S, 52° S and between 5° S and 15° N. Similar peaks were found in the WWRA data set, with maxima centered at 52° N, 10° N, 39° S, and 48° S to 50° S. However, the WWB runoff values were larger, with maximum values two to three times greater than the values estimated by the other two studies in the areas of relative precipitation maxima. (Runoff values as high as 3000 mm in North America and 4900 mm in South America were estimated, values that are much higher than the precipitation averages for the same latitude.) The differences are particularly noteworthy for latitudes between 35° S and 50° S. Runoff values in excess of annual precipitation suggest that either the runoff values are too high or that the precipitation is not accurately represented at the latitudes being considered, or, possibly, both. Underestimates of precipitation are most likely to arise in mountainous terrain where precip-

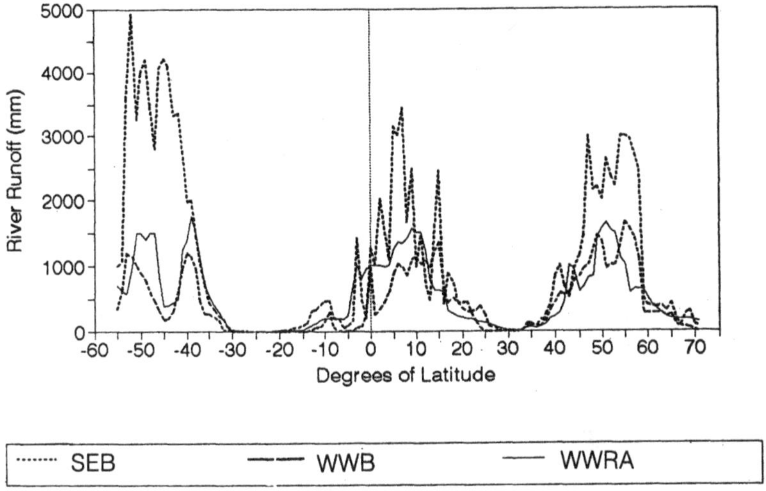

Figure 1.10. The latitudinal changes in the annual runoff computed from a) Surface Heat Budget; b) World Water Balance; and; c) World Water Resources.

itation gauges are located in nonrepresentative areas. The averaging technique used in this study may also have contributed to this problem.

One way of assessing the accuracy of the runoff measurements is to see if the water balance is closed by computing the sum of runoff and evaporation and to compare this total to the precipitation amounts. When the WWRA runoff values are used, the sums of runoff and evaporation compare quite well with precipitation, except for an underestimate of precipitation at latitudes between 35° S and 50° S. A similar pattern is obtained using the SEB estimates. This result gives some confidence in the WWRA and SEB runoff estimates.

Runoff-to-precipitation ratios display these differences very graphically. They also provide some insight into the water yields of different areas. Ratios greater than 1.0 suggest that there may be problems with either the precipitation data or, more likely in this case, the runoff estimates. With SEB runoff values, the ratios reach a maximum of 0.7; with WWB values, they exceed 1.0 for mid-latitudes in both North and South America; the WWRA values tend to be less than 1.0 except in the 10° S to 0° band.

Runoff and streamflow are related but not equivalent. Runoff represents the amount of water which flows over the surface or in the subsurface layer to streams, rivers, and the ocean. Streamflow refers to the water that is carried to the oceans in river channels. As shown in Figure 1.11, the fresh-water discharges carried in individual rivers on the western coast of North America to the Pacific Ocean are generally much larger

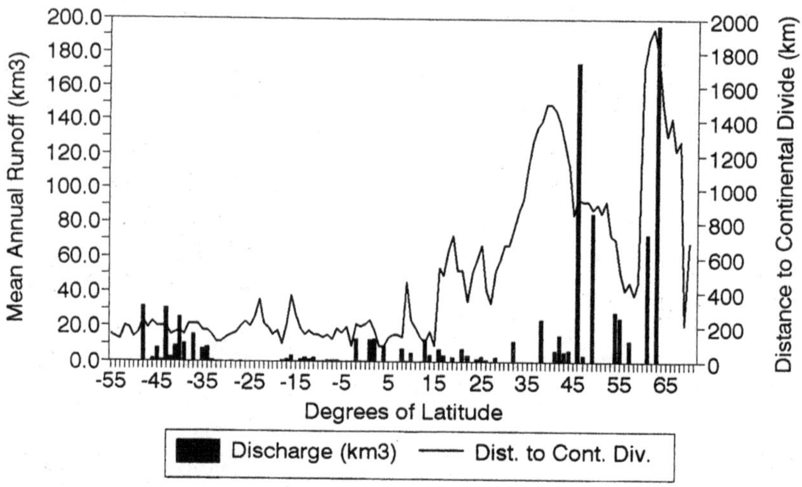

Figure 1.11. The latitudinal distribution of river discharge into the Pacific Ocean and the distance between the shoreline and the Continental Divide.

than river discharges in South America, despite the similar runoff values found at mid-latitudes in both continents.

River discharge is dependent upon the area that the river drains. As a crude first approximation, the drainage area of a dominantly east–west river in the study area can be represented by the distance from the coast to the continental divide, the eastern boundary for water draining west to the Pacific Ocean. This distance is also shown in Figure 1.11. The larger discharges in North America are due primarily to larger watershed sizes along the western coast of the North America rather than to differences in the precipitation and runoff regimes. Specific examples of these larger drainage areas are the Yukon River, which drains an area that extends 2000 km inland, and the Fraser River, which flows north and south as well as east–west, with a river length of 1368 km and a drainage area of approximately 233,000 km^2.

Links Between Hydroclimatology and Biology

The analyses in this paper deal predominantly with average conditions. However, the limits and ranges of many biota are often determined more by extremes of heat or cold or by excessive wetness or aridity than by average values. Along the coasts of North and South America, extremes in temperature and precipitation amounts have been shown to arise from E1 Niño events. The following analysis does not deal with this interannual variability and its causes but is limited to the effects of long time-averaged values of hydroclimatic parameters.

As noted previously, hydrometeorological processes strongly influence the structure and functioning of natural ecosystems. In most areas, biological systems are in a state of dynamic equilibrium with regional and local climates, although in some cases, unusual hydroclimatic environments have maintained ecosystems that are remnants of earlier times (e.g., high-altitude arid-region rainforests maintained by cloud water). When ecosystems are perturbed by human intervention or a natural event, the response of the ecosystem will depend on its state (which, in turn, is determined by abiotic factors such as the hydroclimatology) as well as the natural biological buffers that exist in the ecosystem. One approach that may assist in understanding the complex responses of ecosystems involves the consideration of an ecosystem as a complex yet geographically bounded system embedded in a larger, simpler abiotic system with slower variations, such as land-use changes and water contamination. In turn, these variations are embedded in the global climate, which is subject to interannual variability and change arising from increasing concentrations of greenhouse gases. The following sections summarize some of the linkages that exist between the bulk characteristics of these ecosystems and the larger-scale hydroclimatic factors that affect the development of these ecosystems.

Hydroclimatology and Biodiversity

An important aspect of an ecosystem is its biodiversity. Terborgh (1992) suggested that biodiversity is inversely proportional to the latitudinal temperature gradient and increases near the equator because the latitudinal temperature gradient is less there than it is at more polar latitudes. He indicated that the width of the zone over which the temperature is uniform, or nearly uniform, is a good indicator of the potential for allopatric speciation or biodiversity. According to his analysis, the most uniform temperature gradients exist near the equator and decrease north of 20° N and south of 20° S.

Temperature gradients were analyzed for the study area by computing the value of $\Delta T/\Delta Y$ (where ΔT is the temperature change over a latitude band of width ΔY) and then inverting this number to get the value of $\Delta Y/\Delta T$ (the width of a latitude band over which a 1°C temperature change takes place). Since the values of temperature along the north–south transect undergo considerable variability, the temperature differences were averaged by a double pass of a 5-point triangular filter. The filter smoothed data according to the following relationship:

$$T = (T_{-2} + 2T_{-1} + 3T_0 + 2T_{+1} + T_{+2})/9.0 \qquad (1.3)$$

where T_0 = temperature at the latitude under consideration
T_{-1} = temperature 1° South

T_{-2} = temperature 2° South
T_{+1} = temperature 1° North
T_{+2} = temperature 2° North

The resultant graph, shown in Figure 1.12, has a general trend similar to that reported by Terborgh (1992) north of the equator, where values of $\Delta Y/\Delta T$ are generally large within 25° of the equator. On the other hand, the temperature gradient along the west coast of South America is more variable than the corresponding gradient in North America. In part, this is the result of highly variable elevations in the 10° S to 30° S latitude band that affect surface temperatures. This result suggests that the effect of the elevation of the Andes is sufficiently large to limit the utility of general relationships such as the biodiversity relationship from Terborgh for the west coast of South America. However, the graph does show high values for $\Delta Y/\Delta T$ near 38° S, an area of high biodiversity where the cohabitation of rainforest and Mediterranean-type scrublands occurs (Arroyo et al., this volume).

Controls on Dry Matter Productivity

Temperature and precipitation have been related to biomass productivity. Lieth (1975) derived two equations for biomass productivity: one that related dry matter productivity to mean annual temperature, and one that related it to annual precipitation. According to his methodology, productivity can be computed using Equation (1.4).

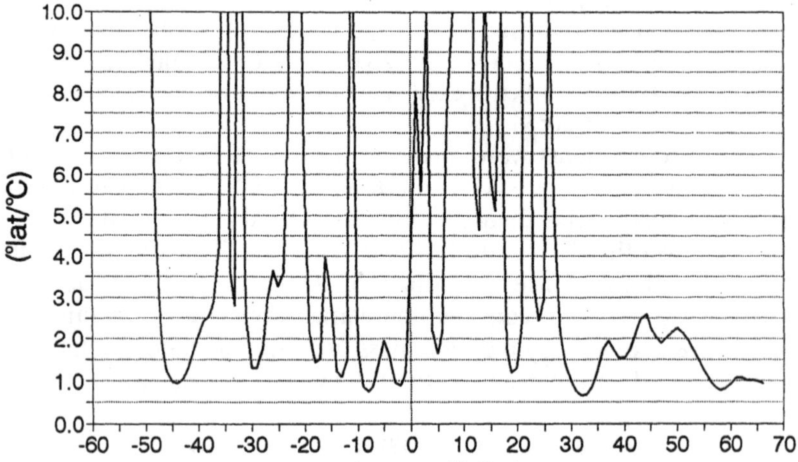

Figure 1.12. Smoothed distribution of $\Delta Y/\Delta T$ with latitude for the west coasts of North and South America.

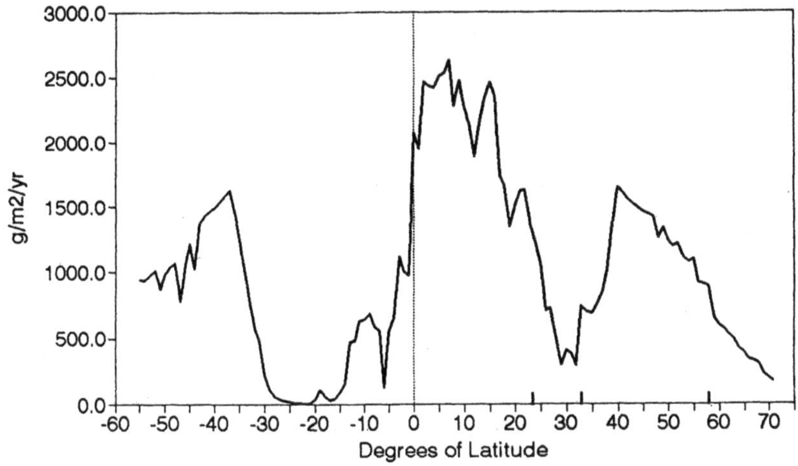

Figure 1.13. North–south transect of dry matter productivity derived from mean annual temperatures and annual precipitation along the west coasts of North and South America using relationships from Lieth (1975).

$$y = \min\left[\frac{3000}{1 + e^{1.315-0.119MAT}}, 3000(1 - e^{-0.000664P})\right] \qquad (1.4)$$

where y = annual productivity ($gm^{-2} \cdot yr^{-1}$)
MAT = mean annual temperature (°C)
P = annual precipitation (mm)

As indicated by Equation 1.4 the lower of the two computed values of y is selected as the best estimate of biomass productivity. Based on the latitudinal profiles of mean annual temperature and annual precipitiation shown in Figures 1.4 and 1.6, values of biomass productivity were computed for the study zone. The resultant curve of biomass productivity is shown in Figure 1.13. At latitudes within 35°C of the equator, except near the ITCZ, precipitation is the limiting factor in biomass productivity. Toward the poles, temperature becomes the limiting factor. Although these biomass productivity values have not been validated using field measurements, the minimum and maximum occur where they should according to the classification shown in Figure 1.3.

Hydroclimatic parameters also provide some insight into the carbon balance and the transport of carbon by rivers. Locally, the carbon balance is maintained by industrial activities that release carbon, plant growth that fixes carbon, soils that act as sinks for carbon, and rivers that transport carbon to the oceans.

In particular, Mulholland and Watts (1982) showed the importance of North American rivers in transporting organic carbon to the ocean. They correlated TOC (Total Organic Carbon) with long-term average streamflows, but they found there was potential for considerable year-to-year variablility in TOC transport because low frequency floods had been found to convey extremely large amounts of carbon to the ocean. However, if it is assumed that TOC transport is correlated with runoff rather than streamflow (Brinson, 1976), then the contributions to the TOC transport are largest from the areas in the study zone between 45°N and 58°N; 5°S and 15°N; 38°S and 42°S, and 47°S and 52°S. Irons et al. (1992) demonstrated that soil type and the presence of permafrost can affect the relationship between TOC transport and runoff. Furthermore, as noted earlier, stream discharge to the ocean is not always uniquely correlated with local runoff, due to differences in the size and orientation of drainage systems; consequently, there is a need to consider watershed characteristics in assessing carbon transport to the ocean.

Conclusions

Hydrometeorological processes and patterns provide a framework for understanding how the large-scale climate influences terrestrially based hydrological and biological patterns. However, not all aspects of the diversity of biological patterns and their temporal and spatial variability can be explained by average values of hydroclimatic parameters. One important factor not considered in this study involves the role of climatic extremes in influencing biological regimes. In cases where the biological patterns are not in equilibrium with their abiotic surroundings, the deviations may be related to the genetic history of biological communities, local industrial activities such as harvesting, biological invasions, or climatic extremes. Both natural and man-induced variations occur at small scales in biological systems. However, the controls on these variations are the larger chemical and physical environmental fluctuations. The complexity inherent in these biological systems is the result of combined variability in the hierarchy of abiotic processes that leads to the distribution of energy and water within the ecosystems and genetic factors that determine the specific characteristics of the ecosystem components.

More specific results from this study include the following:

- The climatological "equator" is displaced by approximately 5° northward from the geographical equator in the study area.
- Ecosystems in the study zone at the same latitude in the two hemispheres do not necessarily experience similar climatic and hydrometeorological conditions.
- Temperatures at higher latitudes and precipitation at lower latitudes limit biomass productivity. These factors also influence other biological characteristics, both on an annual basis and on shorter time scales.

- The west coast of South America experiences higher net radiation, more intensive aridity, and higher sensible heat transfers than does the west coast of North America in the 14° to 31° latitude band.
- Rivers on the west coast of North America have higher river discharges in the 45° N to 60° N latitude band than do rivers at comparable latitudes in South America, due primarily to the larger river-basin sizes.
- Signficant differences exist in the estimates of runoff, depending on the methodology used.

Research Needs

The analysis presented here has provided a preliminary consideration of the effects of hydroclimatic processes on regional biotic patterns. However, this work should be regarded as only preliminary, because a number of gaps remain to be filled.

Given the sparse observational networks and the highly variable terrain in this study area, it is essential to develop more reliable models and techniques for future studies of water and energy budgets on space and time scales appropriate to understanding and monitoring plant and ecosystem responses. Remote sensing offers many new data sets that could enable scientists to monitor a wide range of abiotic factors over a number of time scales on a spatially continuous basis. However, it is necessary to design studies to develop and validate appropriate techniques for estimating the relevant parameters from remote-sensing data. The relationships between climatic and biological factors, such as the range of species, cannot be determined with average values alone.

The role of extreme values and the interannual variability arising from El Niño events in the biology of this region also need to be studied. When using localized relationships dealing with links between hydrometeorological metrics and biological indicators developed in different parts of the world, it is essential to validate them, and it may be necessary to modify or replace these techniques for the complex geographical region considered in this study. Factors such as cloud water need to be considered in studying the water budget of coastal mountain regimes. Many of these issues can only be addressed by combining field observations with models and remote-sensing data to account for the extremes in topography that exist on the west coasts of North and South America.

Acknowledgments. The assistance of Ray Keller in data analysis and in the preparation of diagrams is greatly appreciated. The comments by the reviewers have been helpful in describing the limitations of these analyses. The contributions of Sandra Forsberg in preparing the manuscript are also appreciated.

References

Abreu, M.L. 1991. *On the dynamics of the low-level subtropical flow along the West Coast of South America*. Unpublished doctoral dissertation, Department of Meteorology, Pennsylvania State University, State College, PA.

Brinson, M.M. 1976. Organic matter losses from four watersheds in the humid tropics. *Limn Oceanog*, 21, 572–582.

Bromwell, M. 1987. *The Rand McNally Atlas of the Oceans*. New York: Rand McNally.

Cerededa, P., and Schemenauer, R.S. 1991. The occurrence of fog in Chile. *J Appl Meteorol*, 30(8), 1097–1105.

Henning, D. 1989. *Atlas of the Surface Heat Balance of the Continents: Components and Parameters Estimated from Climatological Data*. Berlin, Stuttgart: Gebrüder Borntraeger.

Irons III, J.G., Oswood, M.W., and Slaughter, C.W. 1992. Carbon dynamics in permafrost-affected stream systems: Implications for global change. In T.D. Prowse, C.S.L. Omonannery, and K. Ulmer (eds.), *Proceedings of the Ninth International Northern Research Basins Symposium/Workshop, 14–22 August 1992*. (Sorry omit parenthesis) NHRI Symposium, No. 10, Vol. 1, pp. 199–215.

Korzun, V.I., Sokolov, A.A., Budyko, M.I., Voskresensky, K.P., Konoplyantsev, A.A., Korotkevich, E.S., Kuzin, P.S., and L'vovich, M.I. 1978. *World Water Balance and Water Resources of the Earth*. Paris: UNESCO Press.

Lawford, R.G. 1993. Regional hydrologic responses to global change in western North America. In H. Mooney, E. Fuentes, and B. Kronberg (eds.), *Earth System Responses to Global Change: Contrasts between North and South America* (pp. 73–99). San Diego: Academic Press.

Lieth, H. 1975. Modelling the primary productivity of the world. In H. Lieth and R.H. Whittaker (eds.), *Primary Productivity of the Biosphere* (pp. 237–263). New York: Springer-Verlag.

L'vovich, M.I. 1979. *World Water Resources and Their Future* (trans.), R.L. Nace (ed.). Chelsea, MI: American Geophysical Union, Lithocrafters, Inc. First published in 1974.

MacMahon, J.A. 1981. Successional processes: comparisons among biomes with special reference to probable roles and influences on animals. In D.C. West, H.H. Shugart, and D.B. Botkin (eds.), *Forest Succession: Concepts and Application* (pp. 277–304). New York: Springer-Verlag.

McBean, G.A. 1994. *Circulation Patterns for Stream Flow and Temperature Predictions in Canadian Surface Climate*. Montreal: Canadian Electrical Association.

Mulholland, P.J., and Watts, J.A. 1982. Transport of organic carbon to the oceans by rivers of North America: A synthesis of existing data. *Tellus*, 34, 176–186.

Peixoto, J.P., and Oort, A.H. 1992. *Physics of Climate*. New York: American Institute of Physics.

Peña, H.T., and Vidal, F.J. 1993. Estimaciőn estadistica de la linea de nieves durante los eventos de precipitaciőn entre las latitudes 28 y 38 grados sur. In *Agua Para el Desarrollo y Protecciőn del Ambiente [Water for Development and Environmental Protection]*. XI Chilean Hydraulics Engineering Congress, Chilean Society of Hydraulics Engineering, University of Concepciőn, Concepciőn, Chile.

Terborgh, J. 1992. *Diversity and the Tropical Rain Forest*. New York: Scientific American Library.

Waring, R.H., and Schlesinger, W.H. 1985. *Forest Ecosystems: Concepts and Management*. New York: Academic Press.

2. Factors Controlling the Climate of the West Coast of North America

G.A. McBean

The west coast of North America extends from the tundra near the Arctic Ocean to the tropical jungles of Panama, with a wide variety of climates in between. Lawford (this volume) discusses the north–south variations throughout the range of the Americas. A special American Geophysical Union monograph, edited by Peterson (1989), is devoted to "Aspects of climate variability in the Pacific and the western Americas" and provides a valuable source of information on this topic. In that same volume, Roden (1989) and Tabata (1989) examine the long-term variability of the west coast and northeast Pacific Ocean climate. Walters and Meier (1989) show that variations in glacial mass balances in western North America are related to the steering effect of the Aleutian/Gulf of Alaska low.

Here the focus will be on the coastal zones between latitudes 60° N and 30° N, which can be divided into two general climate regimes: namely, the mid-latitude climate under the control of the westerly wind circulation; and the areas of the subtropical high pressure zones, where oceanic upwelling is an important factor. The equatorial climate regimes, with the dominance of the intertropical convergence zone, will not be discussed. Important features of the zones under consideration are the seasonal variations as the westerlies and the high-pressure cells move north and south, and the superimposed, longer-term variations due to natural factors in the climate system. Oceans play an important role in determining both the mean climate and its variations. Through better understanding of the factors controlling the climiate, we can gain the possibility of predicting climate variability and change. In this chapter, we cannot comprehensively cover all aspects of the climate of this region, as has been done in works such as Hare and Thomas (1974) and Bryson and Hare (1974). Instead, emphasis will be given here to only two features: 1) the precipitation variability of the British Columbia region (which is representative of the mid-latitude westerlies climate zone); and 2) ocean upwelling, which influences regional climate. Spatial and temporal variations in precipitation are probably the single most important factor in determining vegetation types and ecosystem distributions along the western part of North America. Upwelling is included to highlight the interactions of the oceans and atmosphere in the coastal regions.

The climatic factors discussed here are central to several programs with international global change activities. Variability of precipitation is a key component of the Global Energy and Water Cycle Experiment (GEWEX) of the World Climate Research Programme. Oceanic upwelling will be an important physical process for the Land-Ocean Interactions in the Coastal Zone (LOICZ) Project of the International Geosphere-Biosphere Programme (IGBP). These physical factors will strongly influence regional ecosystems, which are the focus of the Global Change and Terrestrial Ecosystems (GCTE) and Joint Global Ocean Flux Study (JGOFS), both of the IGBP.

Scales of Variation and Predictability

When we deal with natural ecosystems, it is important to remember the interactions of differing spatial scales and response times. There is a full range of natural variability, from millisecond and millimeter scales of turbulence through global and decadal scales of climate variations. For most natural systems, time and space scales are linked. Small features tend to change more rapidly than larger features, but their other characteristics are also important. Generally, we can say that dynamically, the atmosphere adjusts in a few days; it will reestablish its radiative equilibrium in about a month; and water cycles through the atmosphere in about 10 days. Our best estimates of other adjustment times include a few months for soil moisture and about a year for snow and sea ice. The upper layers of the ocean adjust in months through to the seasonal cycle, while the deep ocean may take centuries to respond to a change in external forcing. Connected to this concept of time and space scales is the concept of predictability. It is not possible to predict the evolution of a phenomenon beyond a few of its time scales.

Climate prediction from an initial state of a sequence of future climate states (Lorenz referred to this as prediction of the first kind) (Palmer,

1993), is like weather prediction, except the predicted states are given in a statistical sense (warmer than normal, wetter, etc.) rather than as a sequence of synoptic weather conditions. The emphasis of climate prediction has been on seasonal-to-interannual time scales. Examination of long time series of climatic characteristics reveals a dominant signal associated with the El Niño–Southern Oscillation, the coupling of the tropical Pacific atmosphere and ocean with global climatic implications.

Within the climate system, the statistical impacts of shorter time-scale phenomena may be reflected in slow variations in other characteristics. For example, although precipitation shows large day-to-day variations, its impacts are integrated through river runoff and, subsequently, on even longer time scales on lake levels, as shown in Figure 2.1. Biological phenomena may be affected on any (or none) of these time scales. Terrestrial plants may bloom quickly in response to rain showers, while aquatic plants vary more slowly as the lake level changes.

Mid-Latitude Westerlies Climate Zone

This climate zone extends from the Gulf of Alaska, through British Columbia, to Washington, Oregon, or northern California, depending on the season. The Pacific Ocean provides a moderating influence and an essentially infinite source of moisture along the western boundary. The mountain ranges, from the Coast Mountains through to the Rockies in the north or the western cordillera farther south, alter the climate to strongly reflect topographic characteristics. Isotherms of mean daily temperature run parallel to the coast and along the mountains in the

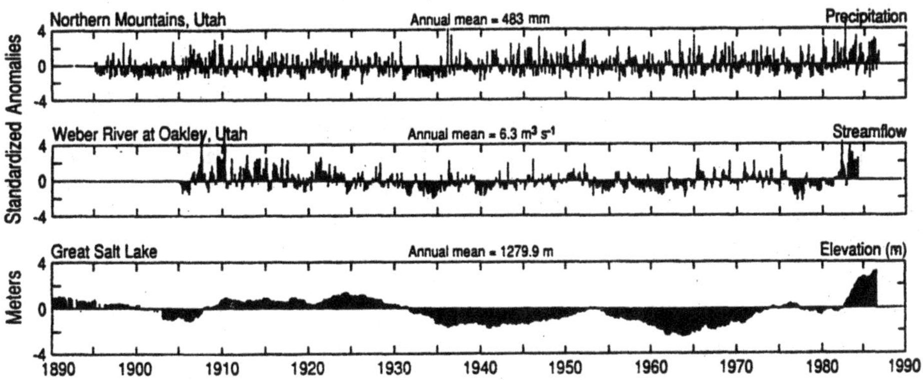

Figure 2.1. Monthly time series of regional mean precipitation in the Northern Mountains region of Utah, stream flow in the Weber River at Oakley, Utah, and level of the Great Salt Lake. Reproduced from Peterson (1989).

coastal zone and generally show more complex patterns (but clearly related to topography) than in other parts of North America.

Precipitation is one of the major features of climate and provides an essential ingredient for terrestrial life forms. Precipitation also varies dramatically in both time and space in mountainous regions such as those that prevail along the mid- and higher-latitude west coasts of the Americas. Studies under way to understand the nature and climatic connections of precipitation in British Columbia will give some insight into what is happening in similar regions of the Americas. Most of British Columbia's precipitation is supplied by North Pacific air masses and is liberated primarily during the cool season by a continual progression of extratropical storms carried in the westerly flow. Mean annual precipitation exceeds 200 mm/month for parts of the Coast Mountain region but decreases to less than 100 mm/month in some regions between Vancouver Island and the mainland mountains. The gradients of precipitation along the slopes of the Coast Mountains are very pronounced. In the interior plateau region of the province, precipitation decreases to 40 to 50 mm/month. Secondary maxima in precipitation, of 100 mm/month^{-1}, occur over the interior mountains and the Rockies.

Despite these complexities, principal component analyses show that large-scale variations in precipitation can be represented in terms of four patterns: north coastal, south coastal, north interior, and south interior regions (McBean & Hourston, 1994). Distinctions among these regions are apparent in the box-plot summaries of the seasonal variation for each regional precipitation series, as shown in Figure 2.2, as well as in other statistics. Both the north coastal and the south coastal regions have maximum monthly precipitation occurring in October–January, when cyclones are most frequent (e.g., Murty, McBean, & McKee, 1983). North coastal region values are higher than those for the south coastal region, except for February. The onset of high winter values (above about 230 mm/month) starts in October for the north coastal region and in November for the south coastal region, corresponding to the southward shift of the westerly storm belt in the fall and winter. The west coast of Washington state also has a pronounced annual precipitation cycle, with maximum in the winter (Finkelstein & Truppi, 1991). There is considerable variability, especially during the winter months. During the summer months, both coastal regions come under the influence of a semipermanent ridge of high pressure, resulting in less cyclonic activity. Precipitation in the months June, July, and August falls to below 80 mm/month. Although standard deviations are largest in the months October to January, the relative variability (as given by the coefficient of variation, defined as the standard deviation divided by the mean) is higher in summer.

Both interior regions lie in the rain shadow of the Coast Mountains. Mean annual precipitation is much lower, generally in the range of 30 to

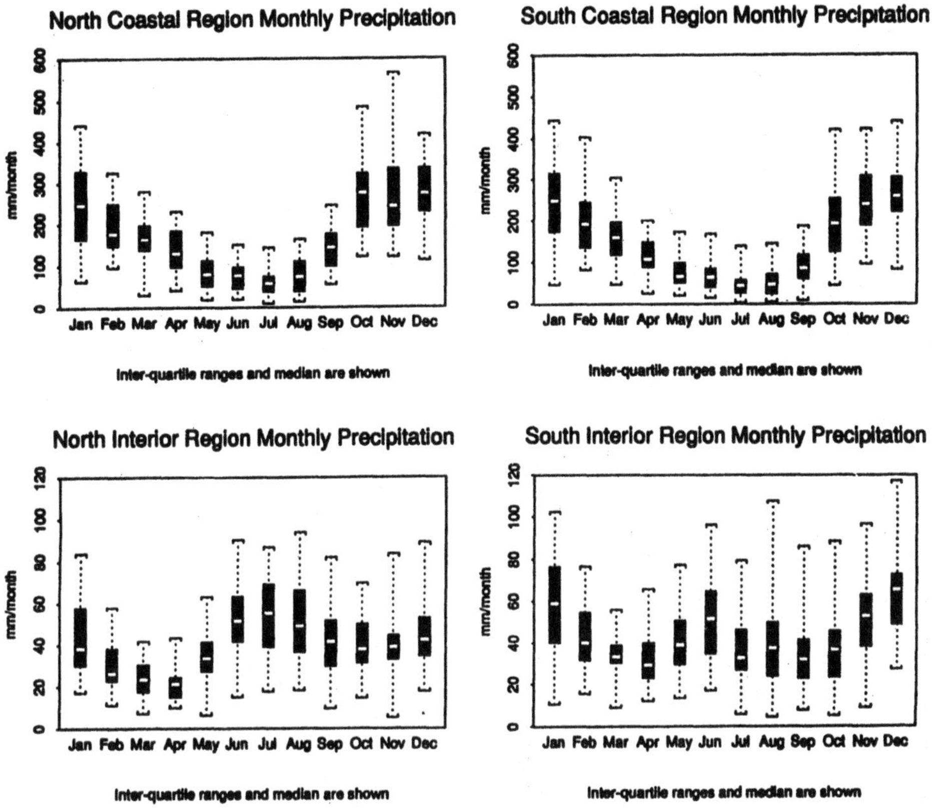

Figure 2.2. Regional monthly precipitation for four areas of British Columbia. Shown are the median and interquartile ranges.

60 mm/month, and the annual cycle has peaks, one in summer and one in winter, with a smaller range to the annual cycle than the coastal regions. Western Montana precipitation shows much less annual cycle, but with maxima in June–July and in winter (Finkelstein & Truppi, 1991). For the north interior region, the summer maximum is higher (54 mm/month August) than the winter maximum (44 mm/month in both December and January), while for the south interior region, the winter maximum (63 mm/month, December) is larger than the summer maximum (50 mm/month, June). Their annual variations of precipitation are much different from the coastal regions. The sumer peak in precipitation is associated with the greater occurrence of convective activity. Monthly mean regional precipitation is a minimum during April (21 mm/month) for the north interior region. The standard deviations for the interior regions are also much smaller in magnitude. However, the coefficients of variation for the north interior region are almost constant (near 0.4)

during the year, about the same as the average value for the coastal regions.

Orographic enhancement is a factor in the south interior region, due to the influence of the Columbia and Rocky Mountains, which run approximately north–northwest/south–sourtheast along the eastern boundary of the province. However, the eastward-flowing air has already precipitated much of its moisture in passing over the Coast Mountains and, consequently, precipitation amounts are not as high as for the coastal regions. Annual mean monthly precipitation ranges from 25 to 69 mm/month. As noted previously, the seasonal pattern shows two peaks, with the winter one being slightly larger. The coefficient of variation is largest in the late summer–early fall (reaching 0.56 in October) and smallest in the spring (0.30 in March).

Time series of regional precipitation show different patterns of decadal and shorter-period fluctuations. North coastal precipitation was consistently below the long-term average until about 1962, when there was a change to consistently above-average conditions. The south coastal region has a trend similar to that of the north coastal region. The south interior region time series exhibits a long period of generally below-mean conditions, with a sudden recovery beginning in 1980.

Several authors have examined relationships between precipitation and atmospheric circulations. Klein and Bloom (1987) related monthly mean precipitation over the United States to the 70 kPa height field and found that the best statistical relationships occurred in winter (December–Febuary) and for the west coast. Station relationships were less successful for regions just east of the continental divide and for summer. Redmond and Koch (1991) correlated the Pacific North America (PNA) pattern (Wallace & Gutzler, 1981) and the Southern Oscillation indices with precipitation, temperature, and stream flow over the western United States. The Pacific Northwest was identified as an area of higher correlation. Cayan and Roads (1984) also examined west coast precipitation in relationship to localized circulation parameters, such as 70 kPa height, winds, etc. Some studies found that precipitation frequency was better correlated with circulation parameters than precipitation amount. Cayan and Peterson (1989) looked at relationships between stream flow in western North America and North Pacific atmospheric circulation. They followed work by Namias (1975) and others who used sea-level pressure variations in the central North Pacific Ocean as a variable for representing atmospheric circulation. Their Central North Pacific index, sea-level pressure anomaly averaged over 170° E–150° W, 35° N–55° N, was highly correlated with the PNA pattern. Cayan and Peterson found that British Columbia was close to the node of the atmospheric long-wave pattern that emanates from the central North Pacific, so that small variations in the position of the remote circulation anomaly center could yield either positive or negative river stream-flow fluctuations. They concluded that British Columbia stream-flow variability did not respond consistently to broad-scale Pacific atmospheric circulation indices but is related to regional atmospheric anomaly features over the eastern North Pacific. Yarnal and Diaz (1986) found that precipitation anomalies were sensitive to minor fluctuations in the general placement and orientation of the PNA pattern. Yarnal and Diaz showed that during warm-event winters, precipitation in California and southeast Alaska were negatively correlated, but northern British Columbia and Vancouver Island precipitation were not significantly correlated with either.

To examine relationships between precipitation over British Columbia and atmospheric circulation, circulation patterns when the precipitation is anomalous were examined, splitting the data into winter and summer half-years. The average 50 kPa heights for all cases when the standardized anomaly index (anomaly-mean divided by standard deviation) was either greater than 1 or less than −1 were computed. For the winter half-year and both north and south coastal regions, high precipitation occurs when the atmospheric flow is toward the coastal mountains, with 50 kPa heights lower in the Gulf of Alaska by up to 60 m and increased over California by 20 m to 30 m, as shown in Figure 2.3 (part A). For low-precipitation events in winter for the coastal regions, a ridge along the coast is dominant, as shown in Figure 2.3 (part B). Low-precipitation anomaly patterns are not opposite of high-precipitation pattern anomalies. For both coastal regions, low precipitation occurs with more than 100 m increases in 50 kPa height along the British Columbia coast, and 40 m to 60 m height falls nearer the dateline and over the Great Lakes. As the t-test shows, the anomaly patterns for low precipitation are a tripole of changes (−, +, −) approximately along the 5400 m contour of the 50 kPa height, with the height increases over British Columbia being about twice as large as the decreases. In comparison, the high-precipitation cases can be described as a dipole pattern with its orientation along the coast (across the flow). Thus, high precipitation occurs with an increase in flow strength as well as some redirection, while low precipitation corresponds to a major flow curvature resulting in ridging over the region. For the interior regions, flow anomalies occurring with high precipitation are similar to their coastal counterparts.

Because the summer gradients are smaller, the height anomalies corresponding to summer precipitation anomalies also tend to be smaller in an absolute sense, but the statistical significance can be as high. Positive summer rainfall anomalies in the coastal regions occur with reduced heights (60 m to 70 m) along the British Columbia coast. Changes elsewhere are less significant. The pattern of significant height changes for summer low-precipitation months for the coastal regions is more widespread. There are large height increases (80 m–90 m) along the

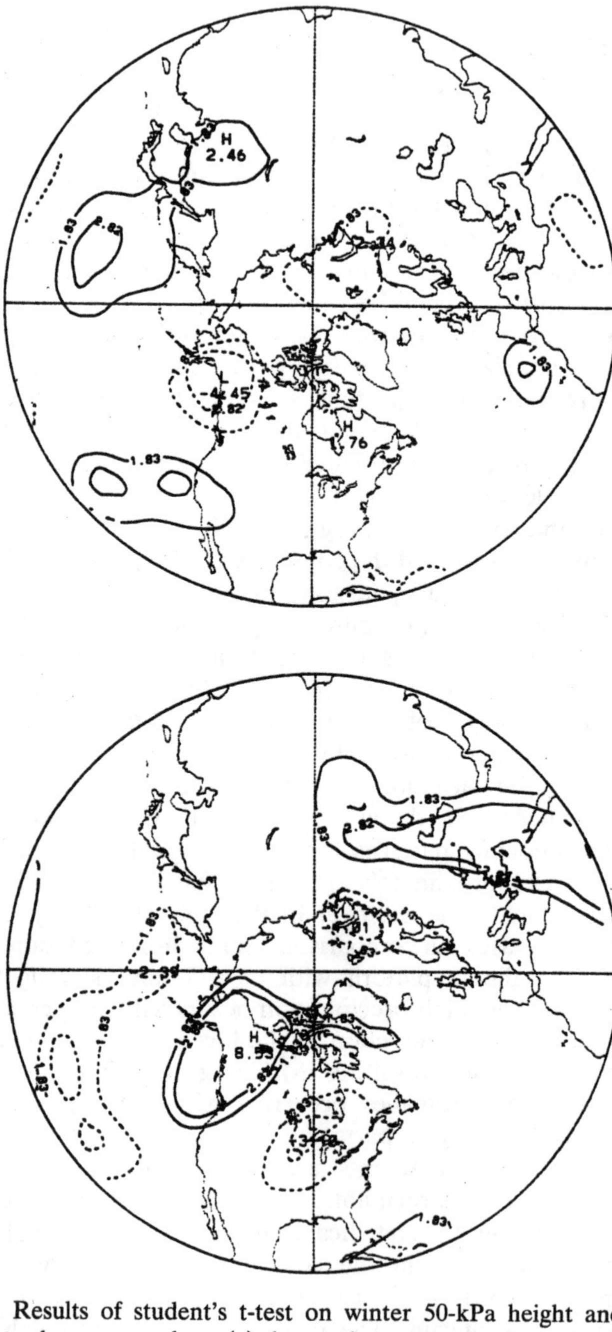

Figure 2.3. Results of student's t-test on winter 50-kPa height anomalies corresponding to those cases where (a) the north coast regional standardized precipitation anomaly indices were greater than 1 and (b) the south coast regional standardized precipitation anomaly indices were less than −1. The contour 1.83 corresponds to 5% significance and 2.82 to 1% significance.

British Columbia coast, height decreases downstream over central North America and other areas (that are different for the north and south coastal regions). For each case, there are significant changes quite distant from British Columbia; these may indicate large-scale flow anomalies or just be statistical noise.

Role of Sea-Surface Temperature Patterns over the North Pacific Ocean

It has been noted that there were some long-period variations to both the precipitation anomalies and the atmospheric flow indices. For example, anomalies of coastal precipitation tended to be negative from the late 1940s until 1958. The question is, what is the cause of these long-period variations in the atmospheric flow and the precipitation? Since the atmosphere has a relatively short memory or adjustment time (a month or less), it seems likely that these long-period variations are linked to variations in the ocean. Correlations between the sea-surface temperature (SST) anomalies and the regional time-series anomalies and standardized anomaly indices for precipitation time series, separating the data into winter and summer half-years, were generally quite small (McBean & Hourston, 1994).

In a review of all composite SST anomaly plots for standardized anomaly indices for precipitation, two anomaly patterns emerge consistently for summer and winter. The summer pattern is characterized by two regions of significant SST anomalies of the same sign in the northwest and northeast Pacific. The winter recurring anomaly pattern is characterized by two regions of significant SST anomalies of different signs in the northeast Pacific near 30° N and 50° N. Persistence of these SST anomaly patterns over the given season for indices based on precipitation over a large area lends support to the idea that these large-scale SST anomalies are associated with large-scale precipitation anomalies over British Columbia. In the case of the second (winter) pattern, the large-scale positive SST anomaly region near 30° N covers most of the North Pacific, and this anomalous heat source may serve to increase the energy and moisture content for cyclones moving east over the Pacific, resulting in larger precipitation amounts over British Columbia. The dynamics of this relation (and for the first, or summer pattern) are not clear, however, but this analysis suggests that they warrant further investigation. Emery and Hamilton (1985) and others have investigated the relationship between the El Niño and the interannual variability of the northeast Pacific Ocean and found that although there were general tendencies for sea-surface temperature to be anomalously high during El Niño events, the lack of a "truly predictable relation" points out that other factors are also important.

Oceanic Upwelling

Linkages between the atmosphere and the oceans take place on a wide range of space and time scales, from basin scale ocean gyres, driven by prevailing wind patterns, to capillary winds on wave crests, generated by local winds. A very important phenomenon is the upwelling of subsurface water to the surface, caused by variations in winds and/or winds interacting with topography (Thomson, 1981). Upwelling occurs due to the basic physics of wind stress on a water body on a rotating earth. Whenever a wind stress is applied to the surface of the ocean, there is a net transport of water to the right (in the Northern Hemisphere) due to the Ekman spiral in the water. For uniform winds over the open ocean, this results in a net horizontal transfer of water. However, if the wind is spatially varying, the horizontal transport will also be spatially varying, and vertical motions in the ocean will be necessary to compensate for converging or diverging surface waters; the result is upwelling (for diverging surface waters) or downwelling. When the wind blows parallel to the coast, there will be either transport toward or away from the coast, resulting in downwelling or upwelling, as shown in Figure 2.4. Upwelling

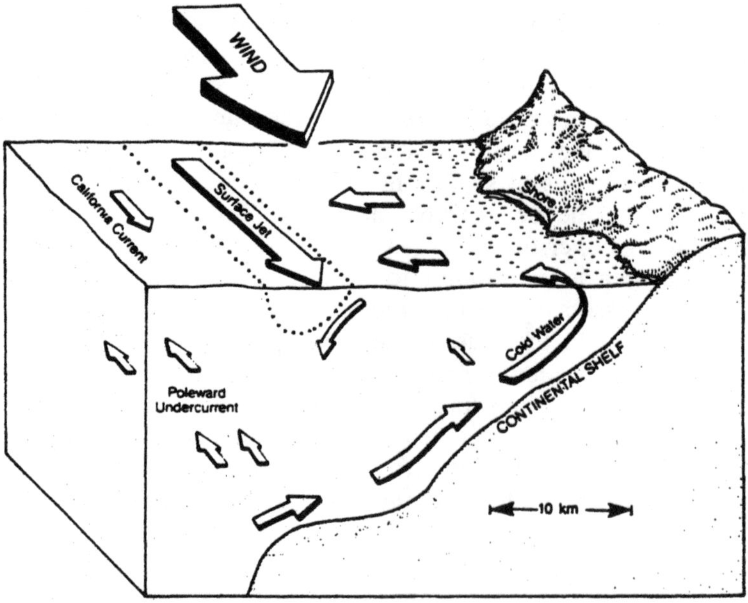

Figure 2.4. Main features of classical wind-induced upwelling in the northern hemisphere. Equatorward winds combined with Coriolis force move nearshore surface water offshore. Colder subsurface water rises to the surface to replace it. Reproduced from Thomson (1981).

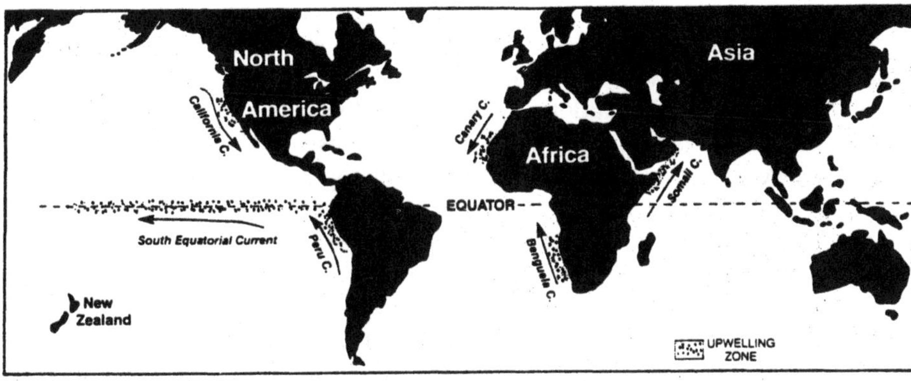

Figure 2.5. Major oceanic upwelling zones and regional coastal currents. Reproduced from Thomson (1981).

of cold, nutrient-rich water from below influences the local climate as well as marine biological processes. Costal upwelling occurs along most coasts at some times but is prevalent along the eastern sides of the subtropical high-pressure systems, as indicated in Figure 2.5. Upwelling may also be caused by ocean currents (including tidal currents) passing over underwater ridges, shoals, or other bumps on the bottom and being forced upward (Thomson, 1981). Although not very important in the open ocean, this type of upwelling is important in many coastal regions, particularly in straits, such as Juan de Fuca Strait and the southern Stait of Georgia on the British Columbia coast. Globally, upwelling regions are most important for fisheries and are areas of strong biomass production.

Variations in Upwelling Along the West Coast of North America

Prevailing northerly winds along the west coast of North America between 30° N and about 45° N cause upwelling of colder subsurface water, the impact of which is seen in colder sea-surface temperatures, marine clouds, and fog. The Coastal Ocean Dynamics Experiment (CODE) (Beardsley & Lentz, 1987) focused on this region. The upwelling strength is determined primarily by the sea-level pressure distribution over the eastern North Pacific Ocean, which is dominated by the Pacific subtropical high-pressure cell. This cell moves northward in summer, and hence the upwelling region extends farther north during the summer months. Variability on time scales of days to months was examined by Beardsley et al. (1987), with emphasis on the atmospheric boundary layer structure, and by Strub et al. (1987). For the California region (at least), the local wind stress that directly drives the upwelling has shown large changes over the past century, being generally lower before the 1950s and then higher until the mid- 1980s (Hsieh & Boer, 1992), as indicated in Figure 2.6. Breaker (1989) examined the variability over the period 1972–1984 and found relations with El Niño.

Although upwelling is a dominant feature along the California coast, it is a much more variable feature along the coast of Vancouver Island, where upwelling may arise due to current–topography interactions as well as wind forcing. Fang and Hsieh (1993) used satellite sea-surface temperature imagery to assess the variability for eight summers (1984–1991). An empirical orthogonal-function analysis of the sea-surface temperatures identified the dominant patterns with river outflows and topographically forced upwelling. The dominant pattern is one of cold waters along the coast and warmer water offshore. The summers of 1986 and 1991 had the coolest coastal water. Variations in temperature of coastal waters have important effects on coastal climate, ecosystems, and economic activities.

It has been speculated by Bakun (1990) that global climate warming would lead to increased coastal upwelling. He reasoned that warmer land temperatures (relative to the oceans, which warm more slowly) would result in lower pressures over land, increased along-shore geostrophic winds, and stronger upwelling. Thus, the cool, foggy summeritime conditions of northern California would become more pronounced and marine ecosystems could be dramatically enhanced. Hsieh and Boer (1992) used the Canadian Climate Centre's global climate model (GCM) to show that although the land areas do warm more than the oceans, the pressures (which are an integration of effects through the atmosphere) do not

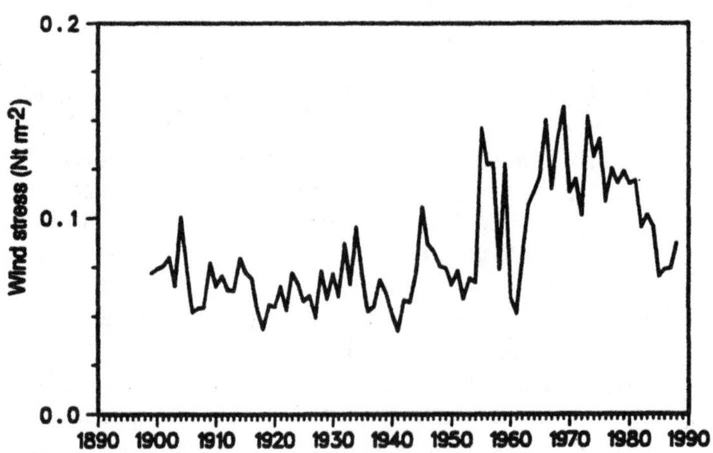

Figure 2.6. Time variations in alongshore wind stress off California (37.5° N), which drives the upwelling along the California coast, based on the geostrophic wind along the coast, computed from pressure fields. Reproduced from Hsieh & Boer (1992).

change as Bakun postulated, and the model actually predicts reduced upwelling along the west coast of North America (and most other upwelling regions as well). In the open ocean, the equatorial and subpolar zonal upwelling bands generally weaken as the winds diminish, due to the reduced equator-to-pole surface temperature gradient. However, the spatial resolution of GCMs is still poor, and the pressure differences are small. The results do point out the need to consider the globally connected climate system in estimating climate response to external forcing.

Conclusions

The climate of the west coast of North America results from the interactions of large-scale atmospheric circulation systems (the westerlies and the subtropical highs) with the ocean and the mountains along the coast. These climates demonstrate variability on a variety of time and space scales. Upwelling is a near-coastal phenomenon related to regional winds that has important effects on the ocean biology as well as the local climate. Variations in large-scale regional precipitation indices over British Columbia also show long-term differences that may be related to sea-surface temperature variations across the Pacific basin. It is important to recognize the interaction of scales of variability and regional differences of the upwelling and all other air–sea interaction phenomena. Studies should also be conducted in cooperation with other projects that are being developed to optimize the use of resources and gain the benefits of multiparticipant and multidisciplinary research.

Acknowledgments. The scientific work reported here was undertaken at the University of British Columbia in cooperation with R.A.S. Hourston and G. Thomas and was supported by grants from the Atmospheric Environment Service of Environment Canada, B.C. Hydro, and the B.C. Ministry of Environment.

References

Bakun, A. 1990. Global climate change and intensification of coastal ocean upwelling. *Science*, 247, 198–201.

Beardsley, R.C., Dorman, C.E., Friehe, C.A., Rosenfeld, L.K., and Winant, C.D. 1987. Local atmospheric forcing during the Coastal Ocean Dynamics Experiment. 1. A description of the marine boundary layer and atmospheric conditions over a northern California upwelling region. *J Geophys Res*, 92, 1467–1488.

Beardsley, R.C., and Lentz, S.J. 1987. The coastal ocean dynamics experiment collection: An introduction. *J Geophys Res*, 92, 1455–1463.

Breaker, L.C. 1989. El Niño and related variability in sea-surface temperature along the central California coast. In D.H. Peterson (ed.), *Aspects of Climate Variability in the Pacific and the Western Americas*. American Geophysical Union, *Geophys Monog*, 55, 133–140.

Bryson, R.A., and Hare, F.K. (eds.) 1974. *Climates of North America*. Vol. 11, *World Survey of Climatology*, H. Landsberg (ed.), New York: Elsevier.
Cayan, D.R., and Peterson, D.H. 1989. The influence of North Pacific atmospheric circulation on streamflow in the west. In D.H. Peterson (ed.), *Aspects of Climate Variability in the Pacific and the Western Americas*. American Geophysical Union, *Geophys Monog*, 55, 375–397.
Cayan, D.R., and Roads, J.O. 1984. Local relationships between United States west coast precipitation and monthly mean circulation parameters. *Mon Wea Rev*, 112, 1276–1282.
Emery, W.J., and Hamilton, K. 1985. Atmospheric forcing of interannual variability of the northeast Pacific Ocean: Connections with El Niño. *J Geophys Res*, 90, 857–868.
Fang W., and Hsieh, W.W. 1993. Summer sea surface temperature variability off Vancouver Island from satellite data. *J Geophys Res*, 98, 14391–14400.
Finkelstein, P.L., and Truppi, L.E. 1991. Spatial distribution of precipitation seasonality in the United States. *J Clim*, 4, 373–385.
Hare, F.K., and Thomas, M.K. 1974. *Climate Canada*. Toronto: Wiley.
Hsieh, W.W., and Boer, G.J. 1992. Global climate change and ocean upwelling. *Fish Oceanog*, 1, 333–338.
Klein, W.H., and Bloom, H.J. 1987. Specification of monthly precipitation over the United States from the surrounding 700 mb height field. *Mon Wea Rev*, 115, 2118–2132.
McBean, G.A. 1994. Observed patterns of precipitation and mid-tropospheric atmospheric circulation. *Atmosphere Ocean* (to be submitted).
McBean, G.A., and Hourston, R.A.S. 1994. Relationships between anomalies in circulation and precipitation and sea-surface temperature patterns over the north Pacific Ocean. *Atmosphere Ocean* (to be submitted).
Murty, T.S., McBean, G.A., and McKee, B. 1983. Explosive cyclogenesis in the northeast Pacific Ocean. *Mon Wea Rev*, 111, 1131–1135.
Namias, J. 1975. Northern Hemisphere seasonal sea level pressure and anomaly charts, 1947–1974. In A. Fleminger (ed.), *CalCOFI Atlas No. 22*. Marine Life Research Program, Scripps Institute of Oceanography, University of California–San Diego, San Diego, CA.
Palmer, T.N. 1993. Extended-range atmospheric prediction and the Lorenz model. *Bull Am Meteorol Soc*, 74, 49–65.
Peterson, D.H. (ed.) 1989. Aspects of climate variability in the Pacific and the western Americas. American Geophysical Union, *Geophys Monog*, 55.
Redmond, K.T., and Koch, R.W. 1991. Surface climate and streamflow variability in the western United States and their relationship to large-scale circulation indices. *Wat Res Res*, 27, 2381–2399.
Roden, G.I. 1989. Analysis and interpretation of long-term climatic variability along the west coast of North America. In D.H. Peterson (ed.), *Aspects of Climate Variability in the Pacific and the Western Americas*. American Geophysical Union, *Geophys Monog*, 55, 93–111.
Strub, P.T., Allen, J.S., Huyer, A, Smith, R.L., and Beardsley, R.C. 1987. Seasonal cycles of currents, temperatures, winds and sea level over the northeast Pacific continental shelf: 35° N to 48° N. *J Geophys Res*, 92, 1507–1526.
Tabata, S. 1989. Trends and long-term variability of ocean properties at Ocean Station P in the northeast Pacific Ocean. In D.H. Peterson (ed.), *Aspects of Climate Variability in the Pacific and the Western Americas*. American Geophysical Union, *Geophys Monog*, 55, 113–132.
Thomson, R.E. 1981. Oceanography of the British Columbia coast. *Can Spec Pub Fish Aquat Sci*, 56.

Wallace, J.M., and Gutzler, D.S. 1981. Teleconnections in the geopotential height field during the Northern Hemispheric winters. *Mon Wea Rev*, 109, 748–812.

Walters, R.A., and Meier, M.F. 1989. Variability of glacier mass balances in western North America. In D.H. Peterson (ed.), *Aspects of Climate Variability in the Pacific and the Western Americas*. American Geophysical Union, *Geophys Monog*, 55, 365–374.

Yarnal, B., and Diaz, H.F. 1986. Relationships between extremes of the Southern Oscillation and the winter climate of the Anglo-American Pacific coast. *J Climatol*, 6, 197–219.

3. Sulfur Cycling in Coastal Upwelling Systems and Its Potential Effects on Climate

S. Mary P. Benbow

Climatic variability and change occur as a result of interaction between the atmosphere and the other components comprising the earth's system (i.e., hydrosphere, lithosphere, biosphere, and cryosphere). The complex relationship between the oceans and the atmosphere is only beginning to be understood, and many of the factors implicated in their interaction include the biogeochemical cycles. Sulfur may not have the same media attention as, for example, carbon dioxide, but it is nevertheless an influential element, not only because of its role in anthropogenic emissions as sulfur dioxide but also in its many forms and consequent roles in the environment. Of particular interest is the role of the marine biosphere, notably marine phytoplankton, in the production of dimethylsulfide and its potential effects on climate.

Upwelling regions are characterized by large primary productivities and may therefore demonstrate the role of DMS not only upon climate but also upon the transfer of sulfur from oceans to land surfaces. The DMS feedback may be envisioned as a subsystem of near-coastal ecosystems, and the resultant input of DMS may be of considerable benefit to resident terrestrial primary producers. The model presented illustrates the relationship between DMS production and resultant cloud cover and also endeavors to illustrate the relationship between terrestrial ecosystems and the production of DMS.

Dimethylsulfide

Dimethylsulfide is a volatile sulfur compound that forms the majority of non-sea-salt sulfates (NSS-SO_4^{2-}), which over ocean areas are almost exclusively responsible for cloud condensation nuclei (CCN) (Charlson, Lovelock, Andreae, & Warren, 1987). In addition to DMS, a wide range of other sulfur compounds is derived from marine biogenic sources (e.g., hydrogen sulfide), but DMS and its precursor, dimethylsulfonium propionate (DMSP), usually predominate (Charlson et al., 1987). Hedin and Hetherington (this volume) note that sea salts are the major route by which near-coastal terrestrial temperate rainforests acquire SO_4, though marine emissions of DMS enhance atmospheric inputs of sulfur.

DMS is produced predominantly by marine phytoplankton; although the first discussion of DMS occurred in 1935 (Haas, 1935) the precise causes are still unclear. The factors affecting the production of DMS and its effects upon the sulfur cycle and the earth's environment are still poorly understood. In evaluating the causes of DMS production, the quantities produced in different environments may be determined. The factors affecting the production of DMS can be classified according to two factors. Abiotic factors such as temperature and salinity indicate the adaptive features of the phytoplankton to their environment. For example, Iida, Nakamura, and Tokunaga (1985) partially purified an enzyme from one phytoplankton species and noted that the optimum temperature for the production of DMS from this enzyme is around 40°C and the maximal pH is between 7.0 and 7.5. Vairavamurthy, Andreae, and Iverson (1985) have shown that DMSP, the substance from which DMS originates, may be associated with internal neutralization of phytoplankton's ionic charges, rendering their salty environment less of a toxic hazard. However, it is unlikely that DMS is a buoyancy control, despite the effect of different salinities on DMS production, as it is such a small proportion of the total amount of the organic matter excreted (Barnard, Andreae, & Iverson, 1984). In addition, as large amounts of sulfur reside in the oceans, marine organisms must find a way to rid themselves of excess amounts; excreting sulfur as a methylated compound ensures quick removal in the form of a gas or vapor.

Biotic factors affecting DMS production include species type and predation; yet information about this is still limited. Dacey (1987) proposed that the rate of DMS release increases when phytoplankton are grazed by zooplankton. However, Chameides and Davis (1982) proposed that DMS may be a defense against predation; this view is supported by observations in the south Bering Sea (Barnard et al., 1984). In addition, different groups of phytoplankton produce different amounts of DMS; for example, coccolithophorids have the highest rate of DMS excretion per unit biomass (Charlson et al., 1987).

The range of factors affecting phytoplankton and their production of DMS is therefore extremely varied, and the combination of factors influencing the production of DMS in different regions at different times will be the result of a complex mixture of these factors. Currently, it is not understood which factors dominate and how they vary spatially or temporally. In order to understand the role of DMS, it is vital to examine not only the broad range of factors outlined above but also to consider their role in otherwise productive regions. Highly productive regions, such as upwelling regions, do not produce significantly higher amounts of DMS (Charlson et al., 1987). However, once DMS has been produced, regardless of conditions, its effects are dependent upon its transfer to the atmosphere and its subsequent deposition.

Increasingly, DMS is being shown to affect climatic variables. Shaw (1983) proposed that the emission of sulfur gases by marine organisms was the most efficient method of climate control. He proposed that global warming could be offset by marine life increasing its output of carbonyl sulfide and carbon disulfide; this would result in an increase in the number of droplets of sulfuric acid in the stratosphere, thereby cooling the earth by increasing the albedo. The DMS feedback, proposed by Charlson et al. (1987), suggests that the DMS produced by marine phytoplankton forms, CCN via oxidation in the atmosphere. These influence the earth's albedo and hence the global climate, in turn affecting the phytoplankton. The DMS feedback has been tested in order to understand its role in climatic regulation. For example, Rampino and Volk (1988) proposed that mass extinctions at the Cretaceous/Tertiary boundary, when the marine calcareous plankton were reduced by 90%, may have produced a warming of up to 6°C or more if the DMS feedback was operational.

Therefore, study of the role of the DMS feedback in upwelling regions and the relationships between some of the factors outlined above needs to be undertaken. The remainder of this chapter focuses upon these problems and addresses the role of DMS in the climate system.

A GIS Model of DMS in the Eastern Pacific Environment

The operation of the DMS feedback can be demonstrated using a simple box model, as shown in Figure 3.1, which provides plausible figures for the production of DMS worldwide (see Table 3.1). The model itself is a compilation of simple population biology models, a sea–air transfer model, and an energy balance model to represent global climate. It is a simple, broad-scale model to provide an indication of the operation of the feedback and is by no means predictive. The population biology models are designed to represent plausible behavior, though they are obviously generalized to conform to the generality of the other model components. To model the DMS feedback loop, observations of the processes inherent

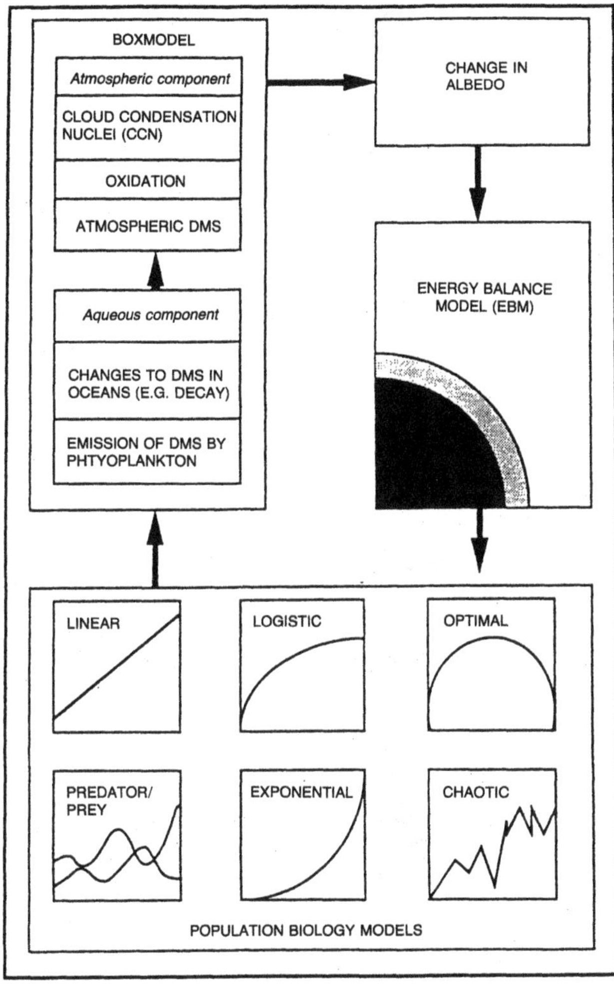

Figure 3.1. Schematic diagram to illustrate the DMS model presented.

in the feedback loop are required as well as more detailed measurements of the amounts of DMS, DMSP, and other components in the atmosphere and oceans.

The model is based on a one-dimensional model that incorporates simple stock/flux calculations. If 1 week is the characteristic residence time of a moderately reactive gas in the atmosphere (Lovelock, Maggs, & Rasmussen, 1972), then the stock of DMS in the atmosphere can be calculated. A zero-dimensional energy balance model (Henderson-Sellers & McGuffie, 1987) considers the earth as a single point, such that:

$$(1 - \alpha) S/4 = \sigma Te^4 \qquad \text{(Eq. 3.1)}$$

Table 3.1. Amounts of Dimethylsulphide Calculated Using the DMS Model and Different Population Functions (Grams per cm^3)

Population Function	Polar	Equatorial
Linear	$7.55 * 10^{-11}$	$1.26 * 10^{-11}$
Exponential—R = 1.5	$2.22 * 10^{-11}$	$2.5 * 10^{-10}$
Exponential—R = 2.0	$3.58 * 10^{-11}$	$3.33 * 10^{-10}$
Logistic—R = 1.6	$3.95 * 10^{-11}$	$2.13 * 10^{-10}$
Logistic—R = 2.0	$3.33 * 10^{-11}$	$2.79 * 10^{-10}$
Optimal	$7.54 * 10^{-11}$	$1.25 * 10^{-11}$
Chaotic—R = 3.00	82.18	$8.76 * 10^{-12}$
	$1.90 * 10^{+7}$	$5.03 * 10^{-12}$
	0.132	$4.88 * 10^{-12}$
Chaotic—R = 3.01	839520.4	$8.51 * 10^{-12}$
	$2.34 * 10^{-11}$	$8.74 * 10^{-12}$
	119494.9	$5.47 * 10^{-12}$
Lotka—Volterra	$3.57 * 10^{-11}$	$1.03 * 10^{-8}$

where $S/4$ = time-averaged energy input rate
$(1 - \alpha)$ = albedo
σ = Stefan–Boltzmann constant
Te = effective temperature

The amount of aqueous DMS (grams/m^3) input to the model and the flux of DMS from sea to air (grams/cm^3/hour) is calculated after Liss and Slater (1974) by the following equation:

$$F = KD \qquad \text{(Eq. 3.2)}$$

where F = flux of DMS through layer (grams/cm^3/hour)
K = exchange constant or flux of DMS per unit concentration gradient (19.2 m^3/$hour^1$)
D = concentration difference across layer (grams/cm^3)

The concentration difference across the layer is essentially identical to the concentration of DMS in the surface ocean layers, as the seas are highly oversaturated with DMS relative to the atmosphere. The flux of DMS per hour is converted to a figure for flux of DMS per year, and then the flux over the whole ocean is calculated using the following equation:

$$FO = FA \qquad \text{(Eq. 3.3)}$$

where FO = flux of DMS over the whole ocean (grams/year)
F = flux of DMS (grams/cm^3/year)
A = area of the earth's oceans ($3.63 * 10^{18}\,cm^2$)

Residence time is the stock over the flux; therefore, the residence time of the DMS in the ocean is expressed by:

$$R = S/F \qquad \text{(Eq. 3.4)}$$

where R = residence time (years)
S = stock (grams/cm^3)
F = flux (grams/year)

Therefore, the stock of DMS in the atmosphere is:

$$DMS = FO * R \qquad \text{(Eq. 3.5)}$$

where DMS = atmospheric DMS (grams)
R = residence time (years)

Residence time of a moderately reactive gas is approximately 1 week (Lovelock et al., 1972). The amount of sulfur dioxide formed from atmospheric DMS is expressed by the following equation (Ferek, Chatfield, & Andreae, 1986):

$$(CH_3)_2S + OH = SO_2 \text{ (40\% yield)} \qquad \text{(Eq. 3.6)}$$

Therefore, the amount of SO_2 produced is expressed by the equation:

$$SO_2 = DMS * 0.4 \qquad \text{(Eq. 3.7)}$$

where SO_2 = atmospheric sulfur dioxide

The quantity of SO_2, expressed as a mass, is then expressed as the density of atmospheric sulfur dioxide of DMS origins as:

$$DS = SO/VAT \qquad \text{(Eq. 3.8)}$$

where DS = density of atmospheric sulfur dioxide of DMS origins
SO = mass of sulfur dioxide of DMS origins
VAT = volume of the atmosphere (m^3)

Average density of cloud condensation nuclei (CCN) is $0.3 * 10^{-3}$; therefore, assuming all SO_2 becomes CCN, then the number of CCN per cubic meter is:

$$CCN = DS/P \qquad \text{(Eq. 3.9)}$$

where P is the average size of a CCN.

Real-world data are still lacking on many aspects of the sulfur cycle, and the variables included here are therefore generally defined. A spatial model is thus a useful way to indicate areas of high and low DMS production and to show how the values relate to other variables such as cloud cover. A small area of the eastern Pacific is used in this study to identify the variation in cloud, primary production and biomass of zooplankton in order to demonstrate the possible variations in DMS. Obviously, many more processes are involved, and this is an exercise designed to indicate how future observations may contribute to future work. The most readily available data were from maps prepared by the USSR Ministry of Defense (Gorshkov, 1979), in the form of an oceanographic atlas. As well as being easily accessible and inexpensive, the entire data set is in the same projection, at the same scale, and from the same source, so reducing input error. The major areas of interest are the impact of zooplankton grazing and the relationship of relative DMS production to cloud cover.

Phytoplankton–Zooplankton Interactions

Zooplankton graze phytoplankton, and both phytoplankton, expressed as the primary productivity, and zooplankton biomass are expressed as grams carbon/meter^2/day. Figure 3.2 shows the relative abundances of phytoplankton and zooplankton for the region according to mass of carbon/meter^2; in most regions, as expected, there is a greater biomass of phytoplankton than zooplankton. In some regions, however, the biomass of zooplankton can equal and in some cases exceed abundances of phytoplankton. It is in these areas where grazing stress upon phytoplankton is likely to be highest. Cooper, Cooper, Saltzman, and Zika (1987) suggest that DMS production is restricted to the photic zone of the oceans and that the subsurface maximum of DMS concentration is a result of grazing by zooplankton. Biomass of zooplankton can appear to exceed primary productivity, as primary productivity is limited to the euphotic zone while zooplankton graze throughout the water column.

Cloud Cover and DMS

Cloud cover is estimated from the available maps indicating the annual average number of clear days (percentage of observations in which the general cloud cover is from 0.0 to 0.3) and the annual average number of cloudy days (percentage of observations in which the general cloud cover is from 0.7 to 1.0). These were averaged to produce annual averages, as shown in Figures 3.3 and 3.4. Figure 3.5, the relative clear and cloudy days in the region, shows a gradual continuum of cloudiness from the offshore to near coastal regions. The Gulf of Mexico was omitted during operations to improve efficiency.

Figure 3.6 shows the combined regions of Figures 3.2 and 3.5 to demonstrate that relative cloudiness coincides with relative biomasses of zooplankton and phytoplankton. The number of cells for each class is included in the legend, and it is immediately apparent that the greater areas where phytoplankton exceed zooplankton are predominantly cloudy (54% of the marine area). None of the regions where zooplankton exceed

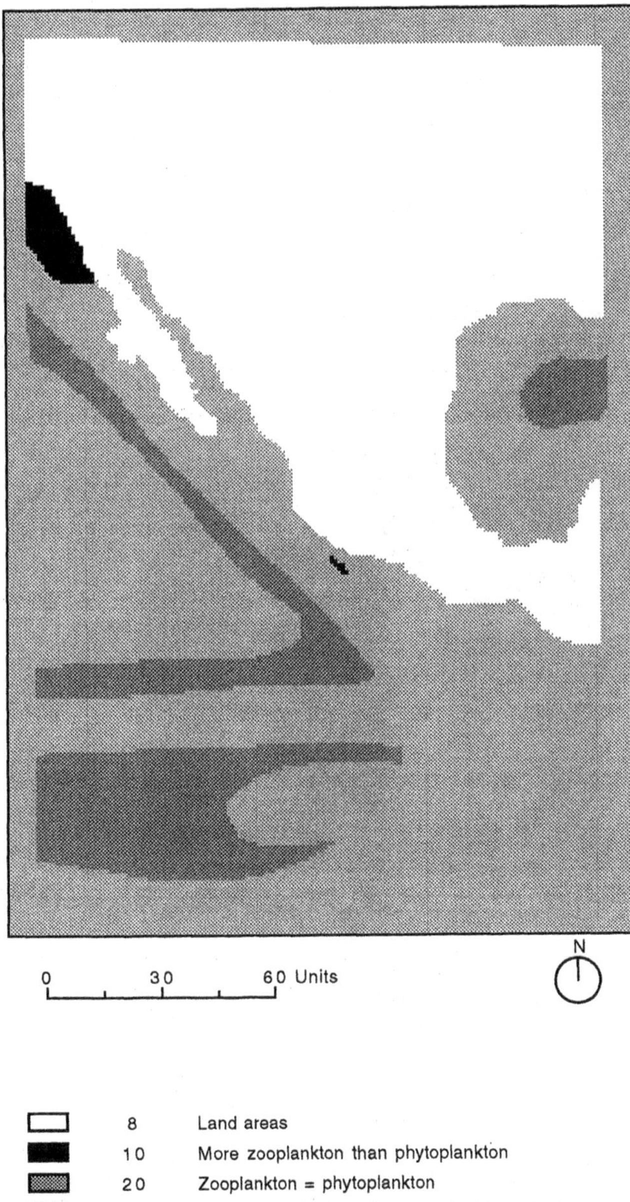

Figure 3.2. Relative phytoplankton and zooplankton.

phytoplankton are classified as cloudy, although where phytoplankton equals zooplankton biomass, the region is predominantly cloudy (17.96% of the marine region). In addition, regions that are predominantly clear also tend to occur in regions where phytoplankton exceed zooplankton.

Figure 3.3 Clerclass; percentage of days clear.

Therefore, it seems that in general, for annual averaged climatic variables and generalized measures of phytoplankton and zooplankton, cloudiness and predation stress do not seem to be related.

The eastern Pacific region has a wide variety of phytoplankton and zooplankton species, each of which will differ according to their tempera-

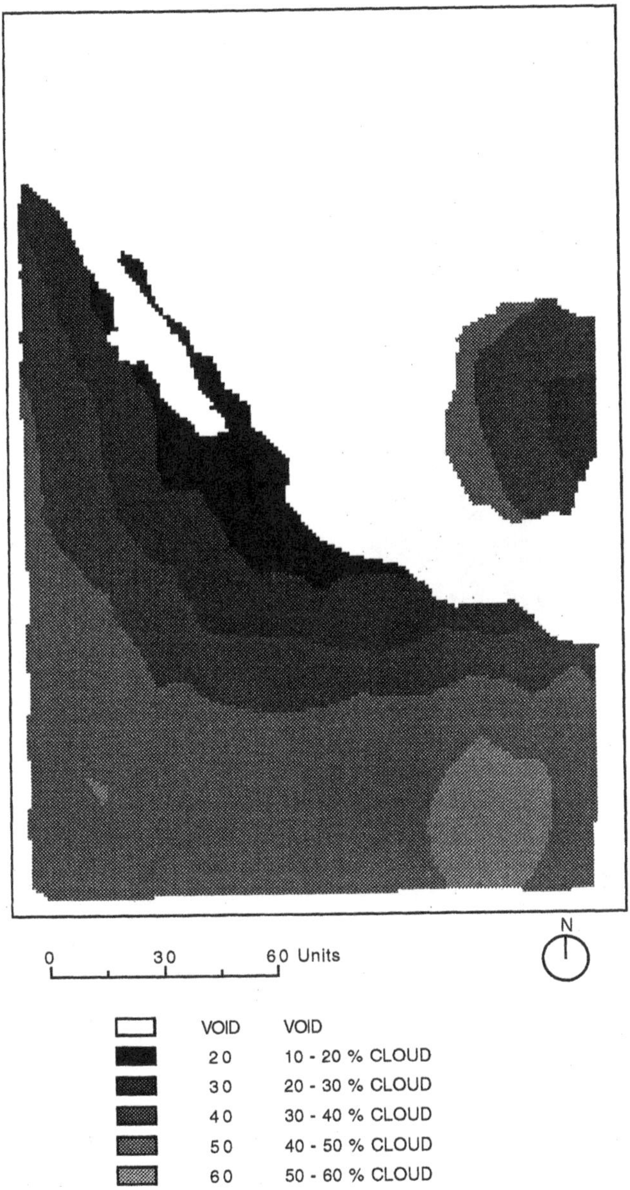

Figure 3.4. Cldclass; percentage of days cloudy.

ture, salinity, pH, and predation niches. Therefore, the amount of DMS produced will be extremely variable. Additional factors not within the auspices of this initial study, such as wind speed that would affect sea-to-air transfer, would also affect results. Therefore, further work is required using a more complex data base, including both biotic and abiotic

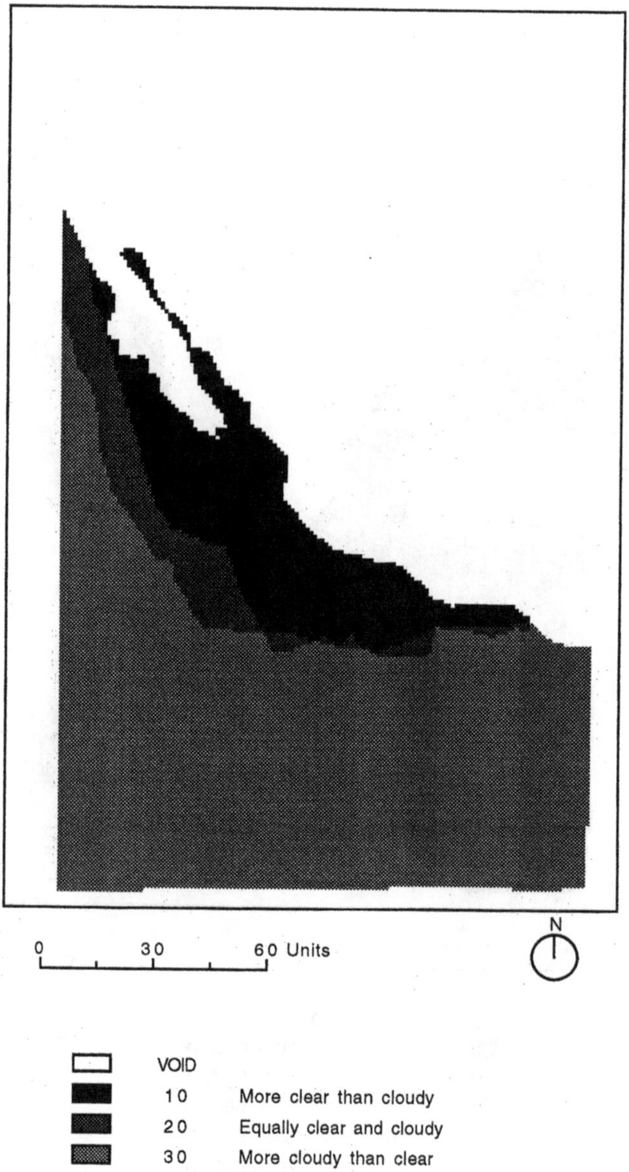

Figure 3.5. Combined clear and cloudy cases.

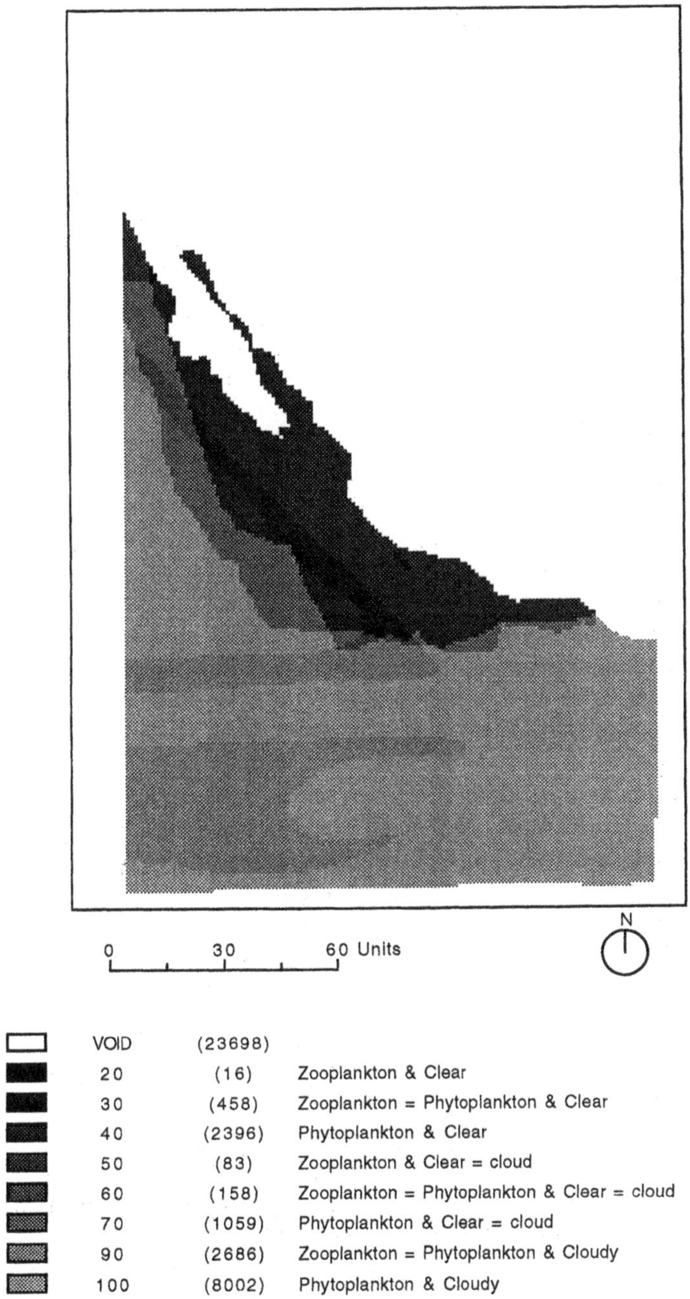

Figure 3.6. Phytoplankton/zooplankton interactions and cloudiness.

components as well as measurements of DMS in the atmosphere and oceans.

Implications for the Terrestrial Environment

Coastal zones are difficult to define and may include areas of influence from certain selected ecological factors that extend landward and seaward from the seashore. The features representing each domain must be characteristic of that domain, such as the primary production of ocean areas or the extent of rainforest in terrestrial regions. For the coastal regions, the influence of land, air, and sea features affect all ecoystems.

Sulfur deposition encompasses both dry deposition and wet deposition, the former as gaseous and particulate deposition and the latter determined by precipitation. Once emitted to the atmosphere, DMS is eventually oxidized to sulfur dioxide, most of which is removed by precipitation, although some is deposited on vegetation and available surfaces. The resolution of the data presented here is not sufficient to permit partitioning between these two processes. The predictions presented emphasize that productive marine environments produce large amounts of DMS. In turn, these enhance terrestrial sulfur deposition and impact upon climatic regimes, in particular cloud and fog physics and chemistry.

The major emphasis of this paper is to examine the role of DMS on climate; however, additional impacts that require further consideration include the influence of DMS on terrestrial surfaces and ecosystems. Sea salts are the major route by which near-coastal terrestrial temperate rainforests acquire SO_4, though marine emissions of DMS enhance atmospheric inputs of sulfur (Hedin & Hetherington, this volume). The enhancement of sulfur deposition by marine-derived DMS has also been noted in other regions; for example, Tarrason (1991) indicated that the seasonal variation of emissions in the North Atlantic modifies the ratio of anthropogenic versus natural emissions and as a result affects deposition values over Western Europe.

Conclusions

The spatial study of the eastern Pacific presented is the first known use of a Geographic Information System (GIS) for the analysis of the DMS feedback. It is likely that the problems associated with using spatial data for the analysis of large-scale environmental and climatic change will become a future feature of scientific endeavor. Spatial models are representations of reality in the same way that climatic models are, and so although they contain and use real-world data, there are a number of ways in which they may be used. Models can be used not only to predict and reconstruct but also to indicate new observation sites, flaws in model assumptions, and inequities in spatial and temporal representation. The results presented unequivocally indicate that to understand the components of the DMS feedback and the many other biogeochemical cycles, the variables involved must be represented in complex detail, and how those details piece together must be exactingly represented. Such data are limited but will form the furture work to elucidate further not only the DMS feedback but the operations of the processes involved in environmental change.

The DMS model aimed to simulate the DMS feedback using numerical modeling techniques and thereby to identify whether such a feedback is operable. The population biology functions incorporated into the model exerted a large influence, although the energy balance model (EBM) used is particularly stable. The most obvious feature of the incorporation of DMS-producing populations into the model is the significant amelioration of the surface temperature, a consequence noted by Charlson et al. (1987). This amelioration of the climate has many repercussions, including the role of DMS in the impact of the greenhouse effect (Charlson & Wigley, 1994).

References

Barnard, W.R., Andreae, M.O., and Iverson, R.L. 1984. Dimethylsulfide and *Phaeocystis pouchetti* in the southern Bering Sea. *Con Sh Res*, 3, 103–113.

Chameides, W.L., and Davis, D.D. 1982. Chemistry in the troposphere. *Chem Eng News*, 60, 39–52.

Charlson, R.J., Lovelock, J.E., Andreae, M.O., and Warren, S.G. 1987. Oceanic phytoplankton, atmospheric sulfur, cloud albedo and climate. *Nature*, 326, 655–661.

Charlson, R.J., and Wigley, T.M.L. 1994. Sulfate aerosols and climatic change. *Sci Am*, 270, 48–57.

Cooper, W.J., Cooper, D.J., Saltzman, E.S., and Zika, R.G. 1987. Distribution of dimethylsulfide in oceanic environments. *Am Chem Soc.*

Dacey, J.W.H. 1987. A role for marine herbivores in the formation of marine dimethylsufide. *Am Chem Soc.*

Ferek, R.J., Chatfield, R.B., and Andreae, M.O. 1986. Vertical distribution of dimethylsulfide in the marine atmosphere. *Nature*, 320, 514–516.

Gorshkov, S.G. 1979. World Ocean Atlas. Vol. 2: *Atlantic and Indian Oceans*. Oxford, England: Pergamon Press.

Haas, P. 1935. The liberation of methyl sulfide by seaweed. *Biochem J*, 29, 1297–1299.

Henderson-Sellers, A., and McGuffie, K. 1987. A climate modeling primer. Chichester, England: Wiley.

Iida, H., Nakamura, K., and Tokunaga, T. 1985. Dimethyl sulfide and dimethyl-beta-propiothetin in sea algae. *Bull Japan Soc Sci Fish*, 51, 1145–1150.

Liss, P.S., and Slater, P.G. 1974. Flux of gases across the air-sea interface. *Nature*, 247, 181–184.

Lovelock, J.E., Maggs, R.J., and Rasmussen, R.A. 1972. Atmospheric dimethyl sulfide and the natural sulfur cycle. *Nature*, 237, 452–453.

Rampino, M.R., and Volk, T. 1988. Mass extinctions, atmospheric sulfur and climatic warming at the K/T boundary. *Nature*, 332, 63–65.
Shaw, G.E. 1983. Biocontrolled thermostasis involving the sulfur cycle. *Clim Ch*, 5, 297–303.
Tarrason, L. 1991. *Biogenic sulfur emissions from the North Atlantic Ocean* (EMEP Report). Co-operative Programme for Monitoring and Evaluation of the Long Range Transmission of Air Pollution in Europe.
Vairavamurthy, A., Andreae, M.O., and Iverson, R.L. 1985. Biosynthesis of dimethylsufide and dimethylpropiothetin by *Hymenomonas carterae* in relation to sulfur source and salinity variations. *Limn Oceanog*, 30, 59–70.

4. Atmospheric and Geologic Constraints on the Biogeochemistry of North and South American Temperate Rainforests

Lars O. Hedin and Eugene D. Hetherington

Terrestrial forest ecosystems develop over time within the constraints of element inputs from atmospheric and geologic sources (e.g., Bormann & Likens, 1979; Gorham, Vitousek, & Reiners, 1979; Likens et al., 1977). Atmospheric and geologic inputs can be defined as "external sources"[1] of elements because they provide long-term supplies of new nutrients to internal ecosystem pools, such as vegetation, soil organic matter, and soil cation or anion exchange complexes. Such external nutrient supplies can generate constraints on the internal pool sizes and processes (e.g., rates of biomass accretion or hydrologic nutrient loss) of terrestrial ecosystems and may affect rates of recovery after disturbance events (resilience). Here we evaluate the role of external element supplies for temperate rainforests of North and South America. Our focus is on atmospheric and

[1] The question of whether geologic weathering should be considered an external or internal source of nutrients is a matter of current debate. However, since both views are equivalent from a practical point of view, the question becomes a matter of operational definition. Because our aim here is to evaluate how geological sources of elements, together with atmospheric inputs, act to constrain the development of forest ecosystems, we have, in accordance with Gorham et al. (1979), defined *weathering* as an external input. Other analyses of ecosystems, however, commonly include weathering as an internal process (e.g., Hedin & Campos, 1991; Likens et al., 1977).

geologic sources of elements, which are of direct importance in defining the biogeochemical environment within which plant communities have developed and evolved.

Atmospheric inputs of elements occur as wet, dry and cloud deposition or as active biological fixation (e.g., of N_2 or CO_2), and can supply important amounts of nitrogen (N), phosphorus (P), sulfur (S), and base cations (sum of Ca^{2+}, Mg^{2+}, Na^{2+}, and K^+) to terrestrial ecosystems over time (Gorham, 1961; Hedin & Campos, 1991; Hedin et al., 1994). Geologic inputs of elements derive from chemical weathering of primary and secondary minerals, and variations in weathering rates can exert considerable control on soil fertility and biotic processes within terrestrial ecosystems (e.g., Kronberg & Melfi, 1987). Internal sources of elements in forest ecosystems derive from inorganic nutrients that are produced during re-mineralization of soil organic matter and from elements that are stored in soil cation or anion exchange pools.

The relative contribution of internal versus external supplies of nutrients is thought to change predictably during ecosystem succession. In early primary succession, plant communities are thought to depend mainly on external nutrient inputs from atmospheric and weathering sources (Bormann & Likens, 1979; Gorham et al., 1979; Vitousek & Reiners, 1975). External nutrient supplies are also thought to be significant after disturbance events that are severe enough to cause significant loss of internal nutrient pools in soil organic matter or on exchange surfaces. Such disturbances may include landslides, deforestation, and especially hot fires. External sources consistently dominate for elements, such as Cl^-, which do not participate significantly in internal ecosystem cycles. In contrast internal supplies of nutrients generally increase as soil organic matter accumulates over successional time (Bormann & Likens, 1979; Gorham et al., 1979; Riley and Vitousek, 1995; Vitousek & Reiners, 1975). Thus, old-growth forest ecosystems are generally thought to be characterized by high internal nutrient supply rates, with external inputs playing a less important role in the supply of nutrients to vegetation and heterotrophs (Robertson & Vitousek, 1981; Tilman, 1986; Vitousek, Matson, & Van Cleve, 1989).

Despite the increasing role of internal nutrient supplies during ecosystem succession, it is not always appreciated that the long-term biogeochemical state of a forest ecosystem continues to be constrained by external element supplies from atmospheric and geologic sources. Barring any significant changes in plant nutrient-use efficiencies, long-term net accumulation of nutrients in internal ecosystem pools can occur only as a function of external inputs of new nutrients to the ecosystem. Thus, external nutrient inputs constrain the long-term accumulation of nutrients and biomass within forest ecosystems. External inputs ultimately also constrain patterns of element loss from watersheds. For example, increased inputs of N from air pollution can, over time, result in increased soil N pools, increased vegetation growth rates, and increased hydrologic N losses from temperate forest ecosystems (Hedin et al., 1994; Aber et al., 1991; Aber, Nadelhoffer, Steudler, & Melillo, 1989).

Here we use the chemistry of watershed streams to understand the importance of external sources of major elements as long-term biogeochemical constraints for temperate rainforests in North and South America. Because stream-water chemistry is also affected by internal watershed processes, such as vegetation nutrient demand or organic matter mineralization (e.g., Hedin & Campos, 1991; Likens et al., 1977), we use several independent approaches to evaluate the role of atmospheric and weathering supplies of individual elements. We focus on elements with significant atmospheric and/or geologic sources (i.e., Ca^{2+}, Mg^{2+}, Na^+, K^+, S, Cl^- and SiO_2). We have not considered N or P; a detailed evaluation of patterns of N loss from temperate rainforests in Chile can be found in Hedin, Armesto, & Johnson (1995).

Study Areas

Temperate rainforests in North and South America are characterized by high rates of precipitation (>1400 mm yr^{-1}; see definition in Alaback, 1991) and stream discharge (see Lawford, this volume), moist soils, a strong influence of marine air masses, moderate temperatures with cool summers and mild winters, and the occurrence of well-developed forest ecosystems with relatively high contents of live and dead biomass (Alaback, 1991).

For North America, we analyzed data from three watershed studies of temperate rain forest in British Columbia, Canada: Carnation Creek (CC) (49° N, 125° W; Scrivener, 1975); the University of British Columbia Research Forest (UBCRC) (49° N, 122° W; Feller, 1987); and Jamieson Creek (JC) (49° N, 123° W; Zeman & Slaymaker, 1978). These sites receive >2500 mm/yr precipitation, as shown in Table 4.1, and are located in the western hemlock biogeoclimatic zone (Krajina, 1969), with forest communities dominated by western hemlock (*Tsuga heterophylla*), western red cedar (*Thuja plicata*), Douglas fir (*Pseudotsuga menziesii*), and varying contributions of amabilis fir (*Abies amabilis*) and Sitka spruce (*Picea sitchensis*). CC and JC contain relatively undisturbed old-growth forests, while UBCRC contains a 70- to 90-year-old second-growth forest subjected to selective logging and fire. UBCRC and JC are located on predominantly acid igneous rock (quartz diorite), while CC is located on metamorphic quartz-dominated substrates, with some basalt and crystalline limestone. We used published annual volume-weighted averages for each element. We also used data from the Dennis Creek (DC) watershed study (49° N, 119° W; Hetherington, 1976) (see Table 4.1) located near the JC and UBCRC watersheds yet not strictly classified as a temperate rainforest because of lower annual precipitation amounts (ca. 700 mm/yr). DC vegetation is dominated by Engelmann spruce (*Picea*

Table 4.1. Summary of Major Ion Chemistry of Watershed Streams Used in This Study

Watershed	Elevation: (m.a.s.l.)	Precipitation Amount (mm)	Stream-water Output (mm)	Ca (mg/l)	Mg (mg/l)	Na (mg/l)	K (mg/l)	SO_4 (mg/l)	Cl (mg/l)	HCO_3 (mg/l)	SiO_2 (mg/l)	TDS (mg/l)
Carnation Creek	0–900	3020	2770	3.0	0.5	2.0	0.1	3.8	2.5	8.4	5.8	26
UBCRC	145–455	2500	1700	1.4	0.2	0.9	0.08	1.8	0.8	7.0	4.3	17
Jamieson Creek	305–1280	4540	3670	1.1	0.3	0.7	0.07	0.3	1.1	1.2	3.0	10
Dennis Creek	1500–2100	520	—	1.7	0.3	1.2	0.4	<1.0	0.7	—	9.2	15[b]
Marmot Creek	1585–2805	964	529	38.7	12.9	0.9	0.8	17.3	0.3	157	4.5	232
Tri-Creeks	1260–1707	688	—	24.3	5.0	5.0	0.5	3.1	0.5	104	9.5	152
Streeter Basin	1317–1660	545	—	54.5	21.5	9.3	2.7	13.0	0.8	280	—	381[b]
Cordillera de Piuchué, Median, 31 sites	0–800	~3000	—	0.4	0.4	2.6	0.09	0.5	4.3	<1[a]	2.3	10
Cordillera Pelada, Median, 8 sites	0–700	—	—	1.7	1.1	2.8	0.2	1.3	4.1	—	5.9	17[b]

[a] Based on incomplete number of analyses.
[b] Value is conservative, since data are missing for at least one major chemical species.

engelmanii) and subalpine fir (*Abies lasiocarpa*) on igneous quartz diorite geology.

As a comparison to these coastal watersheds, we included data from inland sites along a transect from British Columbia to the western part of Alberta. Marmot Creek (MC) (51° N, 115° W; Water Survey of Canada, 1979), Tri-Creeks (TC) (53° N, 117° W; Jablonski, 1980), and Streeter Basin (SB) (50° N, 114° W; Singh & Kalra, 1976) are located on sandstones and shales. Annual precipitation is relatively low (ca. 550–950 mm/yr), and vegetation ranges from coniferous forests at MC (Engelmann spruce and alpine fir) and TC (mainly lodgepole pine, *Pinus contorta*) to mixed aspen (*Populus tremuloides*) and grassland at Streeter Creek. Data from MC, TC, and SB are based on published growing-season averages.

Information on element losses from South American temperate rainforests are scarce to nonexistent. Our own data derive from 31 watersheds at Cordillera de Piuchué (CP) (42°22′ S, 74°03′ W) and 8 watersheds at Cordillera Pelada (CPE) (40°10′ S, 73°35′ W) collected during the summers of 1989, 1992, and 1993 (Hedin & Campos, 1991; Hedin et al., 1995). In both areas, watersheds were located along west–east transects from the ocean to approximately 20 km inland and from between 0 m to approximately 900 m above sea level.

Procedures for sampling, preservation, and analysis are given in Hedin et al., 1995. The two study areas have similar floristic composition, with high-elevation conifer forests dominated by Alerce (*Fitzroya cuppressoides*), mid-elevation mixed forests dominated by *Nothofagus nitida* and *Drimys winteri*, and low-elevation coastal forests generally dominated by Olivillo (*Aextoxicon punctatum*) and *Luma apiculata* (Armesto & Fuentes, 1988; Hedin et al., 1995; Villagrán, 1985). CP watersheds are underlaid by strongly weathered metamorphic feldspar schists and have not been glaciated (Holdgate, 1961; Watters & Fleming, 1972). Soils are moist with infrequent gley horizon at approximately 40 cm depth. While CPE is of similar geological origin as CP, less is known about the mineralogy, soils, extent of glaciation, and disturbance history at CPE.

Element Contributions from Atmospheric Versus Geologic Sources

We use three different approaches to evaluate the relative importance of atmospheric versus geologic sources of elements to watersheds, each subject to different conceptual advantages and limitations: 1) the so-called Gibbs diagram; 2) correlation analysis of Na^+ versus Cl^-; and 3) sea-salt correction of stream-water chemistries.

Gibbs (1970) suggested that the relative importance of major sources of elements to natural waters can be evaluated by graphing the weight ratio of $Na^+:(Na^+ + Ca^{2+})$ against total dissolved solids (TDS). Gibbs used

Na^+ as a tracer of atmospheric sources and Ca^{2+} as a tracer of weathering sources. Specifically, Gibbs (1970, 1971; modified in Gibbs, 1992) suggested that the world's major waters would fall within or near a general area (as shown in the dashed envelope in Figure 4.1), constrained by three end-member conditions: 1) dominance by atmospheric precipitation inputs (high $Na^+:[Na^+ + Ca^{2+}]$ ratio and low TDS); 2) dominance by rock weathering (low $Na^+:[Na^+ + Ca^{2+}]$ ratio and intermediate TDS); or 3) dominance by evaporation–precipitation reactions (high $Na^+:[Na^+ + Ca^{2+}]$ ratio and high TDS). Recent criticisms (Eilers, Brakke, & Henriksen, 1992) of this model have questioned whether the original "boomerang-shaped" envelope drawn by Gibbs (1970, 1971) appropriately

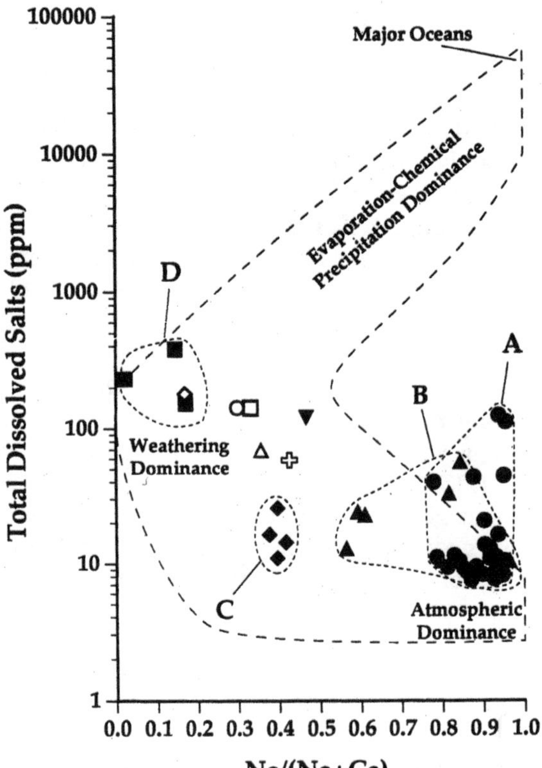

Figure 4.1. Total dissolved salts (calculated as sum of individual ions) as a function of the weight ratio $Na^+:(Na^+ + Ca^{2+})$ in watershed streams draining coastal temperate forests at Cordillera de Piuchué, Chile (●); Cordillera Pelada, Chile (▲); and British Columbia, Canada (◆). ■ = Streams draining inland watersheds in Alberta, Canada. Additional symbols represent average continental river water in North America (○), South America (△), Europe (◇), Asia (□), Africa (▼), and Australia (+), from Livingstone (1963).

describes natural waters of the world and whether Na^+ in dilute waters is, in fact, an exclusive tracer of atmospheric inputs. We have here enlarged this envelope based on Eilers et al. (1992) and based on Gibbs' own revisions (Gibbs, 1992) (Figure 4.1). The assumption that Na^+ reflects atmospheric inputs in dilute waters is generally valid for our analyses, since we focus on near-coastal watersheds with strong inputs of marine aerosols with high Na^+ and Cl^- (Galloway, Likens, Keene, & Moody, 1982).

It is relatively rare to find waters that, according to the Gibbs diagram referred to in our Figure 4.1, are nearly exclusively dominated by atmospheric sources of elements (high $Na^+:[Na^+ + Ca^{2+}]$ and low TDS) (Gibbs, 1970; Kilham, 1990). Such waters are most notably found draining strongly weathered lateritic terranes such as in tropical Amazonia, where soils are depleted of easily weathered Ca-rich minerals (Gibbs, 1970; Kronberg & Melfi, 1987).

When plotted on the Gibbs diagram, the watersheds considered in this study separate into three well-defined groups, as shown in Figure 4.1. Values from inland North American watersheds (Figure 4.1, group D) fall in the area of weathering dominance, with high Ca^{2+} relative to Na^+ and TDS values of approximately 100 to 400 $mg\,l^{-1}$. Values from coastal North American temperate rainforests (group C in Figure 4.1) fall in between weathering and precipitation dominance, with significant Ca^{2+} relative to Na^+ and low levels of TDS (10–40 $mg\,l^{-1}$). In contrast, values from South American temperate rainforests (groups A and B in Figure 4.1) fall in the bottom right-hand area of the graph, indicating a strong dominance of atmospheric aerosol inputs relative to weathering inputs. CP watersheds show particularly high values of $Na^+:(Na^+ + Ca^{2+})$, with TDS ranging from 7.5 to 125 $mg\,l^{-1}$. CP watersheds with high TDS (>100 $mg\,l^{-1}$) were of the coastal forest type and located immediately near the ocean, thus receiving direct sea spray.

Contributions from Sea-Salt Aerosols

We further evaluated the role of atmospheric versus geologic inputs of elements to forest ecosystems by comparing abundances of Na^+ versus Cl^- in watershed streams. Atmospheric inputs (wet and dry deposition) from unpolluted marine air masses are dominated by sea-salt aerosols, in which Na^+ and Cl^- are major elements and occur in a ratio approximating 0.56 by weight (Galloway et al., 1982; Keene, Pszenny, Galloway, & Hawley, 1986; Warneck, 1988). Except for areas with halite deposits, natural nonmarine sources of Cl^- are generally low and associated with the weathering of sedimentary rocks (Berner & Berner, 1987). We thus would expect a relationship between Na^+ and Cl^-, near the ideal sea-salt ratio, in streams that drain igneous or metamorphic terranes and where atmospheric inputs are dominated by sea-salt aerosols. Stream-water $Na^+:Cl^-$ ratios that are significantly elevated relative to sea salt would mainly reflect weathering inputs of Na^+.

We found a strong overall relationship between Na^+ and Cl^- for most watershed streams considered in this study, as shown in Figure 4.2. Two of the inland North American watersheds showed elevated Na^+ relative to Cl^- (see the triangles in Figure 4.2), indicating contributions from weathering to watershed outputs of Na^+. When only coastal watersheds were considered, we found an exceptionally strong correlation ($r^2 = 0.99$; $p < 0.001$) between Na^+ and Cl^- with a slope (0.58; $n = 43$; s.e. = 0.007) nearly identical to the ideal sea-salt ratio of 0.56.

These results indicate a strong link between hydrologic Na^+ and Cl^- losses from coastal temperate rainforest ecosystems and the atmospheric input of marine aerosols. To further explore the biogeochemical constraints caused by this apparent link to atmospheric inputs, we subtracted the contribution of sea-salt aerosols from the observed outputs of elements from watersheds, as shown in Figure 4.3. The sea-salt corrected stream-water chemistry was calculated by using Cl^- as a conservative tracer of sea-salt aerosol inputs with the following element-to-Cl weight ratios: $Mg^{2+} = 0.067$; $SO_4^{2-} = 0.14$; $Ca^{2+} = 0.021$; $K^+ = 0.021$; and $Na^+ = 0.56$ (Keene et al., 1986).

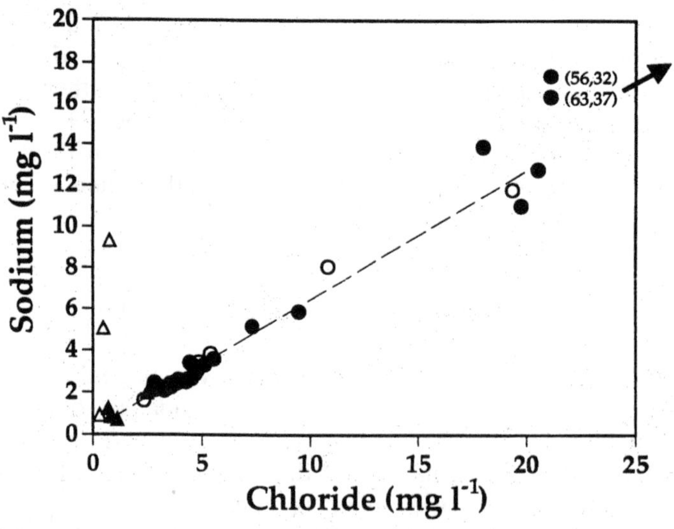

Figure 4.2. Concentrations of Na^+ as a function of Cl^- in watershed streams that drain coastal temperate forests at Cordillera de Piuchué, Chile (●); Cordillera Pelada, Chile (○); and British Columbia, Canada (▲). △ = Streams draining inland watersheds in Alberta, Canada. Line indicates statistically significant regression at $p < 0.05$. Values from two coastal watersheds (indicated by the arrow) were excluded from the correlation.

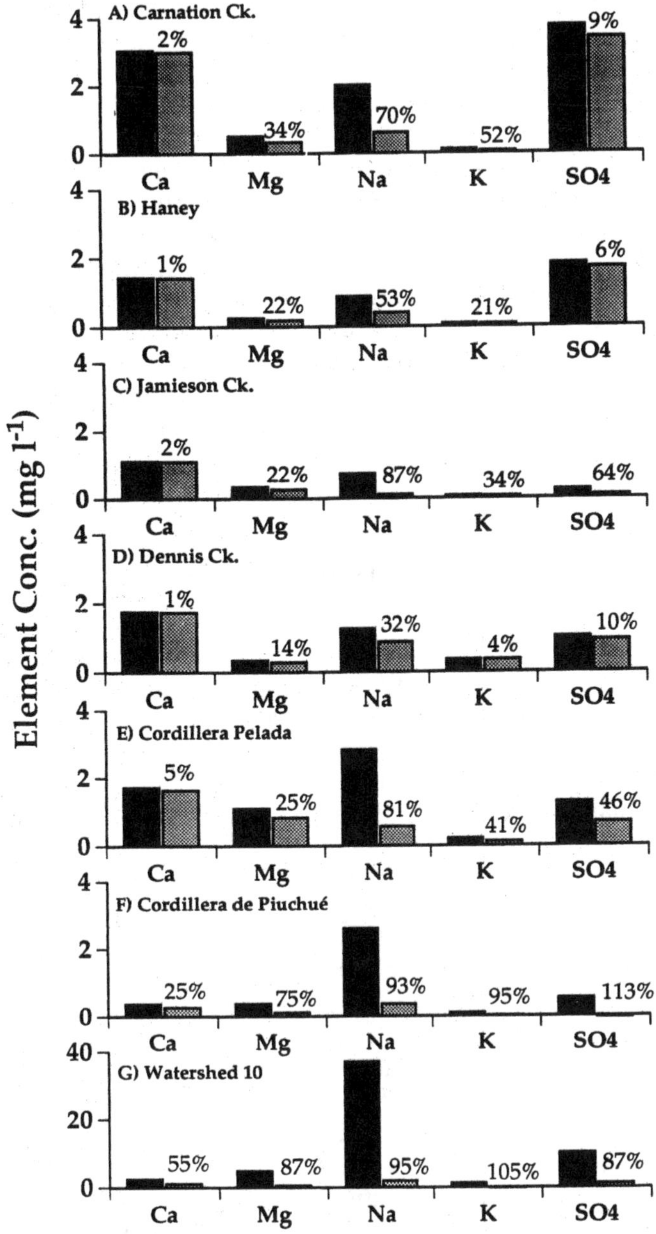

Figure 4.3. Concentrations of major elements in watershed streams draining coastal temperate rainforests in North and South America, before (■) and after (▨) correction for the contribution of sea salt. (See text for methods.) Values on top of bars indicate percent decline of element concentrations from sea-salt correction.

Corrections for sea-salt influence had a relatively small (<32%) effect on concentrations of divalent base cations (Ca^{2+} and Mg^{2+}) in North American temperate rainforests as well as at CPE (Sections A–D in Figure 4.3). Monovalent base cations (Na^+ and K^+) showed greater, yet varying, reductions for the same locations (Na^+:32–81%; K^+: 4–52%). In contrast, sea-salt corrections resulted in strong reductions of both mono- and divalent base cations for South American temperate rainforests at CP (section F in Figure 4.3). Concentrations of Mg^{2+}, Na^+, and K^+ declined by >71% from these corrections, while the already low levels of Ca^{2+} declined by approximately 25% (refer again to section F in Figure 4.3). Sea-salt corrections had varying effects on SO_4^{2-} concentrations, with reductions ranging from 6% to 64% for North American temperate rainforests. In contrast, the South American CP watersheds showed close to 100% reduction in SO_4^{2-} from the sea-salt aerosol correction.

Contributions from Weathering

We evaluated the role of weathering as a source of base cations in coastal North and South American temperate rainforests by examining correlations between residual (after sea-salt correction) concentrations of base cations versus dissolved silica (expressed as SiO_2). If silicate weathering were the major source of residual base cations, we would expect residual cation concentrations to correlate positively with variations in SiO_2.

When data from all coastal watersheds were considered together, residual concentrations of Ca^{2+}, Mg^{2+}, and Na^+ were, in fact, positively correlated with concentrations of SiO_2 (see Figure 4.4). ($Ca^{2+}: r^2 = 0.64$, $p < 0.001$; Mg^{2+}: $r^2 = 0.59$, $p < 0.001$; Na^+: $r^2 = 0.22$; $p < 0.002$; all are log-log regressions.) The slopes of these relationships were consistently less than 1.0 ($Ca^{2+} = 0.5$; $Mg^{2+} = 0.5$; and $Na^+ = 0.4$), indicating that on a linear scale, the non-sea-salt supply of base cations did not increase proportionately as a function of SiO_2. We do not present results for K^+ in Figure 4.4, since some residual K^+ levels were slightly negative and therefore could not be log-transformed. However, evaluation of non-log-transformed data showed a positive and highly significant ($p < 0.001$; $r^2 = 0.30$) relationship between residual K^+ and SiO_2. It is particularly interesting that watersheds with low levels of residual cations were exclusively associated with very low SiO_2 (ca. $0.9–3.0\,mg\,l^{-1}$) (refer to Figure 4.4). Such levels of SiO_2 are low even when compared to waters that drain lateritic terranes of central Amazonia ($3–12\,mg\,l^{-1}$; Kronberg & Melfi, 1987).

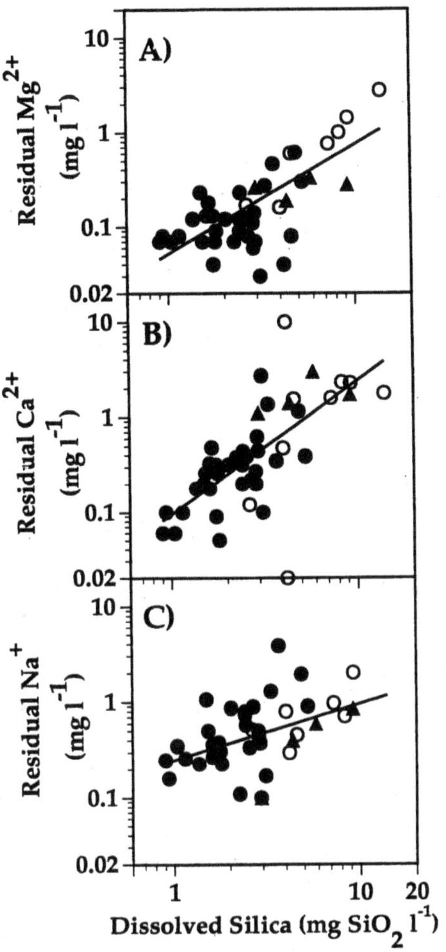

Figure 4.4. Relationships between residual (after sea-salt correction) concentrations of Mg^{2+} (A), Ca^{2+} (B), and Na^{+} (C) versus dissolved SiO_2 in watershed streams draining coastal temperate forests at Cordillera de Piuchué, Chile (●); Cordillera Pelada, Chile (○); and British Columbia, Canada (▲). Lines indicate statistically significant regressions. One extreme value was excluded from both regressions A and B.

Conclusions

Our analyses focus on understanding broad-scale patterns in external inputs of elements to coastal temperate rainforests (Figure 4.5). Our results show important differences, as well as similarities, between forest ecosystems of North and South America. Specifically, our results indicate

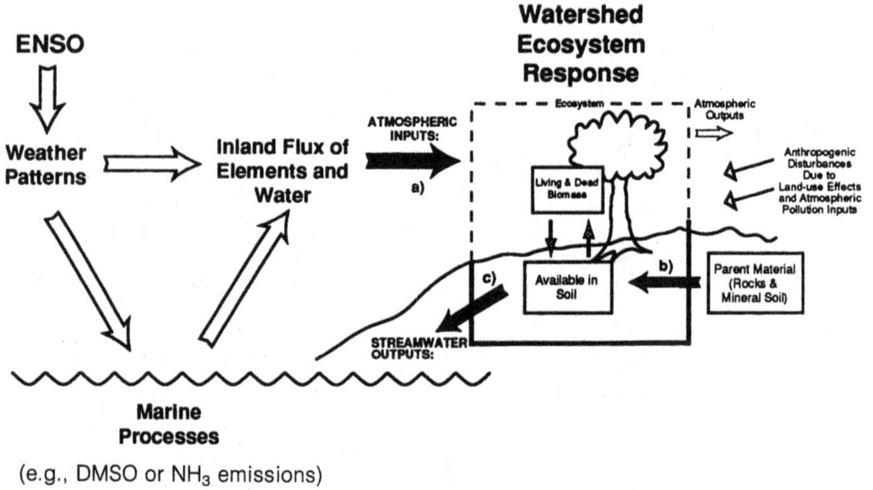

Figure 4.5. Conceptual model of external biogeochemical constraints on temperate rainforest ecosystems in North and South America. External constraints are imposed by atmospheric inputs of marine origin and by generally low rates of weathering supply of elements. Black arrows indicate external sources to the forest ecosystem (atmospheric and geologic inputs) and hydrologic losses in watershed streams. Our analyses have focused on evaluating the link between streamwater outputs (c) and external inputs of elements from atmospheric (a) and geologic (b) sources. Internal element cycles are of great importance for these forests (Hedin et al., 1995). Inputs of elements and water from the atmosphere are, in turn, influenced by factors such as weather patterns, marine emissions of elements (biogenic or as sea salt), and long-term variations in weather and transport patterns associated with El Niño-Southern Oscillation (ENSO) events. Anthropogenic disturbances due to land-use effects or air pollution may modify the natural relationships between external constraints and watershed ecosystem response.

a strong difference in the role of geologic versus atmospheric inputs between coastal temperate rainforests in North versus South America (group C versus groups A and B in Figure 4.1). We propose that this difference reflects differences in glacial history between the areas. The South American watersheds at CP have not been glaciated and contain soils with highly weathered quartz and feldspar schist parent material (Ruthsatz & Villagrán, 1991; Watters & Fleming, 1972). In contrast, soils of the North American temperate rainforests considered here have developed after the exposure to glacial activity during the Pleistocene. Because of the physical disturbance associated with glaciation, recently glaciated soils can show enhanced rates of weathering, particularly for divalent base cations (Kirkwood & Nesbett, 1991; Stauffer, 1990). While CPE and CP belong to the same extension of the Coastal Range Cordillera schist in Chile, the exact glacial history of CPE is less well known.

However, outcrops of soft, strongly weathered, saprolite parent material indicate a lack of glaciation also at CPE.

Inland North American watersheds showed evidence of stronger contributions from weathering processes, with relatively high TDS in conjunction with low $Na^+:(Na^+ + Ca^{2+})$ ratios (see group D in Figure 4.1). We interpret this to reflect two factors: 1) a diminished influence of marine aerosol inputs due to the inland location of these watersheds; and 2) a stronger contribution of geologic sources from the more easily weathered parent materials of sandstone and shale. In fact, except for Marmot Creek, these watersheds showed $Na^+:Cl^-$ ratios that were substantially increased relative to sea-salt expectations (Figure 4.2), indicating a significant weathering source for Na^+.

Sea-salt corrections also differed between CP forests and forests of CPE or North America. Levels of divalent cations (Ca^{2+} and Mg^{2+}) were particularly low in CP streams (Figure 4.3). Except for Ca^{2+}, major cations and anions could be almost exclusively (75%–100%) explained by sea-salt aerosol inputs (sections F and G in Figure 4.3; Hedin et al., 1995). Thus, atmospheric sources appear to dominate over geologic sources for these elements at CP. In contrast, while levels of Ca^{2+} were exceedingly low (median ca. $0.4\,mg\,l^{-1}$) at CP as compared to other areas, only approximately 25% of the observed concentrations could be explained by atmospheric inputs. When compared to other cations the low contribution of atmospheric sources to Ca^{2+} in CP watersheds correlates with the observation that Ca^{2+} is the least abundant base cation in sea-salt aerosol (Keene et al., 1986). Thus, even very low geologic supplies of Ca^{2+} can be significant relative to atmospheric supplies at CP. In fact, soils in unglaciated silicated areas tend to contribute little Ca^{2+}, because Ca-rich feldspars are unstable and weather relatively rapidly on a geological time scale (Berner & Berner, 1987; Stauffer, 1990). The overall lack of weathering sources for most elements at CP is consistent with old soils, lack of glaciation, and highly weathered parent materials.

In contrast, the relatively recently glaciated North American watersheds showed markedly higher Ca^{2+} and Mg^{2+} in stream waters, and corrections for sea-salt aerosols were not as effective as for CP. For example, only 1% to 2% of Ca^{2+}, 14% to 34% of Mg^{2+}, and 4% to 52% of K^+ could be explained by atmospheric inputs of sea-salt aerosols in North American forests (sections A–D, Figure 4.3). Similarly, low fractions of Ca^{2+}, Mg^{2+}, and K^+ could be explained by sea-salt aerosol at CPE (section E, Figure 4.3). Yet, ratios of $Na^+:Cl^-$ in North American coastal forests and at CPE did not differ substantially from expectations based on sea-salt aerosol (Figure 4.2). In contrast to CP, our results from temperate rainforests in North America and at CPE thus indicate that weathering can be a particularly important source of divalent base cations (refer to sections A–D in Figure 4.4 and A and B in Figure 4.5). As for the divalent base cations, sea-salt corrections had limited but varying (6%–64%

reduction) effects on SO_4^{2-} in the North American coastal forests (Figure 4.3).

It should be noted that residual element concentrations after sea-salt correction may not exclusively be due to geologic inputs, but may reflect atmospheric sources other than sea-salt aerosols or element losses associated with internal ecosystem processes (e.g., net biomass depreciation; Bormann & Likens, 1979; Vitousek & Reiners, 1975), or element loss from soil exchange sites. For example, precipitation chemistry in southwestern British Columbia and the northwestern United States is subject to atmospheric pollution that causes increased levels of sulfur and nitrogen in wet and dry deposition (Feller, 1987; Nikleva, 1986). The limited effect of sea-salt corrections on SO_4^{2-} levels in North American watersheds (Figure 4.3) could be due in part to such air-pollution inputs (Feller, 1987). Weathering of sulfur-rich minerals could also supply SO_4^{2-} to watershed streams. While recent studies have indicated a link between human activities and the atmospheric deposition of base cations (Hedin et al., 1994), this effect is likely limited to relatively polluted and arid, non-coastal areas.

For base cations, the strong correlations between residual (non-sea-salt) Mg^{2+}, Ca^{2+}, and K^+ versus SiO_2 (see A and B in Figure 4.5) indicate that silicate weathering processes contribute to variations in supplies of base cation to the coastal rainforests considered here. Feller and Kimmins (1979) found that while watershed budgets showed strong net losses for most elements from a coastal rainforest in British Columbia (UBCRC), losses were especially strong for SiO_2, Ca^{2+}, and Mg^{2+}. Feller and Kimmins interpreted this to indicate a particularly strong supply of these elements from weathering. It is, however, difficult to evaluate Feller and Kimmins' budgets, since atmospheric inputs from cloud water and dry deposition, while potentially significant in such forests (e.g., Borman, Tarrant, McClellan, & Savage, 1989), were not considered. We conclude that external supplies of major elements to coastal rainforests in North America and CPE occur primarily from weathering (Ca^{2+}, Mg^{2+}, and SO_4^{2-}) or as a balance between weathering and atmospheric inputs (Na^+, Cl^-, and K^+). Thus, while marine air masses provide important inputs of several elements, these forests do not appear to be linked as exclusively to marine processes as CP watersheds (Figure 4.5).

General Model

Our results indicate that coastal temperate rainforests in North and South America have developed over time within the biogeochemical constraints of inputs of dilute atmospheric sea-salt aerosols coupled with low rates of weathering. Atmospheric inputs of sea salts constitute the major external biogeochemical constraint on element supplies to CP forests. Inputs of Ca^{2+} are particularly low to these forests due to limited Ca^{2+} in sea-salt aerosol, and the lack of major geologic sources of Ca^{2+} (sections E and F, Figure 4.3). External element supplies to North American temperate rainforests, on the other hand, occur as a balance between atmospheric and geologic sources (Figure 4.3, sections A–D). Inputs of divalent base cations (Ca^{2+} and Mg^{2+}) are dominated by geologic sources, while monovalent base cations are from atmospheric inputs (Na^{+}) or a balance of geologic and atmospheric sources (K^{+}).

Our results provide a view of temperate rainforest ecosystems as relatively poor in many major elements or nutrients. Concentration losses of major elements were consistently dilute in the watershed streams considered here (Figure 4.3). The exact patterns of element loss were, to an important degree, imposed from variations in external atmospheric inputs of elements (refer to Figures 4.2 and 4.3). This overall importance of marine aerosol inputs in defining the biogeochemistry of these forests results from the consistent influence of marine wind patterns in the biogeographical regions of both North and South American temperate rainforests (ca. 40°–50° N and S; see Lawford, this volume). Temperate rainforest ecosystems are therefore naturally closely linked to biogeochemical and meteorological processes that occur within upwind marine ecosystems (Figure 4.5; Benbow, this volume).

Forests at CP are unique in that external supplies of major elements were almost exclusively due to atmospheric sea-salt inputs. This suggests that CP forest communities have developed over time under evolutionary and biogeochemical constraints from dilute sea-salt aerosols. The link between marine processes (including biogeochemical, meteorological, and climatic processes) and the forest ecosystem (Figure 4.5) is thus particularly intimate for CP forests. In Figure 4.5 we outline major factors that affect this link between marine processes and coastal ecosystem response. For example, marine emissions of dimethyl sulfide (DMS) or NH_3 to the atmosphere, from algal productivity and decay (Andreae & Raemdonck, 1983; Benbow, this volume; Quinn, Charlson, & Bates, 1988), will enhance atmospheric inputs of sulfur and nitrogen to coastal watersheds (Figure 4.5). More synoptic influences on coastal temperate forests can be caused by short- or long-term variations in weather patterns, including long-term oscillations associated with the El Niño–Southern Oscillation (refer to Figure 4.5). Temperate rainforests at CP show a potentially strong "downwind" biogeochemical, as well as climatic, link to nearby marine ecosystems.

Because of limited weathering and generally nutrient-poor nature, temperate rainforest ecosystems may be especially sensitive to biogeochemical changes from human activities. For example, air-pollution inputs of strong mineral acids are not likely to be strongly buffered by weathering or base cation exchange reactions.

Acknowledgments. This is a contribution to the program at the W.K. Kellogg Biological Station. Financial support was provided by the A.W. Mellon Foundation. We thank Juan Armesto, Paul Alaback, Michael Velbel, and Michael Brown for comments and assistance.

References

Aber, J.D., Melillo, J.M., Nadelhoffer, K.J., Pastor, J., and Boone, R.D. 1991. Factors controlling nitrogen cycle and nitrogen saturation in northern temperate forest ecosystems. *Ecol Appl*, 1, 303–315.

Aber, J.D., Nadelhoffer, K.J., Steudler, P., and Melillo, J.M. 1989. Nitrogen saturation in northern forest ecosystems. *Bio Science*, 39, 378–386.

Alaback, P. 1991. Comparative ecology of temperate rainforests of the Americas along analogous climatic gradients. *Rev Chil Hist Nat*, 64, 399–412.

Andreae, M.O., and Raemdonck, H. 1983. Dimethyl sulfide in the surface ocean and the marine atmosphere: A global view. *Science*, 221, 744–747.

Armesto, J.J., and Fuentes, E.R. 1988. Tree species regeneration in a mid-elevation, temperate rain forest in Isla de Chiloé, Chile. *Vegetatio*, 74, 151–160.

Berner, K.E., and Berner, R.A. 1987. *The Global Water Cycle: Geochemistry and Environment*. Englewood Cliffs, NJ: Prentice-Hall.

Bormann, F.H., and Likens, G.E. 1979. *Pattern and Process in a Forested Ecosystem*. New York: Springer-Verlag.

Bormann, B.T., Tarrant, R.F., McClellan, M.H., and Savage, T. 1989. Chemistry of rainwater and cloud water at remote sites in Alaska and Oregon. *J Env Qual*, 18, 149–152.

Eilers, J.M., Brakke, D.F., and Henriksen, A. 1992. The inapplicability of the Gibbs model of world water chemistry for dilute lakes. *Limn Oceanog*, 37, 1335–1337.

Feller, M.C. 1987. The influence of acid precipitation on stream chemistry in a small forested basin in southwestern British Columbia. In *Forest Hydrology and Watershed Management*. Wallingford, Oxfordshire: IAHS Press.

Feller, M.C., and Kimmins, J.P. 1979. Chemical characteristics of small streams near Haney in Southwestern British Columbia. *Wat Res Res*, 15, 247–258.

Galloway, J.N., Likens, G.E., Keene, W.C., and Moody J.M., 1982. The composition of precipitation in remote areas of the world. *J Geophys Res*, 87, 8711–8786.

Gibbs, R.J. 1970. Mechanisms controlling world water chemistry. *Science*, 170, 1088–1090.

Gibbs, R.J. 1971. Mechanisms controlling world water chemistry: Evaporation-crystallization process [reply to J.H. Feth]. *Science*, 172, 871–872.

Gibbs, R.J. 1992. A reply to comments of Eilers et al. *Limn Oceanog*, 37, 1338–1339.

Gorham, E. 1961. Factors influencing supply of major ions to inland waters, with special reference to the atmosphere. *Geol Soc Am Bull*, 72, 795–840.

Gorham, E., Vitousek, P.M., and Reiners, W.A. 1979. The regulation of chemical budgets over the course of terrestrial ecosystem sucession. *Ann Rev Ecol Sys*, 10, 53–84.

Hedin, L.O., Armesto, J.J., and Johnson, A.H. 1995. Patterns of nutrient loss from unpolluted, old-growth temperate forests: Evaluation of biogeochemical theory. *Ecology*, 76, 493–509.

Hedin, L.O., and Campos, H. 1991. Importance of small streams in understanding and comparing watershed ecosystem processes. *Rev Chil Hist Nat*, 64, 583–596.

Hedin, L.O., Granat, L., Likens, G., Buishand, T.A., Galloway, J.N., Butler, T.J., and Rodhe, H. 1994. Steep declines in atmospheric base cations in regions of Europe and North Amerîca. *Nature*, 367, 351–354.
Hetherington, E.D. 1976. *Dennis Creek: A Look at Water Quality Following Logging in the Okanagan Basin.* Canadian Forestry Service, Report No. BC-X-147.
Holdgate, M.W. 1961. Vegetation and soils in the south Chilean islands. *J Ecol*, 49, 559–580.
Jablonski, P.D. 1980. *Pretreatment Water Quality of the Tri-Creeks Experimental Watershed.* Alberta Energy and Natural Resources Forest Land Use Branch Watershed Management, Report No. 7.
Keene, W.C., Pszenny, A.A.P., Galloway, J.N., and Hawley, M.E. 1986. Sea-salt corrections and interpretation of constituent ratios in marine precipitation. *J Geophys Res*, 91, 6647–6658.
Kilham, P. 1990. Mechanisms controlling the chemical composition of lakes and rivers: Data from Africa. *Limn Oceanog.*, 3, 80–83.
Kirkwood, D.E., and Nesbitt, H.W. 1991. Formation and evolution of soils from an acidified watershed: Plastic Lake, Ontario, Canada, Geochimica. *Cosmochimica Acta*, 55, 1295–1308.
Krajina, V.J. (ed.). 1969. *Ecology of Western North America*, Vol. 2. Vancover: Department of Botany, University of British Columbia.
Kronberg, B.I., and Melfi, A.J. 1987. The geochemical evolution of lateritic terranes. *Zeits Geomorph*, Supplement band, 64, 25–32.
Likens, G.E., Bormann, F.H., Pierce, R.S., Eaton, J.S., and Johnson, N.M. 1977. *Biogeochemistry of a Forested Ecosystem.* New York: Springer-Verlag.
Livingston, D.A. 1963. *Chemical composition of Rivers and Lakes.* U.S. Geological Survey, Professional Paper 440G.
Nikleva, S. 1986. Acid rain over southeastern British Columbia. In *Acid Precipitation in the Pacific Northwest.* Proc. Victoria, B.C., Vancouver, B.C., Canda.
Quinn, P.K., Charlson, R.J., and Bates, T.S. 1988. Simultaneous observations of ammonia in the atmosphere and ocean. *Nature*, 335, 336–338.
Riley, R.H., and Vitousek, P.M. 1995. Nutrient dynamics and nitrogen trace gas flux during ecosystem development in montane rain forest. *Ecology*, 76, 292–304.
Robertson, G.P., and Vitousek, P.M. 1981. Nitrification potentials in primary and secondary succession. *Ecology*, 62, 376–386.
Ruthsatz, B., and Villagran, C. 1991. Moorland vegetation and soil nutrients in Chiloé island. Revista Chilena de Historia, *Natural*. 64, 461–478.
Scrivener, J.C. 1975. *Water, Water Chemistry and Hydrochemical Balance of Dissolved Ions in Carnation Creek Watershed, Vancouver Island, July 1971–May 1974.* Canadian Fisheries and Marine Service, Technical Report No. 564.
Singh, T., and Kalra, Y.P. 1976. *Water Quality of a Range Watershed in Southwestern Alberta Prior to Aspen Clearing.* Canadian Forestry Service, Report No. NOR-X-168.
Stauffer, R.E. 1990. Granite weathering and the sensistivity of alpine lakes to acid deposition. *Limn Oceanog*, 35, 1112–1134.
Tilman, D. 1986. Nitrogen-limited growth in plants from different successional stages. *Ecology*, 67, 555–563.
Villagrân, C. 1985. Analisis palinologico de los cambios vegetacionales durante el Tardiglacialy Postglacial en Chiloé, Chile. *Rev Chil Hist Nat*, 58, 57–69.
Vitousek, P.M., Matson, P.A., and Van Cleve, K. 1989. Nitrogen availability and nitrification during sucession: Primary, secondary and old-field seres. *Plant Soil*, 115, 229–239.

Vitousek, P.M., and Reiners, W.A. 1975. Ecosystem succession and nutrient retention: A hypothesis. *BioScience*, 25, 376–381.

Warneck, P. 1988. *Chemistry of the Natural Atmosphere.* San Diego: Academic Press.

Water Survey of Canada. 1979. *Water Quality for the Marmot Creek Research Basin.* Calgary, Alberta.

Watters, W.A., and Fleming, C.A. 1972. Contributions to the geology and palaeontology of Chile Island, Southern Chile. *Philosophical Transactions of the Royal Society London*, 263(B), 370–408.

Zeman, L.J., and Slaymaker, O. 1978. Mass balance model for calculation of ionic input loads in atmospheric fallout and discharge from a mountainous basin. *Hydrol Sci*, 23, 103–116.

5. Past Changes in Climate and Tree Growth in the Western Americas

Rosanne D. D'Arrigo and Gordon C. Jacoby

The AMIGO project has been termed "An interhemispheric and interdisciplinary research program to measure and understand ecosystem responses to global change" (Committee on AMIGO). Its goals include the development of comparative studies of ecosystem response in western North and South America, regions that have many parallel geobiospheric features (Fuentes, Kronberg, & Mooney, 1991). It is hoped that these interhemispheric comparisons will help distinguish natural from anthropogenic change and aid in policy making and monitoring of natural systems. Evaluating tree rings are one means by which past climatic change and forest ecosystem response can be compared between the Northern Hemisphere and the Southern Hemisphere.

Dendroclimatic techniques have been employed in numerous studies to reconstruct or extend the instrumental record of climate, based on old-aged trees from selected climatically stressed sites, and to document forest-growth variations. Site-sampling strategies, cross-dating and quality control, standardization (to process raw measurements into final chronologies), and rigorous calibration and verification methodologies for climatic reconstruction are all basic elements of dendrochronology (Cook, 1985; Cook & Kairiukstis, 1990; Fritts, 1976; Holmes, 1983; Stokes & Smiley, 1968). These techniques have been devised in order to minimize errors in dating and measurement and to maximize the signal of interest (e.g., climate).

In remote areas of both hemispheres, instrumental records are often short and sparsely distributed. They can also have missing values as well as changes in station location or methodology. Climatically sensitive tree-ring chronologies, several centuries to millenia in length, can considerably lengthen the limited climatic information available for these regions. Our goal here is to describe some examples of recent dendroclimatic research in western North and South America. We then briefly outline several comparative studies that could be developed to exploit these data within the context of the AMIGO project.

Tree-Ring Width and Maximum-Latewood Density

In addition to the more traditionally used ring-width parameter, maximum-latewood density is increasingly being employed in dendrochronological studies. Yet the similarities and differences in physiological and climatic response reflected in these two variables are still being investigated (e.g., D'Arrigo, Jacoby, & Free, 1992; Hughes, Schweingruber, Cartwright, & Kelly, 1984; Jacoby, Ivanciu, & Ulan, 1988; Schweingruber, 1988). Based on studies thus far, it appears that ring widths respond to year-to-year summer conditions but may also integrate climatic effects throughout the year and on a long-term basis (e.g., D'Arrigo et al., 1992; Jacoby & D'Arrigo, 1989). By comparison, the maximum-latewood density parameter appears most sensitive to high-frequency conditions over a long warm-season interval (Briffa, Jones, & Schweingruber, 1992; Conkey, 1986; D'Arrigo et al., 1992; Jacoby et al., 1988). In our laboratory, a computer-assisted image analysis system (Thetford, D'Arrigo, & Jacoby, 1991) is used for generation of density and also ring-width data.

Tree-Ring Data Bases from Western North America

Tree-ring chronologies have been produced for many regions of western North America. A recent example is an extensive transect of tree-ring width and maximum-latewood density chronologies that has been compiled for the western United States and Canada by Schweingruber of the Swiss Federal Institute of Forest, Snow and Landscape Research (Schweingruber, 1988). This data base includes some 70 sites, ranging from southern Arizona to Alaska, with most of the chronologies extending back before 1750. Density data from this transect were used to reconstruct summer temperature patterns across western North America since 1600 (Briffa et al., 1992). Another tree-ring data base has been developed from an extensive east–west transect of sites across the boreal forest zone of North America sampled by Schweingruber and Jacoby in 1989. Some of these data have also been used for dendroclimatic reconstruction

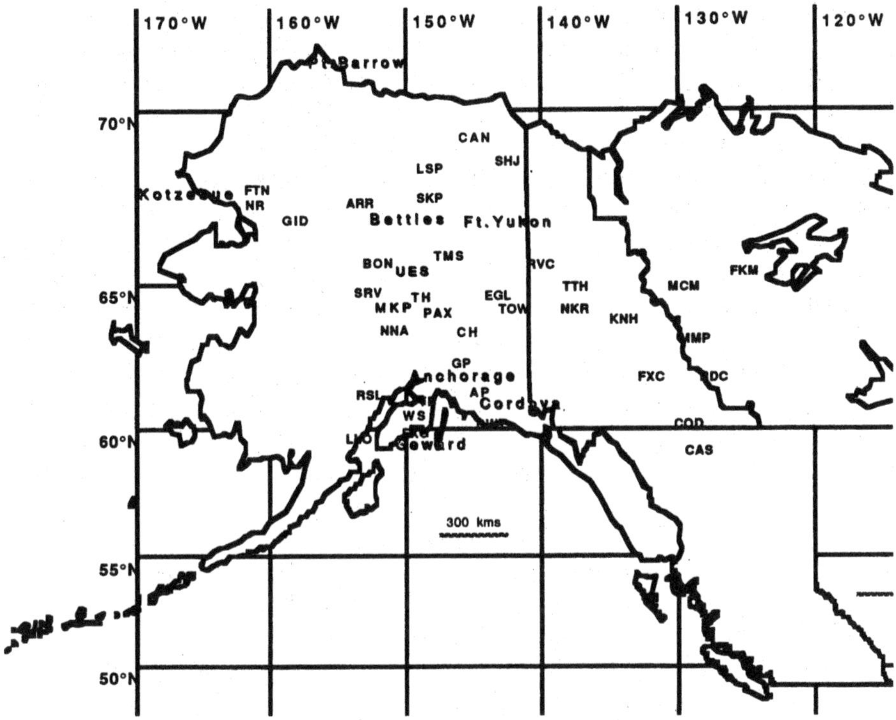

Figure 5.1. Map showing sites in northwestern Canada and Alaska sampled by the TRL at LDEO. Most were sampled in the late 1980s. Some of the Alaskan chronologies, first sampled in the late 1970s–1980s, were updated in the 1990s. ARR is Arrigetch site shown in Figure 5.2.

(Schweingruber, Briffa, & Nogler, 1993). Figure 5.1 shows an additional network of tree-ring chronologies sampled at various intervals within the past two decades by researchers at the Tree-Ring Laboratory (TRL) at the Lamont-Doherty Earth Observatory (LDEO). These three data bases comprise only part of the tree-ring data network that exists for northern/western North America. Below are several examples of the kinds of studies conducted for this region by researchers at the TRL of LDEO based on climatically sensitive chronologies from these and other western North American tree-ring data.

Western North American Tree-Ring Studies: TRL-LDEO

Northern Treeline

The northern limit of tree growth is largely determined by temperature-related factors (e.g., Bryson, 1966). Hence, northern forests may be subject to considerable modification if projected climatic warming effects take place. The expected positive response to warming (i.e., northward and elevational tree-line movement and/or increased rates of tree growth) may be complicated by the more adverse effects on growth of increased insect infestation and disease, changes in stand dynamics, and moisture stress that may accompany warmer conditions.

Figure 5.2 shows tree-ring width and density chronologies for white spruce (*Picea glauca*) near the northern tree-line in Alaska (ARR in Figure 5.1). These trees are inferred to be temperature sensitive, based in part on field observations as well as modeling of tree-growth variations with local and regional temperature data. For reviews of northern tree growth in relation to temperature, see Jacoby, Cook, & Ulan (1985) and Jacoby & D'Arrigo (1989). Tree-ring widths from Alaska and other selected temperature-limited northern sites demonstrate overall enhanced growth since the middle to late 1800s, but with shorter-term growth decreases (cooling) such as during the 1970s (e.g., D'Arrigo & Jacoby,

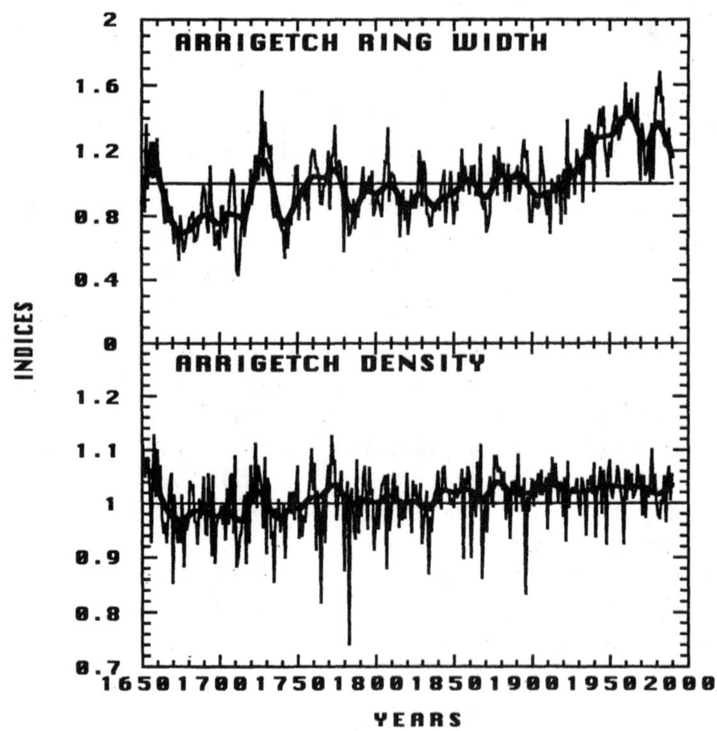

Figure 5.2. Tree-ring width and maximum-latewood density chronologies for a tree-line site in the Arrigetch Peaks, Brooks Range, northern Alaska. Note the overall increase in growth in the ring width series since the turn of the century, with the density series showing primarily high-frequency variability. Also note extremely low density value in 1783 (see text).

1993a; Jacoby & Cook, 1981; Jacoby et al., 1985; Jacoby & D'Arrigo, 1989). The density chronologies for these same sites show primarily high-frequency variations associated with changes in warm-season temperature (D'Arrigo et al., 1992; Jacoby et al., 1988). The extremely low density value in 1783 (Figure 5.2), also present in other density records for Alaska, may be linked to climatic changes following the eruption of Laki in Iceland in that year.

Pacific Northwest

Tree-ring width and density chronologies from the Pacific Northwest (northwest Washington state and near Vancouver Island, British Columbia) were also utilized for dendroclimatic study (Buckley, D'Arrigo, & Jacoby, 1991). Four sets of chronologies were produced (raw measurements obtained from Schweingruber) for these sites for either Douglas fir (*Pseudotsuga menziesii*), Pacific silver fir (*Abies amabilis*), or mountain hemlock (*Tsuga mertensiana*). These data were used to test possible correlations between tree growth and variables such as North Pacific sea-surface temperatures (SST), the El Niño–Southern Oscillation (ENSO), and sea-level pressure (SLP).

The search for such linkages is reasonable since climatic conditions in the Pacific coastal/near-coastal land areas appear to be strongly influenced by ENSO, the Aleutian low-pressure cell, and Pacific North America (PNA) patterns of atmosphere-ocean circulation (e.g., Cayan, 1980; Emery & Hamilton, 1985; Namias, Xiaojun, & Cayan, 1988; Niebauer, 1988; Trenberth, 1990). McBean (this volume) describes some of the processes through which Pacific atmosphere-ocean circulation affects the climate of western North America. Changes in aspects of Pacific circulation can also influence coastal temperate rainforests and watersheds by modifying the composition and concentration of nutrient inputs from marine aerosol sources (Hedin & Hetherington, this volume).

There have been a few studies relating tree-ring widths from western North America locales to SLP and SST (e.g., Blasing & Fritts, 1975; Douglas, 1980) and ENSO (Lough & Fritts, 1985; Michaelson, Haston, & Davis, 1987). Yet there have been relatively few such efforts utilizing maximum-latewood density data. When compared to a grid of monthly SST for the northeast Pacific Ocean, the strongest correlations were found between maximum-latewood density and ring widths and summer (July–August–September) SST (Buckley et al., 1991). Buckley et al. (1991) concluded that with further modeling, there is potential for eventual reconstruction of North Pacific SST based on tree-ring width and density data from this region. Significant correlations were also found between these tree-ring data and winter and spring SLP variations in the general vicinity of the Aleutian low-pressure cell in the northeast Pacific (D'Arrigo, Jacoby, & Wiles, 1993).

The sign and magnitude of ENSO influence on temperature and precipitation vary along a gradient from coastal Alaska down the coast to British Columbia and south to California (Yarnal & Diaz, 1986). Correlation analysis of the four sets of Pacific Northwest chronologies with ENSO indices (e.g., the Southern Oscillation Index or pressure differential between Tahiti and Darwin in the equatorial Pacific) indicated that tree growth tends to be significantly above average during warm ENSO events. These warm events are often associated with a deepened Aleutian low and warmer conditions near Vancouver Island and along the British Columbia and Washington state coastlines (Yarnal & Diaz, 1986). The tree-ring data from the Pacific Northwest could thus prove useful in a comparative study of past ENSO variability along the western coastlines of North and South America. Such a study would also complement transects proposed for AMIGO that aim to compare Vancouver Island with a comparable site in Chile.

Winter Precipitation Reconstruction for Northwestern New Mexico

A 1000-year reconstruction of November–May winter precipitation for northwestern New Mexico was developed by D'Arrigo and Jacoby (1991). This reconstruction was based on six tree-ring width chronologies, including both living and archaeological data for several coniferous species (Dean, 1988; Dean & Robinson, 1978). Other researchers have also related tree rings to precipitation variations in the southwest United States (e.g., Meko, Stockton, & Boggess, 1980), a semiarid region with significant precipitation inputs from the Pacific Ocean.

The southwestern United States is another region of known ENSO teleconnection (Ropelewski & Halpert, 1987), with above- or below-average winter precipitation associated with warm or cold events respectively. Features of the New Mexico reconstruction (D'Arrigo & Jacoby, 1991) include two prolonged drought periods during the broadly defined cold period known as the Little Ice Age (Grove, 1988). One might speculate that these dry periods reflect more pronounced cold-event (La Niña) conditions, although additional paleoclimatic data are needed to confirm this possibility.

Related Studies

Briffa et al. (1992) reconstructed patterns of summer temperature (April to September) since 1600 based on maximum-latewood density chronologies from the Schweingruber data base for western North America. A principal-components-regression technique was employed to compare 53 high-elevation density series to temperature. Briffa et al. (1992) found that the temperature reconstructions were of best quality between 35° N and 55° N, but with uncertainty north of 55° N and before around 1750. Four regional time series were described: British Columbia and the Pacific Northwest; California; the eastern Rockies and the northern High Plains and the southwest U.S. deserts; as well as a more extensive western U.S. series. Other density-based research in western North America include modeling studies of climate response in old-growth Douglas Fir (*Pseudotsuga menziesii*) from Banff (Robertson & Jozsa, 1988) and Englemann spruce (*Picea engelmanii*) from Alberta, western Canada (Parker & Henoch, 1971).

Luckman and his associates have carried out extensive research in western Canada, particularly for tree-line sites in the Canadian Rockies (Luckman, 1992). In addition to direct climate studies (based on Engelmann spruce, alpine larch [*Larix lyalli*], and other species), these records have been applied to the dating of glacial moraine sequences. Some of the high-elevation chronologies from the northern Rockies approach 1 millenium in length (Luckman, 1992). Graumlich and Brubaker (1988) reconstructed mean annual temperatures for Longmire, Washington, based on ring-width series of alpine larch and mountain hemlock (*Tsuga mertensiana*) from the Cascade Mountains. Fritts (1991) describes a number of climatic reconstructions for the period from 1601 to 1961 that he has developed over the past 2 decades. These include reconstructions of large-scale anomalies of sea-level pressure, temperature, and precipitation for the North Pacific, the contiguous United States, and southwestern Canada. All these studies provide valuable information on the paleoclimate of western North America that could be compared with similar South American tree-ring studies within the context of AMIGO.

Tree Rings from Western Southern America

Dendroclimatic analyses in South America and the Southern Hemisphere as a whole are relatively sparse compared to northern latitudes, although this situation has recently improved (e.g., Boninsegna et al., 1990; Lara & Villalba, 1993).

The earliest South American collections were obtained by Schulman (1956) and LaMarche for Argentina and Chile (LaMarche, Holmes, Dunwiddie, & Drew, 1979). More recent dendroclimatic research efforts in southern and western South America include studies by Villalba, Boninsegna, Roig, and others. Winter precipitation has been reconstructed for Santiago, Chile (Boninsegna, 1988; updated from a LaMarche study). Villalba (1989) used tree rings to infer information on the latitude of the subtropical high-pressure belt off western South America over the past 500 years. Villalba, Boninsegna, and Cobbs (1989) also employed chronologies of *Araucaria* species to reconstruct summer temperatures in central-western Argentina back to A.D. 1491. On Chiloe Island, which has been proposed as a site for possible AMIGO transects, Roig and Boninsegna (1989) developed a preliminary reconstruction of precipitation using *Pilgerodendron uviferum*, an austral conifer. Villalba (1990) modeled climate fluctuations over the past millenium in the Rio Alerce region of northern Patagonia based on alerce (*Fitzroya cupressoides*), the longest-lived conifer in South America (Boninsegna, 1992). Lara and Villalba (1993) reconstructed summer temperatures for the past 3620 years for southern South America based on alerce tree-ring width data.

Boninsegna et al. (1990) describe a data base of 21 southern beech (*Nothofagus*) ring width chronologies developed for Tierra del Fuego, Argentina, by Argentinian and U.S. scientists. These chronologies are derived from the southernmost forests on the globe. Most are a few hundred years in length, although a few date back to the 1500s. The chronologies separate into three regions: a northern region, which is drought-sensitive, and two areas to the south (see Figure 5.3) which are primarily sensitive to temperature. This data base is currently being used in studies of subantarctic climate.

Discussion

In this section, we outline three topics of current interest that could be addressed through comparisons of tree-ring records from the west coasts of the two Americas.

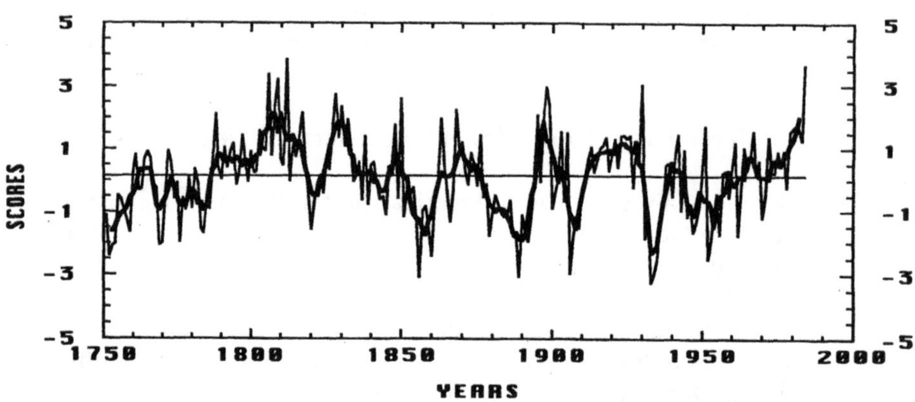

Figure 5.3. Dominant mode of variation (based on principal components analysis) for a subset of four temperature-sensitive tree-ring width chronologies of southern beech (*Nothofagus*) from Tierra del Fuego, Argentina (based on data discussed in Boninsegna et al., 1989). These chronologies have been produced from some of the southernmost trees in South America and the world. The chronologies from Arrigetch (Figure 5.2), by comparison represent some of the northernmost trees in western North America.

The El Niño–Southern Oscillation

ENSO is a key forcing factor for climate variability along the west coasts of the Americas. Because each warm/cold event can differ in phasing, amplitude, and spatial signature, proxy data coverage from different ENSO-sensitive regions around the globe is essential in order to determine how each paleoevent has varied in time and space. Historical records of regional-scale El Niños from along the west coast of South America, for example, show differences when compared to broader-scale ENSO listings for the equatorial Pacific (Quinn & Neal, 1992).

Tree-ring records from western North America have been shown to reflect past ENSO variations (Cleaveland, Cook, & Stahle, 1992; D'Arrigo & Jacoby, 1991; Lough & Fritts, 1985; Swetnam & Betancourt, 1990). Much of South America is also under the strong influence of ENSO (e.g., Aceituno, 1988; Quinn & Neal, 1992). Although chronologies have been produced from Chile and Argentina, there is virtually no coverage in Ecuador and Peru, where some of the strongest ENSO effects are found. Lough and Fritts (1985) attempted a reconstruction of ENSO based on Southern Hemisphere chronologies, although the results were less encouraging than reconstructions based on western North American chronologies. However, additional research is needed to more fully evaluate the ENSO signal in the Southern Hemisphere tree-ring data. Comparison of the chronologies from western North and South America could be useful for studying the differential response of tree growth to drought and ENSO variables in these two settings.

High-Latitude Climatic Response

It is anticipated that the AMIGO project can aid the detection of global-warming effects on the biosphere (Fuentes et al., 1991). Climate models project that increasing levels of atmospheric CO_2 will result in pronounced warming at high latitudes of both hemispheres. Yet due to the smaller Southern Hemisphere land–sea ratio, the Southern Hemisphere warming might be delayed due to the buffering effect of oceans and ice (Fuentes et al., 1991; Gordon, 1992). The instrumental record does not support this projection, however, since SH temperatures have recently been increasing at a faster rate (Gordon, 1992).

Tree rings are one source of high-resolution data that can help determine whether climate and forest growth during the period of anthropogenic change exceed levels of past natural variability. One possible AMIGO project would be to compare tree-ring records from the high latitudes of western North and South America to determine if recent growth exceeds prior levels and to compare the trends for the two regions. The tree-ring records shown in Figures 5.2 and 5.3 represent examples of relevant data. Comparison of the standard errors of recent levels of tree growth or inferred warming as compared to the long-term reconstructed mean is one way to assess whether recent trends are exceeding past natural variation (e.g., Cook et al., 1992). Other techniques include tests for nonstationarity (long-term changes in the mean) and t-tests to evaluate the significance of differences in the means between two time periods (Graumlich, 1991).

CO_2 Fertilization

The issue of whether natural vegetation is currently responding to direct CO_2 fertilization (i.e., enhanced growth due to the increased concentration of CO_2 for photosynthesis) is a matter of considerable debate. In particular, this subject is being examined for high-elevation sites in the southwestern United States (e.g., Graumlich, 1991; Graybill & Idso, 1993; LaMarche, Graybill, Fritts, & Rose, 1984). These sites are of interest because it has been suggested that alpine vegetation may be particularly sensitive to increasing levels of atmospheric CO_2, due to decreased CO_2 partial pressure with altitude (Cooper et al., 1986; Gale, 1972; Korner & Diemer, 1987). This effect may decrease the diffusion gradient and hence the availability of CO_2 into the leaf, resulting in a greater CO_2 limitation (and, presumably, response) than at sea level.

LaMarche et al. (1984) used graphical comparisons between high-elevation western U.S. tree-ring chronologies and temperature and precipitation data to conclude that CO_2 growth enhancement may be taking place. Graumlich (1991) determined that CO_2-induced growth enhancement is not occurring in subalpine conifers in the Sierra Nevada. She based this conclusion in part on residual diagnostic statistics to evaluate trends in the residuals following response surface and regression analyses of tree growth and climate. Graybill and Idso (1993) suggested that CO_2 fertilization in the western United States may be most detectable in trees with a stripbark morphology, in which most new biomass is allocated to cambial production. D'Arrigo and Jacoby (1993a, and 1993b; Jacoby & D'Arrigo, 1989) evaluated residual estimates from regression models relating tree growth to climate at sites in northern North America. No significant positive residual trends (based on the Kendall Rank correlation coefficient test and other statistical tests) were detected that were not modeled by climate and might thus be caused directly by increasing levels of atmospheric CO_2 (D'Arrigo & Jacoby, 1993b). Similar studies can be done to evaluate whether a CO_2 fertilization effect is detectable in high-elevation and high-latitude South American tree-ring records.

Conclusions

A brief overview of some recent topics in dendroclimatic research has been presented, with an emphasis on regions of potential interest to the AMIGO project. We have also suggested several possible topics for comparative study of tree rings from North and South America. These include ENSO, high-latitude forest response to climatic change, and CO_2-induced growth enhancement. Tree-ring records from both Americas provide high-resolution, precisely dated information on past climate and tree-growth changes over prior centuries to millenia. As such, they are a valuable resource that could be exploited for AMIGO's comparative studies of forest ecosystems and their responses to global change.

Acknowledgments. We thank two anonymous reviewers for comments that improved the manuscript. This chapter was written with support from the Climate Dynamics Program of the National Science Foundation, Grant No. ATM89-15353. It is based on a presentation given at the AMIGO Meeting, September 21–26, 1992, Vancouver Island, British Columbia, Canada. Lamont-Doherty Earth Observatory Contribution No. 5346.

References

Aceituno, P. 1988. On the functioning of the Southern Oscillation in the South American sector. Part I: Surface climate. *Mon Wea Rev*, 116, 505–524.

Blasing, T.J., and Fritts, H.C. 1975. Past climate of Alaska and northwestern Canada as reconstructed from tree-rings. In G. Weller and S.A. Bowling (eds.), *Climate of the Arctic* (pp. 48–58). Fairbanks, AK: Geophysical Institute, University of Alaska.

Boninsegna, J.A. 1988. Santiago de Chile winter rainfall since 1200 as reconstructed by tree rings. *Quat J S Am Antarc Pen.*

Boninsegna, J.A. South American dendroclimatological records. In R.S. Bradley and P.D. Jones (eds.), *Climate since A.D. 1500* (pp. 446–462). London: Routledge.

Boninsegna, J.A., Keegan, J., Jacoby, G.C., D'Arrigo, R.D., and Holmes, R.L. 1990. Dendrochronological studies in Tierra del Fuego, Argentina. *Quat J S Am Antarc Pen*, 7, 305–326.

Briffa, K.R., Jones, P.D., and Schweingruber, F.H. 1992. Tree-ring density reconstructions of summer temperature patterns across western North America since 1600. *J Clim*, 5, 735–754.

Bryson, R.A. 1966. Air masses, streamlines, and the boreal forest. *Geog Bull*, 8, 228–269.

Buckley, B.M., D'Arrigo, R.D., and Jacoby, G.C. 1991. Tree-ring records as indicators of air–sea interaction in the northeast Pacific sector. In K.T. Redmond (ed.), *Proceedings of the Eighth Annual Meeting on Pacific Climate* (PACLIM), March 1991 (pp. 35–45). Monterey, CA: Interagency Ecological Studies Program, Technical Report 31.

Cayan, D.R. 1980. Large-scale relationships between sea surface temperature and surface air temperature. *Mon Weath Rev*, 108, 1293–1301.

Cleaveland, M.K., Cook, E.R., and Stahle, D.W. 1992. In H. Diaz and V. Markgraf (eds.), *El Niño. Historical and Paleoclimatic Aspects of the Southern Oscillation* (pp. 271–291). Cambridge, England: Cambridge University Press.

Conkey, L.E. 1986. Red spruce tree-ring widths and densities in eastern North America as indicators of past climate. *Quat Res*, 26, 232–243.

Cook, E.R. 1985. *A time series analysis approach to tree-ring standardization.* Unpublished doctoral dissertation, University of Arizona, Tucson, AZ.

Cook, E.R., and Kairiukstis, L.A. (eds.). 1990. *Methods of Dendrochronology*. Hingham, MA: Kluwer Academic Press.

Cook, E., Bird, T., Peterson, M., Barbetti, M., Buckley, B., D'Arrigo, R., and Francey, R. 1992. Climatic change over the last millenium in Tasmania reconstructed from tree rings. *Holocene*, 2, 205–217.

Cooper, C.F., Gale, J., LaMarche, V.C. Jr., Graybill, D.A., Fritts, H.C., and Rose, M.R. 1986. Carbon dioxide enhancement of tree growth at high elevations. Technical comment. *Science*, 231, 859–860.

D'Arrigo, R.D., and Jacoby, G.C. 1991. A 1000-year record of winter precipitation from northwestern New Mexico, USA: A reconstruction from tree-rings and its relation to El Niño and the Southern Oscillation. *Holocene*, 1, 95–101.

D'Arrigo, R.D., and Jacoby, G.C. 1993a. Secular trends in high northern latitude temperature reconstructions based on tree rings. *Clim Ch*, 25, 163–177.

D'Arrigo, R.D., and Jacoby, G.C. 1993b. Tree growth-climate relationships at the northern boreal forest treeline of North America: evaluation of potential response to increasing carbon dioxide. *Glob Biogeochem Cyc*, 7, 525–535.

D'Arrigo, R.D., Jacoby, G.C., and Free, R. 1992. Tree-ring width and maximum latewood density at the North American treeline: Parameters of climatic change. *Can J For Res*, 22, 1290–1296.

D'Arrigo, R.D., Jacoby, G.C., and Wiles, G. 1993. High-resolution tree-ring records from coastal Alaska and British Columbia: Associations with northeast Pacific climate. *Proceedings of the North Pacific Marine Science Organization Second Annual Meeting*, October 1993, Seattle, WA.

Dean, J.S. 1988. Dendrochronology and paleoenvironmental reconstruction on the Colorado Plateaus. In G.J. Gumerman (ed.), *The Anasazi in a Changing Environment* (pp. 119–167). New York: Cambridge University Press.

Dean, J.S., and Robinson, W.J. 1978. *Expanded Tree-Ring Chronologies for the Southwestern United States*. Chronology Series III. Tucson: University of Arizona.

Douglas, A.V. 1980. *Geophysical Estimates of Sea Surface Temperatures off Western North America since 1671*, CALCOFI (Calif. Coop. Oceanic Fish.), Report No. 21.

Emery, W.J., and Hamilton, K. 1985. Atmospheric forcing of interannual variability in the northeast Pacific ocean: connections with El Niño. *J Geophys Res*, 90, 857–868.

Fritts, H.C. 1976. *Tree Rings and Climate*. New York: Academic Press.

Fritts, H.C. 1991. *Reconstructing Large-Scale Climatic Patterns from Tree-Ring Data: A Diagnostic Analysis*. Tucson: University of Arizona Press.

Fuentes, E.R., Kronberg, B., and Mooney, H.A. 1991. The west coasts of the Americas as indicators of global change. *Ecol Evol*, 6, 203–204.

Gale, J. 1972. Availability of carbon dioxide for photosynthesis at high altitudes: Theoretical considerations. *Ecology*, 53, 494–497.

Gordon, A.H. 1992. Interhemispheric contrasts of mean global temperature anomalies. *Int J Climatol*, 12, 1–9.

Graumlich, L.J. 1991. Subalpine tree growth, climate, and increasing CO_2: An assessment of recent growth trends. *Ecology*, 72, 1–11.

Graumlich, L.J., and Brubaker, L.B. 1988. Reconstruction of annual temperature (1590–1979) for Longmire, Washington, derived from tree rings. *Quat Res*, 25, 223–234.

Graybill, D.A., and Idso, S.B. 1993. Detecting the aerial fertilization effect of atmospheric CO_2 enrichment in tree-ring chronologies. *Glob Biogeochem Cyc*, 7, 81–95.

Grove, J.M. 1988. *The Little Ice Age*. London: Methuen Press.

Holmes, R.L. 1983. Computer-assisted quality control in tree-ring dating and measurement. *Tree-Ring Bull*, 43, 69–78.
Hughes, M.K., Schweingruber, F.H., Cartwright, D., and Kelly, P.M. 1984. July–August temperature at Edinburgh between 1721 and 1975 from tree-ring density and width data. *Nature*, 308, 341–344.
Jacoby, G.C., and Cook, E.R. 1981. Past temperature variations inferred from a 400-year tree-ring chronology from Yukon Territory, Can. *Arct Alp Res*, 13, 409–418.
Jacoby, G.C., Cook, E.R., and Ulan, L.D. 1985. Reconstructed summer degree days in Alaska and northwestern Canada since 1524. *Quat Res*, 23, 18–26.
Jacoby, G.C., and D'Arrigo, R.D. 1989. Reconstructed Northern Hemisphere annual temperature since 1671 based on high latitude tree-ring data from North America. *Clim Ch*, 14, 39–59.
Jacoby, G.C., Ivanciu, I.S., and Ulan, L.D. 1988. A 263-year record of summer temperature for northern Quebec reconstructed from tree-ring data and evidence of a major climatic shift in the early 1800s. *Palaeogeog Palaeoclimatol Palaeoecol*, 64, 69–78.
Korner, C., and Diemer, M. 1987. In situ photosynthetic responses to light, temperature and carbon dioxide in herbaceous plants from low and high altitude. *Funct Ecol*, 1, 179–194.
LaMarche, V.C. Jr., Graybill, D.A., Fritts, H.C., and Rose, M.R. 1984. Increasing atmospheric carbon dioxide: Tree-ring evidence for growth enhancement in natural vegetation. *Science*, 225, 1019–1021.
LaMarche, V.C., Holmes, R.L., Dunwiddie, P.W., and Drew, L.G. 1979. *Tree-Ring Chronologies of the Southern Hemisphere*. Chronology Series V. Tucson: University of Arizona.
Lara, A., and Villalba, R. 1993. A 3620-year temperature record from *Fitzroya cupressoides* tree rings in southern South America. *Science*, 260, 1104–1106.
Lough, J., and Fritts, H.C. 1985. The Southern Oscillation and tree-rings: 1601–1961. *J Clim Appl Met*, 24, 952–965.
Luckman, B.H. 1992. Glacier and dendrochronological records for the Little Ice Age in the Canadian Rocky Mountains. In T. Mikami (ed.), *Proceedings of the International Symposium on the Little Ice Age Climate* (pp. 75–80). Tokyo, Japan.
Meko, D.M., Stockton, C.W., and Boggess, W.R. 1980. A tree-ring reconstruction of drought in southern California. *Wat Res Bull*, 16, 594–600.
Michaelson, J., Haston, L., and Davis, F.W. 1987. Four hundred years of central California precipitation variability reconstructed from tree-rings. *Wat Res Bull*, 23, 809–818.
Namias, J., Xiaojun, Y., and Cayan, D.R. 1988. Persistence of North Pacific sea surface temperature and atmospheric flow patterns. *J Clim*, 1, 682–703.
Niebauer, H.J. 1988. Effects of El Niño-Southern Oscillation and North Pacific weather patterns on interannual variability in the southern Bering Sea. *J Geophys Res*, 93, 5051–5068.
Parker, M.L., and Henoch, W.E.S. 1971. The use of Engelmann spruce latewood density for dendrochronological purposes. *Can J For Res*, 1, 90–98.
Quinn, W.H., and Neal, V.T. 1992. The historical record of El Niño events. In R.S. Bradley and P.D. Jones (eds.), *Climate since a.d. 1500* (pp. 623–648). London: Routledge.
Robertson, E.O., and Jozsa, L.A. 1988. Climate reconstruction from tree-rings at Banff. *Can J For Res*, 18, 888–900.
Roig, F., and Boninsegna, J.A. 1989. *Chiloe Island (Chile) Summer Precipitation since 1556 as Reconstructed by a Tree-Ring Chronology of Pilgerodendron Uviferum*. Unpublished Manuscript.

Ropelewski, C.F., and Halpert, M.S. 1987. Global and regional scale precipitation patterns associated with the El Niño/Southern Oscillation. *Mon Wea Rev*, 115, 1606–1626.

Schulman, E. 1956. *Dendroclimatic Changes in Semiarid America*. Tucson: University of Arizona Press.

Schweingruber, F.H. 1988. *Tree Rings. Basics and Applications of Dendrochronology*. Dordrecht, The Netherlands: Reidel.

Schweingruber, F.H., Briffa, K.R., and Nogler, P. 1993. A tree-ring densitometric transect from Alaska to Labrador: Comparison of ring-width and maximum-latewood-density chronologies in the conifer belt of northern North America. *Int J Biometeorol*, 37, 151–169.

Stokes, M.A., and Smiley, T.L. 1968. *An Introduction to Tree-Ring Dating*. Chicago: University of Chicago Press.

Swetnam, T.W., and Betancourt, J.L. 1990. Fire-Southern Oscillation relations in the southwestern United States. *Science*, 249, 1017–1020.

Thetford, R.D., D'Arrigo, R.D., and Jacoby, G.C. 1991. An image-analysis system for determining densitometric and ring-width time series. *Can J For Res*, 21, 1544–1549.

Trenberth, K.E. 1990. Recent observed interdecadal climate changes. *Symposium on Global Change Systems*, Feb. 1990. American Meteorological Society, Anaheim, CA.

Villalba, R. 1989. Latitude of the surface high-pressure belt over western South America during the last 500 years as inferred from tree-ring analysis. *Quat SAm Antarc Pen*, 7, 273–303.

Villalba, R. 1990. Climatic fluctuations in northern Patagonia during the last 1000 years as inferred from tree-ring records. *Quat Res*, 34, 346–360.

Villalba, R., Boninsegna, J.A., and Cobos, D.R. 1989. A tree-ring reconstruction of summer temperature between a.d. 1500 and 1974 in western Argentina. *Third International Conference of Southern Hemisphere Meteorology and Oceanography*, Buenos Aires, Argentina.

Yarnal, B., and Diaz, H.F. 1986. Relationships between extremes of the Southern Oscillation and the winter climate of the Anglo-American Pacific coast. *J Climatol*, 6, 197–219.

6. Constraints on Terrestrial Primary Productivity in Temperate Forests Along the Pacific Coast of North and South America

R.H. Waring and W.E. Winner

When residents of the temperate forest regions along the Pacific Ocean of North and South America exchange places, their first impression is that the climates of the two continents are similar but out of phase. The second impression is that the forests, although predominantly evergreen in both places, differ in form. Broadleaf, broad-crowned trees are common in the south; whereas narrowleaf, narrow-crowned trees occur in the north. In both regions, the climates are moderated by the Pacific Ocean. This maritime influence allows considerable photosynthesis by evergreen species beyond the normal growing season (Alaback, 1991; Waring & Franklin, 1979).

To assess the implications of current or changing climate in either region requires a model that accounts for how various climatic factors affect growth-related processes. To generalize for the myriad combinations of topography, climate, and soils present, we attempt to interpret environmental factors in an operational sense (Mason & Lagenheim, 1957) as they directly initiate a response that impairs or improves a plant's ability to: 1) capture light and carbon dioxide for photosynthesis; 2) maintain favorable water relations; 3) sequester nutrients; and 4) grow and maintain structure.

A general ecosystem model developed by Running and colleagues at the University of Montana incorporates these operational principles and allows reliable predictions of the growth of coniferous forests throughout much of the western United States (Running, 1994; Running & Coughlan, 1988; Running & Gower, 1991). We extracted basic relationships from this model that couple photosynthesis and growth to the availability of light, water, and nutrients and tested the ability of a simplified model to predict forest growth across a steep climatic gradient in Oregon (Runyon, Waring, Goward, & Welles, 1994; Waring et al., 1993).

In this chapter, we present the logic behind the simplified model and summarize the results of field comparisons. Extrapolation of some components of the simplified growth model provides insights into how a changing climate associated with continued accumulation of CO_2 and other greenhouse gases in the earth's atmosphere might alter forest productivity in the Pacific Northwest and in the temperate forest region of Chile.

General Model Principles

The forest-growth component of the FOREST-BGC ecosystem model accumulates daily estimates of photosynthesis and subtracts the cost of maintenance respiration accumulated throughout a year. The remaining carbon resource is distributed above and below ground, based on the relative availability of nitrogen for leaf growth and an integrated measure of drought stress (Running & Gower, 1991).

Photosynthesis

The absorption of visible light by leaves, branches, and whole canopies has received much study in the last two decades. A number of models estimate light interception and maximum rates of photosynthesis for individual strata of leaves in canopies (Wang & Jarvis, 1990). In mild climates, photosynthesis is a linear function of light absorption by the canopy (Wang, McMurtrie, & Landsberg, 1992).

More complex simulation models allow for variation in the rates of photosynthesis as a function of leaf nitrogen content (or the primary photosynthetic enzyme), soil-water availability, humidity deficits, ambient temperature, and carbon dioxide concentrations (McMurtrie, Comins, Kirschbaum, & Warg, 1992; Raupach, 1989). The data requirements for such models, however, are rather demanding and thus limit their application. We sought a simplified model that generalized responses for whole canopies and assumed thresholds common for a wide range of species (Running & Coughlan, 1988).

Water Relations

Water relations of plants are a function of the supply of water in the soil and the demand from transpiration, balanced against limitations in the rate of transport through the vascular system. Under cloudy and cool conditions, plants have little difficulty in maintaining good water relations. Increasing transpiration beyond certain rates, however, causes some portions of a plant's vascular system to fail and lose hydraulic efficiency. When this happens, stomata begin to close, limiting both transpiration and photosynthesis (Tyree & Sperry, 1988). The threshold when stomata begin to close differs among species. Most temperate forest species in the Pacific Northwest close their stomata completely when the humidity deficit of the atmosphere exceeds 2.5 kPa, as shown in Figure 6.1.

Soil drought can also lead to partial or complete stomatal closure. Under drying conditions, roots extract water from the surface soil downward. Once tree roots have extracted most of the available soil water, tensions increase in the plant's vascular system during the night, when no transpiration is occurring (Waring & Cleary, 1967). As predawn tensions increase, the ability of stomata to open during the day is progressively restricted until complete closure occurs, as shown in Figure 6.2.

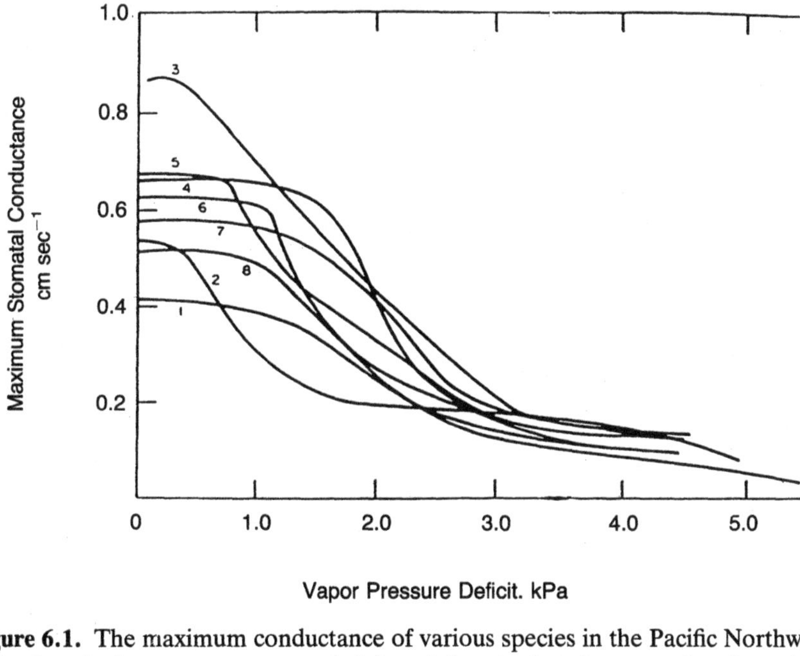

Figure 6.1. The maximum conductance of various species in the Pacific Northwest differ in relation to water-vapor deficits of the air, but most species close their stomata at high deficits. Conifers: 1) *Pseudotsuga menziesii*, 2) *Tsuga heterophylla*; deciduous trees: 3) *Cornus nutallii*, 4) *Acer macrophyllum*; evergreen broadleaf trees; 5) *Castanopsis chrysophylla*; deciduous shrubs: 6) *Acer circinatum*; evergreen broadleaf shrubs 7) *Rhododendron macrophyllum*, 8) *Gaultheria shallon*. From Waring and Schlesinger (1985).

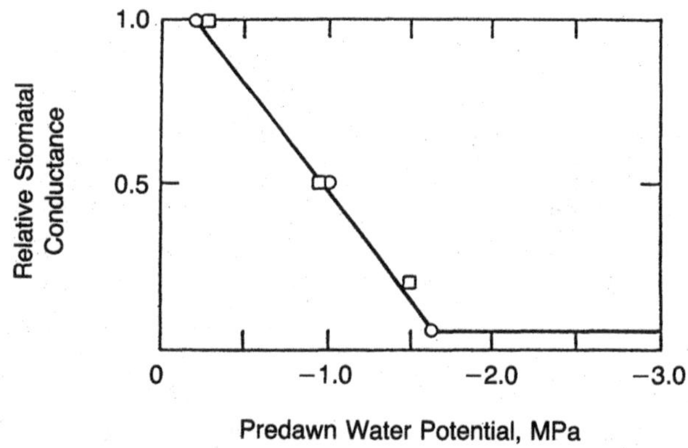

Figure 6.2. As soils dry, predawn water potentials in Douglas fir decrease. Maximum relative stomatal conductance during the day also decreases until complete closure is attained at −1.6 MPa. From Waring and Schlesinger (1985).

Below-freezing air temperatures and cold soil temperatures also restrict the uptake of water and the degree that stomata open. Conifers in the Pacific Northwest show some differences in sensitivity to cold soil temperatures only up to about 10°C, as shown in Figure 6.3. All the

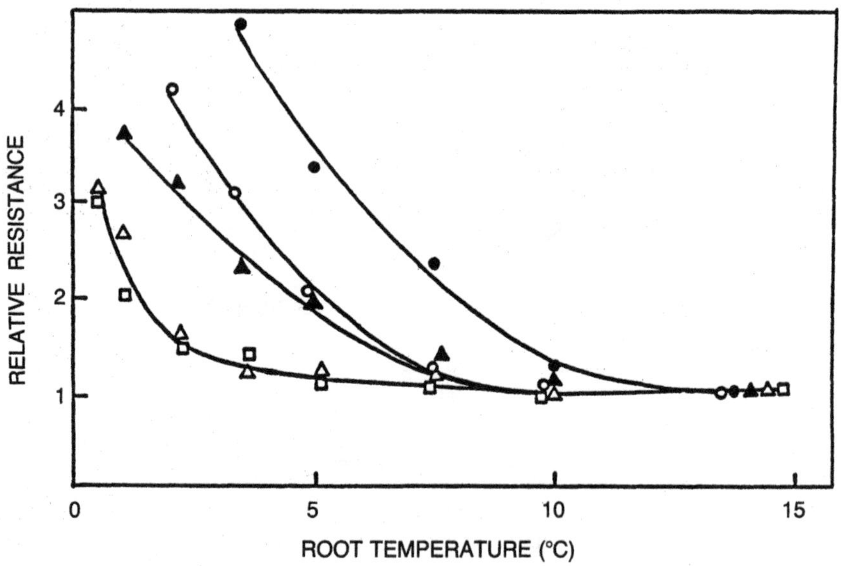

Figure 6.3. Five species of Northwest conifers—*Abies amabilis* (□), *A. procera* (△), *Thuja plicata* (▲), *Tsuga heterophylla* (○), and *Pseudotsuga menziesii* (■)—show differences in their relative resistance to water uptake when root temperatures are only below 10°C.

interacting factors that limit stomata opening were combined into a simulation model by Running, Waring, and Rydell (1975) to predict daily transpiration. Models with both hourly and daily resolution have been compared and proved in close agreement (Running, 1984). The equations and defined thresholds that apply for this water balance model underlie the curent FOREST-BGC ecosystem simulator.

Nutrition

In the Pacific Northwest, nitrogen is generally the most common element limiting growth. Because of its organic origin, nitrogen is concentrated in the upper soil horizons. When drought occurs, roots must extract water from lower horizons, causing the concentration of nitrogen in the sap stream to decrease significantly even if water uptake is not affected (Goulden, 1991; Stark, Essig, & Baker, 1989). The resulting decrease in the flux of nitrogen to leaves limits further growth of foliage much more than it does photosynthesis. When nutrients are not readily available, canopy development is restricted and the ability to absorb solar radiation is reduced. Generally, there is also an increase in the proportion of carbon allocated to root growth (Mooney, 1972; Running & Gower, 1991; Waring & Schlesinger, 1985).

Once physiological thresholds are established, as shown in Table 6.1, constraints on the utilization of intercepted photosynthetically active radiation (IPAR) can be calculated in units of MJ m^{-2} for each climatic variable when requisite data are available. Such an analysis was performed across a steep climatic gradient in Oregon at 44° N latitude; see Figure 6.4. In the coastal zone, where deciduous and evergreen forests both grow (sites 1 and 1A, Figure 6.4), climatic constraints other than cloud

Table 6.1. Criteria for Sequentially Reducing Intercepted PAR Based on Physiological Thresholds Applicable to All Major Tree Species in Oregon. From Runyon et al., 1994

Freezing Temperatures
• If less than −2°C, assume no radiation utilized for 24-hour period
Soil Drought
• If predawn water potential less than −1.5 MPa, assume no radiation utilized for 24-hours • If predawn water potential is between −1.0 and −1.5 MPa, assume half radiation utilized
Vapor-Pressure Deficits (VPD)
• If VPD exceeds 2.5 kPa, assume no radiation utilized for hour period • If VPD is between 1.5 to 2.5 kPa, assume half radiation utilized for hour period

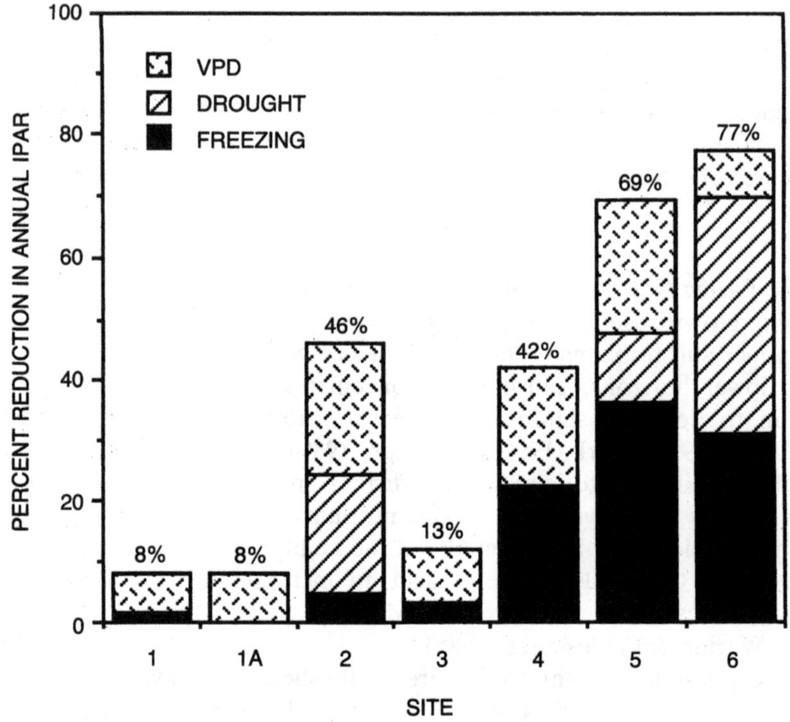

Figure 6.4. The fraction of intercepted radiation not utilized by various forests aligned along a transect (44° N latitude) from coastal rainforests (site 1) across a central valley (site 2) up and over the Cascade Range (sites 3–6) indicate that freezing temperatures, drought, and excessive vapor-pressure deficits constrain photosynthesis to different extents. From Runyon et al. (1994).

cover limit the utilization of intercepted light by less than 10% annually. In the central valley (site 2), a dense coniferous forest mixed with oak uses only about 50% of the annual intercepted PAR, due to limitations from frost, drought, and extreme humidity deficits. To the east of the valley, in the foothills of the Cascade Mountains, precipitation increases and climatic constraints limit IPAR less than 15% annually. In the subalpine zone (site 4) and other sites (sites 5 and 6) in the rainshadow of the Cascade Mountains, 50% to more than 75% of the intercepted radiation is not utilized, due to unfavorable climatic conditions during the year.

Respiration

As trees age, the amount of respiring tissue accumulates. The sapwood, which contains living cells, converts completely to inactive heartwood after a specific number of years (Pothier, Margolis, & Waring, 1989). For young stands of conifers, the maintenance cost of sapwood as a percentage of total annual carbon uptake is generally small, ranging from 5% to 20%, as shown in Figure 6.5. Once height growth slows or ceases, new limb production stops. Additional above-ground growth is restricted to the elongation of previously formed limbs, diameter growth, and the replacement of foliage. The total stand leaf area decreases as the forest opens (Ryan & Waring, 1992). The reduction in leaf area counterbalances to a large extent the increase in sapwood volume associated with increasing tree height. Branch length increases as a consequence of slowed height growth and canopy opening. The hydraulic conductivity for water movement is reduced in such branches and makes stomata more sensitive to humidity deficits (Waring & Silvester, 1994; Yoder et al., 1994). For these reasons, old-growth forests exhibit photosynthetic rates 20% to 30% less than younger stands (Ryan, 1991; Ryan & Waring, 1992; Yoder et al., 1994).

Above-Ground Growth and Utilized Light

When the resource-constrained light-absorption model is applied to forests still growing in height, a linear correlation with estimated annual

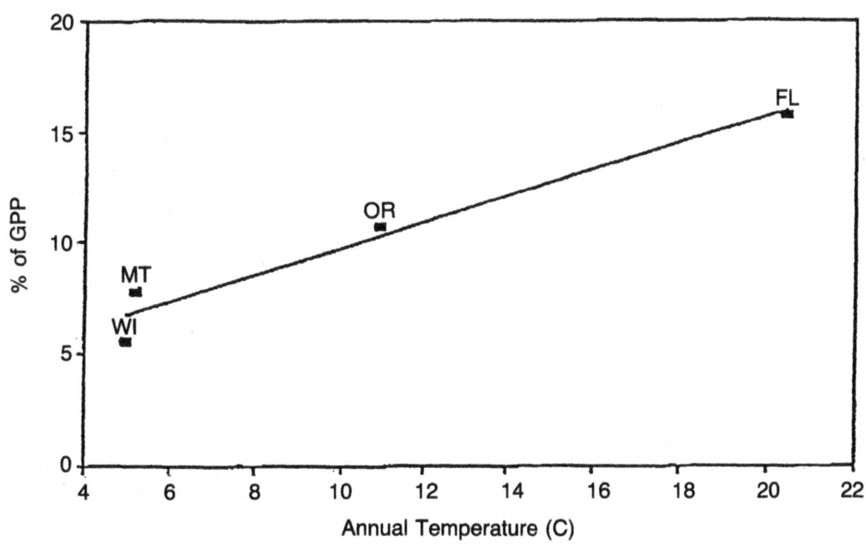

Figure 6.5. The percentage of total gross primary production (GPP) required for maintaining live cells in sapwood increases linearly with annual temperature according to measurements made on pine forests in the States of Wisconsin (WI), Montana (MT), and Florida (FT) and in Douglas-fir and western hemlock forests in Oregon (OR). From M.G. Ryan et al. (1995).

utilized light and above-ground growth was observed across a broad range of environments, as shown in Figure 6.6. No specific threshold for soil-temperature responses was established eliminate because most species in the Pacific Northwest have considerable tolerance for cool soils, perhaps reflecting their boreal origins (Waring & Franklin, 1979). Old-growth forests (1, Figure 6.6), because of reduced photosynthetic capacity and increased maintenance respiration, grow much less than younger stands with similar leaf area (Waring & Franklin, 1979).

Effects of Climatic Change on Forest Growth in the Pacific Northwest

From a sequential analysis of the factors constraining photosynthesis (Table 6.1), climatic warming combined with increased drought should

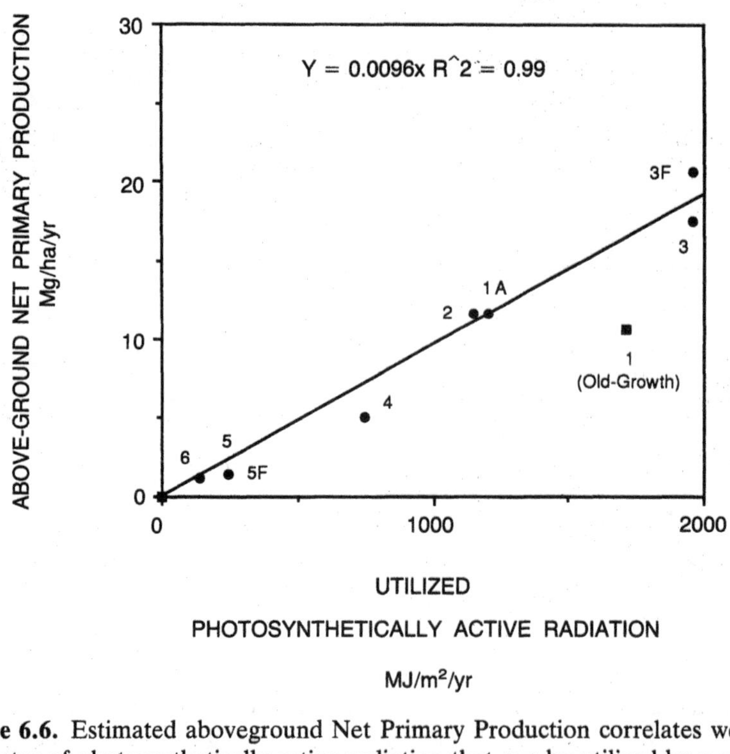

Figure 6.6. Estimated aboveground Net Primary Production correlates well with estimates of photosynthetically active radiation that can be utilized by a range of vegetation across a steep climatic gradient in western Oregon. Letters (A) refer to a deciduous forest of alder, and (F) indicate stands that were fertilized with nitrogen. The old-growth stand (1) was excluded from the regression. It utilized 30% less PAR than indicated. Deducting an additional 100 MJ m^{-2} for increased maintenance respiration compared to an adjacent 120 year-old stand brings the calculated production in line with that observed. From Runyon et al. (1994).

lead to a reduced fraction of IPAP being utilized by forests in the Pacific Northwest. Slower-growing species better adapted to drought should invade and replace more mesic types of forests as the total area occupied by forests contracts in comparison to the current distribution (Franklin & Dyrness, 1973). The transition could be abrupt, spurred by outbreaks of insects, disease, and fire across much of the region (Waring, Savage, Cromack, Jr., & Rose, 1992; Waring & Schlesinger, 1985).

Application of the Light Utilization Model in Chile

In the last 2 decades, much of the native forests of Chile have been logged and the land replanted with introduced species such as *Pinus radiata* (Lara, 1985). The pine forests are planted very densely and absorb >95% of all visible light, as do coniferous species composing forests on the west side of the Cascade Range in the states of Oregon and Washington (Runyon et al., 1994; Waring & Franklin, 1979). Read and Hill (1985) reported that in 21 sampled native Chilean forests, >30% of the light penetrated to the forest floor. The temperate forest region in Chile has a milder climate, with more uniform distribution of rainfall, than the Pacific Northwest (Alaback, 1991; Lawford, 1995). The milder climates result in an extra month of growing season (Goward, Dye, Kerber, & Kalb, 1987). During the growing season, however, the climate is much cooler than at a comparable latitude in the Pacific Northwest. The mild and moist climate in the mid-coastal zone of Chile permits plantations of *Pinus radiata* to produce wood volumes from 22–24 $m^3 ha^{-1} yr^{-1}$. On the other hand, second-growth stands of the fastest-growing native species (*Nothofagus obliqua* and *N. alpina*) produce only 12–14 $m^3 ha^{-1} yr^{-1}$ (Schmidt & Lara, 1985).

Why might native species in Chile form more open canopies than introduced species and generally grow more slowly? One possibility is that most of the native flora is less than perfectly adapted to the current climate as a result of being derived from subtropical stock (Axelrod, Kalin-Arroyo, & Raven, 1991; Raven & Axelrod, 1974). Related genera in New Zealand derived from subtropical origin exhibit a temperature optimal for photosynthesis of 27°C, as compared to 20°C for Pacific Northwest conifers (refer to Table 6.2). Even presumably well-adapted, relatively fast-growing species from Chile such as *Nothofagus obliqua* and *N. alpina* have temperature optima of 23°C, which is well above mean monthly temperatures during the growing season recorded within most of the current range of these species (Alaback, 1991; Read & Busby, 1990; Read & Hill, 1985).

We infer that cool soil temperatures provide a much greater constraint on the growth of most Chilean tree species than equivalent temperatures do on Northwest conifers (Figure 6.3). Unfavorably cool soil temperatures can restrict stomatal opening, photosynthesis, the growth of roots,

Table 6.2. Optimum Temperature for Photosynthesis Compared with Actual Mid-summer Temperatures for Five Genera of Native New Zealand and North American Tree Species. From Hawkins and Sweet, 1989

Species	Optimum Temperature, °C	Mid-Summer Temperature, °C	Difference, °C
New Zealand			
Agathis australis	27.0	22.2	4.8
Dacrycarpus dacrydiodes	27.0	17.9	9.1
Dacrydium cupressinum	27.0	16.8	10.2
Nothofagus solandri	27.0	17.0	10.0
Podocarpus totara	27.0	21.5	5.5
Mean	27.0	19.1	+7.9
North America			
Larix decudua	17.0	19.0	−2.0
Pinus radiata	23.0	21.0	2.0
Pseudotsuga menziesii	21.0	20.2	0.8
Sequoia sempervirens	19.0	17.0	2.0
Tsuga heterophylla	18.0	20.2	−2.2
Mean	19.6	19.4	+0.2

and the uptake of nutrients (Dulucia, 1987; Kramer, 1983). The largest species of *Agathis* (*A. australis*) is very sensitive in its stomatal response to modest humidity deficits when soil temperatures are as high as 15°C. When soils are artificially warmed to 25°C, stomata remain open to much increased humidity deficits (2.5 kPa versus 1.5 kPa; Silvester, University of Waikato, personal communication). Seasonal variations in the maximum observed stomatal conductance also mirror changes in soil temperature (Silvester, personal communication). *Pinus radiata* is sensitive to soil temperatures up to 15°C (Babalola, Boersma, & Youngberg, 1968; Kaufmann, 1977). We might expect many Chilean tree species, given their subtropical origin, to be physiologically constrained at soil temperatures below 20°C.

If native tree species of Chile exhibit similar sensitivity to humidity deficits as *Agathis* in New Zealand at current soil temperatures, then the thresholds at which photosynthesis should be set in our model would be reduced by more than half that assigned for species in the Pacific Northwest (Table 6.1). Other processes, such as maintenance respiration and root growth initiation at the start of the growing season, may also be restricted by cooler temperatures, but these effects are likely to be less important than reductions in photosynthesis, given the generally cool temperatures observed throughout most of the year (Alaback, 1991).

The strong maritime influence in Chile buffers against extremely cold temperatures, allowing plantations of eucalyptus and Monterey pine to be established at the lower elevations in the Andes. In the Pacific Northwest, subfreezing winter temperatures occur sufficiently often to prevent successful plantation of these species, even along the Pacific coast. One might expect in the temperate-forest region of Chile that Northwest conifers such as Douglas fir would not experience sufficient cold temperatures to meet their chilling requirements. This is not the case, however, because young Douglas fir can be observed growing more than a meter in height a year, side by side with eucalyptus and Monterey pine in the Andes at 41° S latitude (Waring, personal observation with photographic evidence available).

Effects of Climatic Change on the Temperate Forests of Chile

If global warming were to increase the growing-season temperature in Chile, native tree species might be expected to respond far differently from species in the Pacific Northwest. We speculate that an increase in temperature would improve the ability of Chilean trees to take up water and nutrients from soils and increase their tolerance to atmospheric humidity deficits. Although a decrease in summer precipitation in Chile is not predicted in climate models (Revkin, 1988), if drought were to occur, it should be less extreme than in the Pacific Northwest, because Chilean forests receive at present from 10% to 25% of their annual precipitation during summer months, whereas forests in the Pacific Northwest receive only 3% to 15% of annual precipitation during the same interval (Alaback, 1991). Other factors may contribute to changes in forest composition, but to assess these requires different kinds of models that account for competition, reproduction, dispersal, introduction, and extinction of species.

Conclusions

With an environmentally constrained light-absorption model, we quantified the influence of frost, drought, and humidity deficits upon photosynthesis and growth across a steep climatic gradient in Oregon. When account was made for reduced utilization of intercepted light due to partial or complete stomatal closure, the solar conversion efficiency was similar for all types of Pacific Northwest coniferous forests. Because Northwest conifers are derived from boreal stock, they show little restriction on water uptake or growth once soil temperatures exceed 10°C. In contrast, the native species of Chilean forests are derived mainly from subtropical stock and thus are likely to be restricted in their performance at soil temperatures below 20°C. If climatic warming occurs in Chile, improved growth of native species is predicted from our model. The opposite prediction applies to the Pacific Northwest forests, because drought is expected to have more impact.

References

Alaback, P.B. 1991. Comparative ecology of temperate rainforests of the Americas along analogous climatic gradients. *Rev Chil Hist Nat*, 64, 399–412.

Axelrod, D.I., Kalin-Arroyo, M.T., and Raven, P.H. 1991. History of temeprate rainforests of the Americas. *Rev Chil Hist Nat*, 64, 413–436.

Babalola, O., Boersma, L., and Youngberg, C.T. 1968. Photosynthesis and transpiration of Monterey pine seedlings as a function of soil water suction and soil temperature. *Plant Physiol*, 43, 515–521.

Dulucia, E.H. 1987. Air and soil temperature limitations on photosynthesis in Engelmann spruce during summer. *Can J For Res*, 17, 527–533.

Franklin, J.F., and Dyrness, C.T. 1973. *Natural vegetation of Oregon and Washington*. U.S. Department of Agriculture, Forest Service, Pacific Northwest Forest and Range Experiment Station, General Techical Report PNW-8, Portland, OR.

Goulden, M.L. 1991. Nutrient and water utilization by evergreen oaks that differ in rooting depth. Unpublished doctoral dissertation, Stanford University, Palo Alto, CA.

Goward, S.N., Dye, D., Kerber, A., and Kalb, V. 1987. Comparison of North and South American Biomes from AVHRR observations. *Geocarto Int*, 2, 27–39.

Hawkins, B.J., and Sweet, G.B. 1989. Photosynthesis and growth of present New Zealand forest trees related to ancient climates. *Ann Sci For*, 512S–514S.

Kaufmann, M.R. 1977. Soil temperature and drought effects on growth of Monterey pine. *For Sci*, 23, 317–325.

Kramer, P.J. 1983. *Water Relations of Plants* New York: Academic Press.

Lara, A.A. 1985. Los ecosistemas forestales en el desarrollo de Chile. *Amb Des*, 1, 81–99.

Lawford, R.G. 1995. In R.G. Lawford, P. Alaback, and E.R. Fuentes (eds.), *High Latitude Rainforests and Associated Ecosystems of the West Coast of the Americas: Climate, Hydrology, Ecology, and Conservation*. New York: Springer Verlag.

Mason, H.L., and Lagenheim, J.H. 1957. Language analysis and the concept of environment. *Ecology*, 38, 325–339.

McMurtrie, R.E., Comins, H.N., Kirschbaum, M.U.F., and Wang, Y.-P. 1992. Modifying existing forest growth models to take account of effects of elevated CO_2. *Aust J Bot*, 40, 657–677.

Mooney, H.A. 1972. The carbon balance of plants. *Ann Rev Ecol Sys*, 3, 315–346.

Pothier, D., Margolis, H.A., and Waring, R.H. 1989. Patterns of change of saturated sapwood permeability and sapwood conductance with stand development. *Can J For Res*, 19, 432–439.

Raupach, M.R. 1989. Turbulent transfer in plant canopies. In G. Russell, B. Marshall, and P.G. Jarvis (eds.), *Plant Canopies: Their Growth, Form and Function* (pp. 41–61). Cambridge, England: Cambridge Universtiy Press.

Raven, P.H., and, Axelrod, D.I. 1974. Angiosperm biogeography and past continental movements. *Ann Miss Bot Gard*, 61, 539–673.

Read, J., and Busby, J.R. 1990. Comparative responses to temperature of the major canopy species of Tasmanian cool temperate rainforest and their ecological significance. II. Net photosynthesis and climate analysis. *Aust J Bot*, 38, 185–305.

Read, J., and Hill, R.S. 1985. Photosynthetic responses to light of Australian and Chilean species of *Nothofagus* and their relevance to rainforest dynamics. *New Phytol*, 101, 731–742.

Revkin, A.C. 1988. Endless summer: living with the greenhouse effect. *Discover*, 9, 50–60.
Running, S.W. 1984. Fort Collins, CO: U.S.D.A., Forest Service, Rocky Mountain Forest and Range Experiment Station, Research Paper RM-225.
Running, S.W. 1994. Testing forest-BGC ecosystem process simulations across a climatic gradient in Oregon. *Ecol Appl*, 4, 238–247.
Running, S.W., and Coughlan, J.C. 1988. A general model of forest ecosystem processes for regional applications. I. Hydrologic balance, canopy gas exchange and primary production processes. *Ecol Model*, 42, 125–154.
Running, S.W., and Gower, S.T. 1991. FOREST-BGC, a general model of forest ecosystem processes for regional applications. II. Dynamic carbon allocation and nitrogen budgets. *Tree Physiol*, 9, 147–160.
Running, S.W., Waring, R.H., and Rydell, R.A. 1975. Physiological control of water flux in conifers. *Oecologia*, 18, 1–16.
Runyon, J., Waring, R.H., Goward, S.N. and Welles, J.M. 1994. Environmental limits on above-ground production: observations from the Oregon transect. *Ecol Appl*, 4, 226–237.
Ryan, M.G. 1991. A simple method for estimating gross carbon budgets for vegetation in forest ecosystems. *Tree Physiol*, 9, 255–266.
Ryan, M.G., and Waring, R.H. 1992. Maintenance respiration and stand development in a subalpine lodgepole pine forest. *Ecology*, 73, 734–763.
Schmidt, H., and Lara, A. 1985. Descripcion y potencialidad de los bosques nativos de Chile. *Amb Des*, 2, 91–108.
Stark, N., Essig, D., and Baker, S. 1989. Nutrient concentration in *Pinus ponderosa* and *Pseudotsuga menziesii* xylem sap from acid and alkaline soils. *Soil Sci*, 148, 124–131.
Teskey, R.O. 1982. *Acclimation of Abies amabilis* to Water and Temperature in a Natural Environment. Unpublished doctoral dissertation, University of Washington., Seattle, WA.
Tyree, M.L., and Sperry, J.S. 1988. Do woody plants operate near the point of catastrophic xylem dysfunction by dynamic water stress? *Plant Physiol*, 88, 574–580.
Wang, Y.-P., and Jarvis, P.G. 1990. Description and validation of an array model—MAESTRO. *Agri For Meteorol*, 51, 257–280.
Wang, Y.-P., McMurtrie, R.E., and Landsberg, J.J. 1992. Modelling canopy photosynthetic productivity. In N.R. Baker and H. Thomas (eds.), *Crop Photosynthesis, Topics in Photosynthesis* (pp. 43–67), Vol. 12. Amsterdam: Elsevier.
Waring, R.H., and Cleary, B.D. 1967. Plant moisture stress: Evaluation by pressure bomb. *Science*, 155, 1248–1254.
Waring, R.H., and Franklin, J.F. 1979. Evergreen forests of the Pacific Northwest. *Science*, 204, 1380–1386.
Waring, R.H., Runyon, J., Goward, S.N., McCreight, R., Yoder, B., and Ryan, M.G. 1993. Developing remote sensing techniques to estimate photosynthesis and annual forest growth across a steep climatic gradient in western Oregon, U.S.A. *Stud For Suec*, 191, 33–42.
Waring, R.H., Savage, T., Cromack, K., Jr., and Rose, C. 1992. Thinning and nitrogen fertilization in a grand fir stand infested with western spruce budworm. Part IV. An ecosystem management perspective. *For Sci*, 38, 275–286.
Waring, R.H., and Schlesinger, W.H. 1985. *Forest Ecosystems: Concepts and Management*. Orlando, FL: Academic Press.
Waring, R.H. and Silvester, W.B. 1994. Variation in foliar $\delta\ ^{13}C$ values within the crowns of *Pinus radiata* trees. *Tree Physiol*, 14, 1203–1213.

Yoder, B.J., Ryan, M.G., Waring, R.H., Schoettle, A.W. and Kaufmann, M.R. 1994. Evidence of reduced photosynthetic rates in old trees. *For Sci*, 40, 513–527.

2. Biotic Patterns

7. Biodiversity Patterns in Relation to Climate: The Coastal Temperate Rainforests of North America

Paul B. Alaback

The west coast of North America presents a rich array of climatic, geologic, and historical environments in which a relatively rich flora and fauna have developed. In contrast to South America, the west coast of North America is part of a large temperate zone and as such shares serveral species with continental and in particular the montane biota of inland portions of the continent. Many species are confined to the coastal region, suggesting an adaptation to the relatively unique, mild marine climate, especially that of the temperate rainforest region. Subtleties in climatic differences between North America and other temperate rainforest types are hypothesized as a critical factor in explaining differences in structure and composition of vegetation (Alaback, 1991).

Although much work has been done on describing the biogeography of many parts of this large region, little work has been done on the broad climatic and biological patterns that occur within the rainforest regions (Alaback, 1991; Cordes, 1972; Franklin et al., 1991; Juday, 1976; Pojar, Klinka, & Meidinger, 1987; Quay, 1982). This broad-scale approach is particularly appropriate for understanding linkages between history, evolution, genetics, and contemporary species interactions. These relationships are of key importance in developing realistic predictions of the effects of global climate change on these ecosystems. Since the temperature and precipitation patterns are consistent across this broad latitudinal gradient, it is a particularly tempting analog for climate change, further underscoring the need to understand the region in a broad-scale sense.

The west coast is one of the last major forest regions to be exploited in North America and as such contains some of the greatest remaining examples of pristine forest on the continent. This region provides one of the best areas for testing ideas about how natural landscapes are structured and how changes in landscape structure influence ecological processes, since both managed and pristine landscapes are present here. The temperate rainforest region is of particular interest for biodiversity studies, since it encompasses a broad range of landscape types both within and between latitudes and includes both Pleistocene and recent glacial environments.

In this chapter, we present an overview and synthesis on vegetation biodiversity patterns in coastal-forest vegetation from the southern limit of continuous forest in the redwood region in northern California north to the limit of coastal forest in central Alaska. The key focus of this paper is on the temperate rainforest region of the coastline, which includes most of the coastal forest in the north but only a narrow belt of forest in the south. Basic patterns of forest biodiversity and forest genetics are discussed to provide a basis for prioritizing research needs on how better to predict the effects of global climate change across this broad region and to identify data gaps. This synthesis should also provide guidance to resource managers on the kinds of information needed to understand how best to adapt ecosystems or landscapes to the uncertain future presented with climate change.

Physical Environment

The gradient from 38° N to 61° N across western North America extends from fog-drenched, almost Mediterranean climates to perpetually wet rainforests and tundra. For our purposes, *rainforest* is defined to include maritime climates with 1400 mm annual precipitation or greater, and with cool temperatures year-round. For convenience of discussion, North American rainforests are divided into four principal zones: 1) the *subpolar rainforest* zone (59° N–61° N), where heavy precipitation occurs during most months (summer rainfall >20% of annual), summers are very cool, and sea-level snow is common; 2) the *perhumid rainforest* zone (50° N–58° N), where summer precipitation is common (>10% of annual), summers are cool, and snow is transient in the winter; 3) the *seasonal rainforest* zone (43° N–50° N), where summer droughts and fire are common and less than 10% of annual precipitation occurs during the summer; and 4) the *warm temperate rainforest* zone (38° N–43° N), where extended droughts and fire can occur at any time of the year and 5% or less of annual rainfall occurs during summer (Alaback, 1991; Franklin &

Dyrness, 1973; Pojar et al., 1987). These zones correspond to mean annual temperatures of 4°C, 7°C, 10°C, and 12°C and growing-degree days (>5°C) from less than 800 to more than 2000 (Farr & Harris, 1979; Lawford, this volume).

Temperate rainforest regions are distinguished from other western temperate forest types by cool annual temperatures and the equability of climate over the year, as shown in Figure 7.1. Whereas a high degree of similarity occurs among the climates of temperate rainforest regions over a wide range in latitude, the climates of interior continental areas vary widely with latitude. In the southern extent of coastal forest, adjacent climatic zones include Mediterranean and semidesert as well as montane zones. In the seasonal zone, interior forests include high-elevation xeric pine and mesophytic types. In the subpolar zone, rainforest rapidly gives way to boreal forest and tundra. The boreal zone is the most distinctive, with the highest degree of annual variation and lowest temperatures. During the growing season, however, the boreal zone has similar temperatures to the subpolar rainforest zone. Xeric and mesophytic forests (e.g., *Pinus ponderosa* and *Pseudotsuga menziesii* respectively) are distinguished from temperate rainforests mostly by the greater degree of annual variation in temperature, but, except at the highest latitudes or at high elevations, also tend to be significantly warmer, especially during the

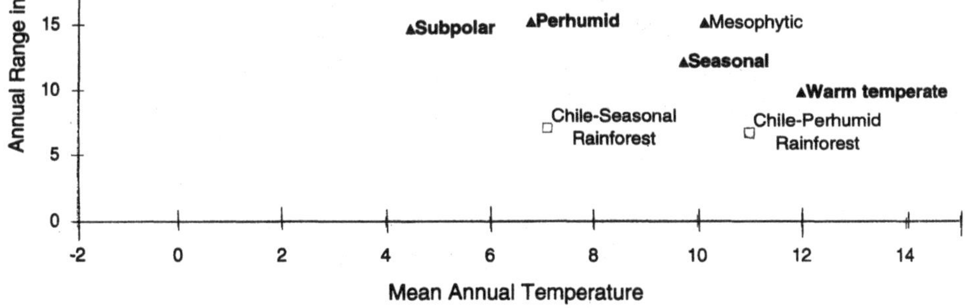

Figure 7.1. Climates of Pacific forest types. Means based on 20–30 year climate normals from U.S. Weather Service and Environment Canada Atmospheric Environment Service from the Pacific coast east to the eastern foothills of the Cascade–Sierra Range and the British Columbia Coast Range crest. Xeric forests represent those dominated by *Pinus ponderosa*. Mesic forests represent those dominated by *Pseudotsuga menziesii* in the western Cascades.

summer. The analogous temperate rainforest types in southern Chile are closely matched with those in North America, except that their climate is even more equitable year-round and they are warmer, particularly during the winter.

One of the key gradients within the temperate rainforest zones of North America is precipitation, which steadily increases northward both in equability of distribution and in yearly accumulation (Alaback, 1991; Alaback & McClellan, 1992). Along this transect, moisture changes from a scarce resource limiting plant productivity to an overabundant resource which is associated with declines in soils fertility (Ugolini & Mann, 1979). In the warm temperate and seasonal rainforests, precipitation primarily occurs in the winter months. In the warm temperate zone, rainfall over 1400 mm occurs only within 30 km of the ocean or at elevations greater than 500 m. The Klamath Mountain region provides a striking anomaly, in that harsh dry climates persist almost to the coast because of the unique physiography of the region. This geographic area also corresponds with an extremely rich and diverse flora. This diversity is due not only to the diversity in climates that occur there but also to a long history and diverse geology (Whittaker, 1960). While the whole Klamath Mountain region is sometimes considered to be in the Mediterranean zone, the wetter coastal portions of this region are included here, due to their floristic and climatic similarity to seasonal rainforests to the north.

Within the northern portions of the temperate rainforest region, the most ecologically important effects of precipitation relate more to its distribution than to its magnitude (Alaback, 1991). Throughout the perhumid and subpolar rainforest regions, semipermanent low-pressure systems bring high annual precipitation during most of the year. Upwelling off the Pacific Ocean also increases moisture by producing fog during late spring and early summer. Fog-drip interception is an important yet seldom quantified element of the hydrological budget (Harr, 1982) and is also of key importance in defining the geographic limit of the most productive forest types, especially along the coastline of northern California and the Pacific Northwest. Fog is also a key climatic factor on the outer coast of British Columbia and Alaska.

Dry interior valleys extend their climatic influence on coastal vegetation patterns by allowing local domination by fire- or drought-resistant species. This pattern is most significant in the warm temperate rainforest, in particular the Klamath Mountain region, but also occurs intermittently in the seasonal rainforest zone and in a few large valleys in the perhumid zone. The only significant occurrence of *Pseudotsuga* in the northern seasonal zone or perhumid rainforest, for example, is in these major interior-influenced valleys (e.g., Nimkish, Bella Coola in British Columbia).

An equally significant but more consistent gradient in temperature also characterizes the north–south transect. The temperature gradient is closely related to forest yield and productivity (Farr & Harris, 1979). Because of persistent westerly winds and warm Kuriosho currents, the entire west coast has a distinctly maritime climate, with moderate temperatures in both winter and summer. Variation both seasonally and diurnally is greatest in the north (see Figure 7.1). Fog plays a key role in depressing spring and summer temperatures, particularly in the warm temperate and seasonal zones. The moderating influence of the ocean is most constricted along the southern coast, often extending to within as little as 1 km, and is most pervasive along the northern coast, where moderated temperatures may extend 150 km inland. Mean annual temperatures decline sharply with latitude, while seasonal range in temperature increases. Thermal energy appears to be a critical limiting resource in the perhumid and subpolar zones for both plant growth, decomposition, nutrient release, and animal activity (Alaback & McClellan, 1992; Parker, 1988; Wallmo & Schoen, 1980).

History

The pollen record is extensive and relatively well known along the Pacific coast of North America, especially along the northern coast and interior (e.g., Axelrod, 1976; Axelrod, Arroyor, & Baven, 1991; Brubaker, 1988; Heusser, 1960). Vegetation has been in a continual state of flux during the entire pollen record (ca. 14,000 years). As with other regions, these fluctuations demonstrate that plant species tend to vary in an individualistic fashion. They respond differentially to stress and vary in their speed of migration to suitable habitat areas. Life-history strategies, reproductive mechanisms, and physiological tolerances vary among the principal dominant species and determine migration rates. For example, many tree species in Alaska are physiologically adapted to grow much farther north or west than the limit of their present-day distribution (e.g., *A amabilis*, *Picea glauca*, *P. sitchensis*). In the Southern Hemisphere rainforests, even more dramatic asymmetries between local populations and exotics occur. Exotic tree species such as *Pinus contorta*, because of their adaptation to the current cool temperature regime, can extend timberline as much as 500 m or dramatically increase wood productivity over that of any native species (Hawkins & Sweet, 1989a). Other Southern Hemisphere endemics appear to grow best at temperatures much higher than current climatic conditions (Hill, Read, & Busby, 1988; Read & Hill, 1985). In the coastal temperate rainforest of southern Scandinavia, *Picea abies* grows well much farther west and north than its natural distribution, suggesting a large time lag in its migration and equilibration to current climatic conditions.

Many historical plant assemblages have no modern counterparts. The current forest species mix of the perhumid zone has been in existence for little more than 2000 years (Brubaker, 1988). Western hemlock and Sitka spruce have existed in Prince William Sound for only 4000 years and have yet to reach a stable species mixture. Subtle changes in global climate correspond to dramatic changes in northern areas, including glacial surges or rapid recessions and rapid timberline movement. For example, during the Little Ice Age approximately 200 years ago, the entirety of Glacier Bay was filled with ice. Currently, ice remains only in the head of the bay, more than 95 km from the mouth of the bay. Following deglaciation, a cool, dry climate developed that favored the development of an extensive *Pinus contorta* forest, which now only occurs in isolated patches in dry local climates usually associated with linkages to dry interior valleys. Contemporary revegetation patterns following glacial retreat is dramatically different in rate, composition, and structure than it was in the past (Chapin, Walker, Fastie, & Sharman, 1994; Fastie, 1990). Because of this recent history of dramatic climate change, the northern flora is considered more resilient to changes in climate than more southerly areas (e.g., Peters, 1990).

Historical linkages to other regions and broad migration corridors, as well as continuous changes in climate from subtropical to cool temperate, help explain much of the richness and adaptation of the current forest flora (e.g., Axelrod et al., 1991). *Pinus* was the first conifer to appear in the region (90 million years ago). *Picea, Chamaecyparis, Pseudotsuga, Tsuga, Abies*, and *Thuja* first appeared 55 million–65 million years ago. During the past 25 million years, the climate rapidly cooled, leading to a major impoverishment and isolation of the flora. During the mid-Miocene (ca. 25 million years ago), coastal forest was separated from eastern mesophytic forest by the dry interior. The coastal forest was still continuous with Beringia and other parts of Asia. Late in the Miocene (15 million years ago) summer temperatures dropped by 7°C, causing deciduous elements in the flora to retreat to central British Columbia. This is also when one of the dominant ericaceous understory shrub genera appeared (*Vaccinium*), which is closely associated with acid soils. The southern portion of coastal forest became more isolated during the Pliocene, when the Cascades and Coast range rose, eliminating forest in the interior. Palynological data suggest that when summers became cooler and drier approximately 7 million years ago, heat became inadequate for most deciduous trees to complete their normal growth and reproductive cycle before autumn frost (Stephenson, 1990; Waring & Franklin, 1979). As a consequence, most deciduous elements were lost in the region at that time. All these reductions in flora were completed well before the Pleistocene glaciations. This elimination of hardwood trees distinguishes the flora of rainforests in North America from virtually all other coastal temperate rainforests in the world.

The *Picea-Tsuga* coastal forest first appeared 2 million to 5 million years ago in the Cook Inlet region and extended northward throughout

Alaska (an area now occupied mostly by boreal forest or tundra and by subpolar rainforest on the coast) (Axelrod, Arroyo & Raven, 1991). This long period of relatively stable climate along the coast (ca. 5 million years) and the existence of a temperate rainforestlike climate over most of Alaska and other high latitudes over 20 million years or more appears to have allowed for the evolution of trees closely adapted to current climatic conditions with minimal gene depletion during glaciations. The Pacific coast forest is often considered the most luxuriant conifer forest in world, reflecting this high level of physiological adaptation (Waring & Franklin, 1979). Ideal growth conditions occur in this climatic zone because of abundant moisture, which is available year-round, and moderate temperatures. Intense selection for height growth and biomass accumulation has occurred over a long period of time in this zone. As a consequence, the world's largest and longest-lived species occur in the Pacific coastal forests of North America for virtually every dominant genus of conifer: *Abies*, *Chamaecyparis Larix*, *Libocedrus Picea*, *Pseudotsuga*, *Sequoia*, *Thuja*, and *Tsuga*. The greatest phytomass accumulations in the world also occur here (>2000 mt/ha in *Sequoia*, >1000 mt/ha in *Picea*, *Tsuga*, and *Abies* forests). This rich genetic diversity in dominant trees suggests that there is much potential for adaptation for climate change. Changes in disturbance regime wrought by changes in storm frequency and intensity in the perhumid zone, or by changes in frequency and intensity of fire in the warm temperate and seasonal zones, may provide the most challenging constraints for trees' adaptation to climate change in this climatic zone (Alaback & McClellan, 1992;Franklin et al., 1991).

There is little information on the genetic structure or the history of understory shrub and herb species that comprise the center of biological diversity of these forests in terms of plant species richness and plant–animal interactions. Most species are clonal and may have limited recruitment of new genetic individuals (Bierzychudek, 1982; Tappeiner & Alaback, 1989). Shrubs and herbs also have a much greater dependence on animal vectors for seed dispersal than dominant trees in these rainforests, making them more sensitive to climate change (Willson, 1991). Much more information is needed on the genetic architecture and physiology of these shrub and herb species to be able to predict how they might respond to climate change and to changes wrought by land-use changes.

Genetic Adaptation of Dominant Tree Species

Although it is clear that the northern rainforest climate often provides ideal growth conditions for trees, climate alone cannot explain the high level of biomass accumulation and productivity that distinguishes the Pacific coastal forests from virtually all others in the world (Waring &

Franklin, 1979). It is generally thought that the conifers in this forest zone have better genetic potential for growth under current climatic conditions than trees from other geographic areas, because of historical and biogeographic factors. As described earlier, coastal temperate rainforest has occurred along the northern coastline for at least 5 million years (and are composed of many genera 70 million or more years old) and has had minimal impacts from glaciations, relative to other high-latitude regions, Genetic selection has apparently favored tree-height growth for many genera, leading to ideal genetic characteristics for rapid biomass accumulation and high wood product values under current climatic conditions. Forests of the North Pacific are among the tallest in the world (50 m–70 m in the perhumid zone, 70 m–100 m in the seasonal zone, and up to 112 m in the warm temperate rainforest zone).

Relatively little work has been done on the genetic structure of dominant tree species of the coastal rainforest zone, with the exception of *Pseudotsuga* and *Populus*. In general, conifers are considered genetically diverse, reflecting a strong tendency toward heterozygosity and outbreeding (Hamrick, Godt, & Sherman-Broyles, 1992). Large production of pollen that is efficiently distributed over long distances helps maintain this high level of natural genetic diversity (Adams, 1992; Adams & Birkes, 1990). Because of its economic importance and widespread distribution, much work has been done on the genetics of *Pseudotsuga*. From this work it is clear that this conifer has an extremely diverse and complex genetic structure both throughout and within parts of its range (Campbell, 1979; Conkle, 1981). In one small watershed in the western Cascades of Oregon, for example, isoenzyme analysis suggests a high degree of genetic differentiation according to microtopography and microclimate (Campbell, 1979).

In general, woody plant species have a high degree of genetic variation relative to other life forms (Hamrick et al., 1992). Species with wide distributions or formerly wide distributions and with outcrossing breeding systems tend to have the highest levels of diversity. Animal-dispersed species tend to have higher levels of diversity than wind-dispersed species. Endemics typically have 30% or less the genetic variation of widespread species (Hamrick et al., 1992). The forest tree species of the Pacific forest zone appear to follow this general trend. *Picea sitchensis*, *Pseudotsuga menziesii*, and *Tsuga heterophylla* are the dominant widespread species of the region and have the greatest levels of genetic diversity. These species would be expected to have the greatest potential to adapt to climate change or to be improved for the broadest range of environmental conditions within this biome. Specialized species with limited distributions such as Cupressids in the warm temperate rainforest, by contrast, would be expected to have the lowest levels of genetic diversity and at the greatest risk with rapid climate change.

Sitka spruce (*Picea sitchensis*) is the principal species (along with western hemlock) that defines the rainforest type throughout the entire coastal Pacific forest zone. I will use *Picea* as the principal species to illustrate how characteristics of history, ecology, and physiology influence the genetic architecture of dominant species in this climatic zone. *Picea* is an old genus, first appearing in the fossil record about 70 million years ago. It has changed relatively little since first radiating from eastern Asia (Wright, 1955). Most of the species of *Picea* developed as a result of geographic isolation and can freely interhybridize. *P. sitchensis* is most closely related to *P. glauca* and often hybridizes with it whenever their geographic ranges overlap. It also freely interhybridizes with *P. jezoensis*, *P. omorica*, and *P. engelmannii* (Wright, 1955). During the Pleistocene, *P. sitchensis* maintained populations about 35 km south of San Francisco Bay, on nunataks 1000 m–1200 m in elevation in coastal Alaska, and had a major refugium south of Puget Sound (Daubenmire, 1968; Heusser, 1960; Hulten, 1937). *P. sitchensis* expanded to southern British Columbia by 8600 years ago and to Southern Alaska by 8000 years ago, to Prince William Sound 4000 ago, and to Kodiak 3000 years ago. It is still expanding its range westward of Kodiak, approximately 1.6 km per century (Griggs, 1934; Heusser, 1960).

Picea sitchensis is extremely diverse throughout its range and appears to be closely adapted to current climatic conditions (Falkenhagen, 1978; Hamrick et al., 1992). *P. sitchensis* has a limited tolerance for moisture stress, so is restricted to fog-influenced coastline in the southern part of its range, and only reaches timberline in the perhumid and subpolar rainforest types (Beadle & Jarvis, 1977; Watt, Nelson, & Jarvis, 1976). *P. sitchensis* also has a relatively high salt tolerance, which explains in part its dominance along beaches, particularly from British Columbia northward (Cordes, 1972). Geographic or microclimatic location of seed source can be used to predict cone and seed morphology. Cone scale length:width can be used to determine position on inland–coastal gradient (including introgression with *P. glauca*) (Daubenmire, 1968; Falkenhagen, 1977, 1978). Introgression between *P. sitchensis* and *P. glauca* can also be identified using isoenzyme techniques (Copes & Beckwith, 1977).

Genetically based differences in seedling growth appear to be related to timing of bud burst and the onset of dormancy, not photosynthetic efficiency (Burley, 1965a, 1966). Bud burst is generally proportional to latitude of the origin of the seed, as is nuclear volume (Burley, 1965b; Roche 1969). Growth is also genetically controlled in relation to both temperature regime and day length, with northern varieties growing and reaching bud burst at lower temperatures (growing degree days). In northern areas, southerly seed sources will produce faster-growing trees but with increased risk of injury by frost (since bud burst is earlier and dormancy is later).

In England, for example, *Picea sitchensis* is one of the primary species used for reforestation or afforestation (Cannell, 1984). At low elevations, seeds from Washington state (40° N) grow the best, but they are frost susceptible. At higher elevations or a farther north, seed from the Queen Charlotte Islands in British Columbia (54° N) appear to be better adapted (greater frost tolerance). In Oregon, Washington, and southern British Columbia, *P. sitchensis* is not normally planted, because of its susceptibility to the white pine weevil (Ruth & Harris, 1979). The narrow distribution of *Picea* along the coast (often 1 km or less) in the seasonal and warm temperate rainforests may be because it is the only microclimate in those regions with cool-enough summer temperatures to deter weevil development. From the Queen Charlotte Islands north, there are insufficient growing degree days for weevils to complete their life cycle, so they do not affect leader growth of *P. sitchensis* (McMullen, 1976).

Tsuga heterophylla also appears to have a high level of genetic diversity, suggesting significant potential for improvement through breeding or for adaptation to climate change (Foster, Campbell, & Adams, 1985; Meager, 1976). *Tsuga* has a strong tendency for outcrossing and is in the same general range of genetic diversity as *Pseudotsuga* and *Picea*. *Tsuga* is considered most closely related to *Picea* but rarely hybridizes with other species. Hybrids between the closely related. *T. mertensiana* are very rare, even though they often occur in mixed stands together. Coastal populations of *Tsuga* in Oregon form a north–south cline of variation; however, interior populations have a more complex pattern that cannot be predicted by latitude (Kuser, 1980; Kuser & Ching, 1980). Significant variation in frost hardiness, date of bud burst and growth characteristics occur along this gradient as well as along the whole latitudinal gradient from Alaska to Oregon.

Cedars appear to have less genetic diversity than *Tsuga* or *Picea* as reflected by their relatively narrow distributions and narrower ecological amplitude. Cupressids also tend to grow slower but have much better defenses against wood decay, allowing for them to grow to great ages (1000 years or more; Hennon & Loopstra, 1991; Waring & Franklin, 1979). Cupressids tend to be stress tolerators and specialize in colonizing extreme sites, such as wet microsites, steep rocky terrain, limestone substrates, or riparian areas. Because of this, they appear to be particularly sensitive to changes in climatic and soil conditions (Demeo, Martin, & West, 1992; Hennon, Hansen, & Shaw, 1990; Martin, 1989). A major anomaly is *Thuja plicata*, which has a relatively wide distribution yet appears to have low genetic diversity both in terms of isoenzyme analysis and leaf oil terpenes (Copes, 1981; Von Rudloff & Lapp, 1979). Tests of *Thuja* in Scandinavia suggest that frost tolerance and leaf blight are genetically controlled (Søegaard, 1966). In plantation trials in Poland and Denmark, Alaskan seed sources show much poorer growth but generally show more resistance to frost than southerly seed sources. In the northern limit of seasonal rainforests on Vancouver Island and on the adjacent mainland coast of British Columbia, *Thuja* achieves its most extensive distribution, particularly in riparian and moist to wet sites (Pojar et al., 1987). In the perhumid zone, *Thuja* is mostly restricted to wet, nutrient-poor sites (DeMeo et al., 1992; Ver Hoef, 1985).

Chamaecyparis nootkatensis, a native of the seasonal and perhumid rainforest types, also appears to have a relatively low level of genetic diversity (Russell and Cartwright, 1991). Its highly fragmented and scattered populations may reflect effects of recent glacial events on the northern Pacific coast. In the perhumid rainforest zone, *Chamaecyparis* has very inconsistent seed output and often fails to regenerate naturally. In old-growth forests, it may maintain itself through layering. It appears that seed is often shed when immature (Colangeli, 1991; Owens & Molder, 1975). Seeds are produced in warm climates within one season, whereas in cool climates it may take two more seasons, with more variability in quality. Physiological studies of *Chamaecyparis* also suggest a narrow environmental tolerance. Photosynthesis appears to be very sensitive to moisture stress, with a 50% reduction occurring from 1 kPa to 2 kPa (Grossnickle & Russell, 1991). *Chamaecyparis* is also extremely sensitive to root temperature (more so than other native conifers in the Pacific Northwest region), so it may benefit significantly from higher temperatures, as predicted under a global-warming scenario (Grossnickle, 1991).

Only limited information is available on the genetic architecture of *Chamaecyparis*. Preliminary provenance studies in British Columbia suggest that *Chamaecyparis* may have more potential for genetic improvement than *Thuja*, but not as much as *Pseudotsuga*, *Tsuga*, or *Picea* (Akasi & Kawamura, 1977; Russell & Cartwright, 1991). A weak relationship exists between seedling frost tolerance and latitude of provenance. Seedling height growth appears to have a significant additive genetic component, mostly at the family level, rather than the provenance level. The closely related analog for this species in Japan, *C. obtusa*, also appears to have low levels of genetic diversity and has a limited natural distribution (Shiraishi, Kaminaka, & Ohyama, 1987). In addition, tree plantations appear to have significantly less diversity than natural ones (as contrasted with those of *Pseudotsuga* and *Picea*, where this relationship does not appear to exist). *C. lawsoniana* appears to be significantly more diverse than the other *Chamaecyparis* species but is still only half or less the expected heterozygosity of dominant widespread conifers such as *Picea* and *Pseudotsuga*. *C. Lawsoniana* occurs only in the warm temperate rainforest zone and nearby mesophytic climatic zones. Moderate levels of variation occur in the coastal zone, and low diversity exists in interior populations (Millar & Marshall, 1991).

Several major native forest trees from Tasmania and New Zealand also appear to have relatively low levels of genetic diversity and limited distributions (Hawkins & Sweet, 1989b, 1889c), perhaps because of "evolutionary bottlenecks" during Pleistocene glaciations. Perhaps similar evolutionary bottlenecks occurred for some of the North American Cupressid species. These less diverse species would be expected to have less capability of adapting to rapid changes in climate as compared to dominant and widespread conifers such as *Tsuga heterophylla* and *Picea sitchensis*, both because of narrower genetic variation and more restrictive environmental requirements.

Tree Species Distributions

The northern Pacific coast, because of its cool temperatures and influence of nearby cold continental climates, provides ideal growing conditions for conifers. Conifers are overwhelmingly dominant throughout the Pacific forest belt, especially in the northern extension of the rainforest belt, as shown in Figure 7.2. In the southern extension of rainforest (seasonal and warm temperate), conifers are physiologically the best adapted to current climatic conditions, because of the prevalence of summer drought and cool growing conditions year-round (Waring & Franklin, 1979).

Most hardwoods in the region are deciduous and are therefore poorly adapted to the climate, since they cannot photosynthesize during the fall, winter, and spring, when mild temperatures and high moisture availability often occur. Leaf flushing of hardwoods usually does not occur until May or June, when the summer drought often begins. At high latitudes, the growing season becomes even shorter and less suitable for deciduous species, except as ephemeral disturbance-dependent species. In the

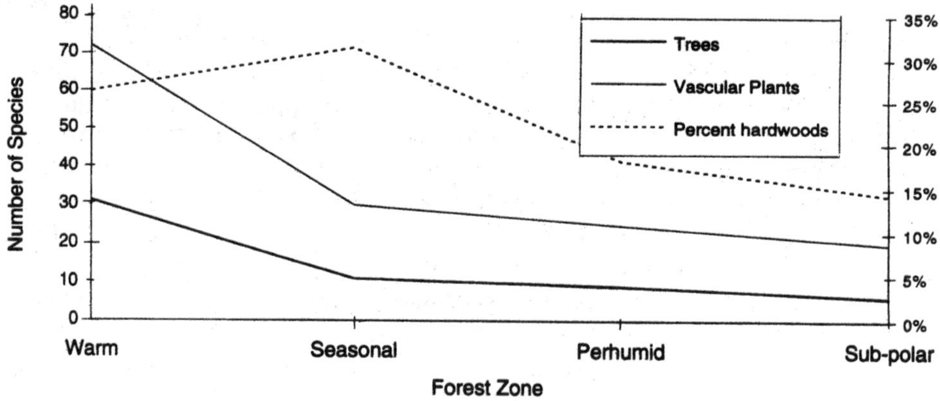

Figure 7.2. Forest diversity by rainforest zone. Tree data derived from distributional maps (Little, 1971, 1976); vascular plant data derived from 0.04 ha plot data on productive *Tsuga* forests (Eck, 1984; Hawk, 1977; Judav, 1976; Martis, 1989; Ver Hoef, 1985).

northern extension of rainforest, conifers are even more dominant than in the south, perhaps because conifers are more efficient at photosynthesis at low temperatures (Sprugel, 1989) and are better able to tolerate low nutrient availability and acid soils. At high latitudes, the multilayered canopy of conifers may be more efficient at capturing solar radiation, because of the low angle of sunlight (Kuuluvainen, 1992). The high density of leaf area may also be more efficient at intercepting low levels of light resulting from persistently cloudy skies combined with a shallow sun angle (Alaback & Tappeiner II, 1991; Gholz, Fitz, & Waring, 1976). The lack of hardwoods in the narrow belt of rainforest on the Pacific coast may also be attributed to the lack of cold-tolerant evergreen hardwood species in adjacent regions from which they might have migrated. Although evergreen hardwoods can tolerate the cool winters that typify the northern rainforest region, few can tolerate the harsh winters that characterize much of the interior forests of North America. This lack of hardwoods relative to South America may also reflect the greater development of aridity in the interior of the continent, leading to the early loss of evergreen broadleaf species in North America.

A high degree of similarity exists between the floras of each of the major forest zones on the Pacific coast. Most of the flora evolved from the arcto-tertiary flora and from refugia in the north (Hulten, 1937). Only in the warm temperate forest zone is there any significant influence from southern refugia (Whittaker, 1960). In the warm coastal temperate forest, more than 80% of the species occur only in that zone or farther east or south in drier and more continental climates. This high level of endemism mostly reflects the intrusion of dry local climates and the complex interplay of oceanic and continental influences in the Klamath Mountain region near the Oregon–California border. Many species there occur throughout the mountainous west in xeric to mesic habitats. In addition, the warm coastal temperate forest also includes several species from the adjacent Mediterranean zone and closely allied fire-adapted species (e.g., *Pinus ponderosa*, *P. lambertiana*, *Lithocarpus densiflorus*).

As has been frequently noted with other terrestrial communities, tree species diversity steadily declines with latitude across the entire Pacific rainforest zone (see Figure 7.2). (Adams & Woodward, 1989; Rapoport, 1982). A decline occurs most consistently with both conifer and hardwood tree species. As contrasted with South America, this decline in tree species is much more gradual in the perhumid and subpolar rainforest zones, further illustrating the minimal effects of Pleistocene glaciations (Alaback, 1991; Arroyo et al., this volume).

Partly because of recent fluctuations in glaciers and climate, species in the north tend to be habitat generalists and are the least likely to be uniquely adapted to current climatic conditions (Kellison & Weir, 1987). The latitudinal range in distribution of tree species reflects these patterns (Little, 1971, 1976, see Figure 7.3). Most species in the subpolar zone

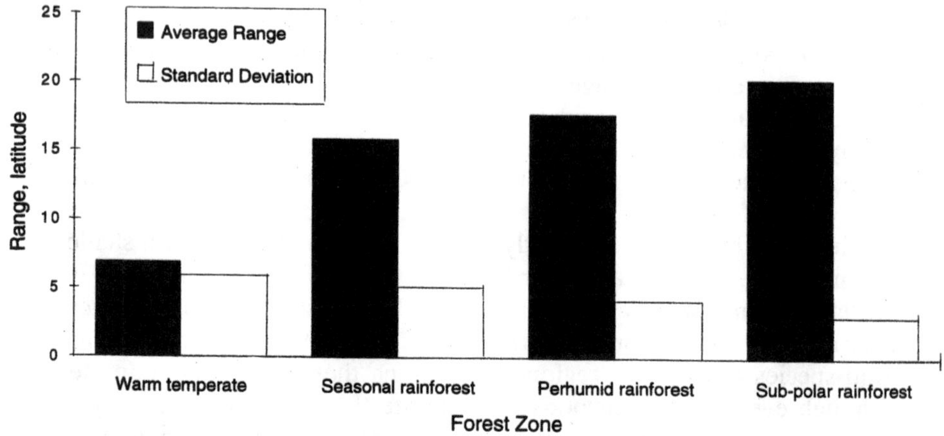

Figure 7.3. Tree species range by rainforest zone. Data derived from distributional maps of Little (1971, 1976).

reside in both the perhumid and seasonal zones. In the warm temperate zone, by contrast, the highest degree of endemism occurs (more than 80% occur only in this rainforest zone or associated interior zones). No species are endemic to the seasonal rainforest, since most of these species also occur in the warm temperate zone. In the perhumid, 9% of the species occur only in this zone (although all these species occur in interior boreal forest zones as well). In the subpolar zone, 28% of the species also occur in the interior boreal forest zone, further confirming its status as a transitional forest zone. The deviation in mean latitudinal range of tree species also declines with latitude, suggesting that the more southerly zones include a broad range of both habitat specialists and generalists.

Vascular Plant Species Richness

Vascular plant species richness patterns reflect a complex interplay between local climatic, historical, biogeographical, and geological factors. Including interior floras, total vascular plant species richness ranges from more than 5000 taxa in California to 1735 in Alaska (Hitchcock & Cronquist, 1973; Hulten, 1968; Munz & Keck, 1959; Taylor & MacBryde, 1977). In the warm temperate and seasonal rainforest zones, the rainforest flora is relatively impoverished as compared to the interior. In the northern coast, however, the rainforest region has a much higher diversity than the interior. In Alaska, 70% of the species occur on less than 20% of the land area in the rainforest region. In the subpolar region, however, the total flora is only 550 taxa (Farr & Hard, 1987; Hulten, 1968). Recent glacial history plays a key role in constraining diversity in these northern regions. For example, in a 496,000 ha region at the southern end of the subpolar type (the Yakutat forelands), only 350 taxa have been recorded (Sheppard, USDA Forest Service Chatham Area, Sitka, Alaska, personal communication). A maximum of 329 species are estimated to occur on the recently glaciated coast from Yakutat north to Prince William Sound, with more than two thirds found in rainforest habitats (Borchers & Wertheim, 1992; Farr & Hard, 1987; Hulten, 1968). Within the perhumid zone of southeast Alaska, approximately 1200 taxa are known to occur (Hulten, 1968; Muller, 1977; USDA Forest Service, unpublished manuscript). Species richness varies substantially by subregion (Farr & Hard, 1987). One of the small northern subregions, Lynn Canal fjord, has the richest flora (804 species) of any subregion, because of its linkage to a wide range of interior climates in British Columbia and the Yukon (it is close to both) and because of the pronounced rainshadow within the fjord, allowing for the development in close proximity of boreal, fire-dominated, and rainforest communities as well as a broad range of alpine and other nonforest communities. The other coastal subzones vary from 497 to 520 species, with no obvious latitudinal trend.

Species richness within forest habitats in the temperate rainforest zone is generally a small fraction of that which occurs over the whole landscape. In the perhumid zone, for example, only 205 species out of a flora of more than 1000 species are considered forest species (Martin, 1989; Muller, 1977). Surprisingly little variation occurs in the proportion of plant species belonging to different structural groups that grow on forest habitats. Whereas approximately 60% or more of the flora is composed of woody species in tropical rainforests and Southern Hemisphere temperate rainforests, in all the central to northern temperate rainforests on the Pacific coast, 55%–75% of the species are herbaceous, with only 13%–17% of the species being woody (compare with Arroyo et al., this volume). Graminoid species are particularly poorly represented in rainforest regions in general and in forest habitats in particular. The slight increase in proportion of graminoid species with latitude thus represents not an increase in graminoid species (it is actually a substantial decrease) but a substantial impoverishment of herbaceous species.

Within the perhumid zone, species diversity varies substantially by forest type and productivity class. On a per plot basis (0.04 ha), the greatest species richness in the perhumid zone is attained on sites with the lowest productivity (DeMeo et al., 1992). This is primarily because the least productive sites have the most open overstory canopy, so more solar radiation is available for the most diverse vegetation strata—the shrubs, herbs, and bryophytes. Owing to the diversity of microsites and hydrological conditions, peatland forests have the highest stand level species diversity, including many forest, alpine, and riparian species as well as endemics to this habitat. Unproductive forests are the next most diverse on this basis, and productive forests are the least diverse (DeMeo et al.,

1992; Martin, 1989). Less productive sites also have greater development of a woody shrub stratum, whereas the more productive sites have a better developed fern and herb flora. This pattern of the relative abundance of shrub and herb species is also found in southern regions where the primary environmental gradient is a moisture gradient rather than a nutrient availability gradient (Gagnon & Bradfield, 1987; Kojima & Krajina, 1975; Zobel, McKee, & Hawk, 1976). It is important to note however, that many of the species that occur in productive forests, especially riparian forests, are habitat specialists, particularly the bryophytes and lichens, whereas unproductive sites are dominated by generalist species, many of which have circumpolar distributions. From a regional biodiversity standpoint, these productive sites are therefore of high value even though their species richness is often low.

Rainforest understory shrubs and herbs provide a more complex pattern with latitude. Since these plants are responding as much to microclimatic conditions created by overstory species structure as to macroclimate, high species diversity can occur at relatively high latitudes, as shown in Figure 7.4. The average number of shrub and herb species in productive *Tsuga*-dominated forests on 0.04 ha study plots, for example, varies between 20 to 30 species across the entire rainforest zone. The total number of species in productive *Tsuga* forests, however, declines dramatically from more than 60 species in the Klamath Mountain area to 20 or fewer in the subpolar zone in Alaska (Borchers, Wattenbarger, & Ament, 1989; Cordes, 1972; Eck, 1984; Hawk, 1977; Juday, 1976; Kratz, 1975; Quay, 1982; Ver Hoef, Neiland, & Glenn-Lewin, 1988). A significant leap in

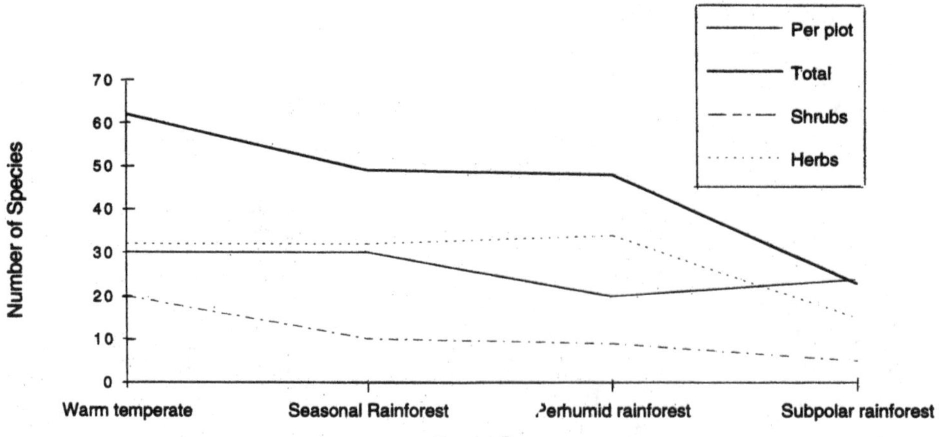

Figure 7.4. Understory species diversity. Data derived by pooling together all species that occurred on 0.04 ha plots on productive *Tsuga* forests (Eck, 1984; Hawk, 1977; Juday, 1976; Ver Hoef, 1985).

species richness occurs in the warm temperate zone where the flora of interior climates intergrades with the rainforest flora and broadleaf evergreens reach their northern limit of distribution. There is relatively little published information available on community structure and its relation to microclimate and soils in the most diverse rainforest regions (Hawk, 1977; Waring & Major, 1964; Whittaker, 1960).

A major exception to this general trend of species diversity declines with latitude appears to be the bryophytes. High levels of species diversity exist in high-latitude rainforest regions, with no obvious decline with latitude (Jarman, Kantrilas, & Brown, 1991; Worley, 1970). It appears that northern areas provide a greater array of recently deglaciated and poorly drained sites, which offer minimal competition from vascular plant species. The cool, persistently wet climate provides optimal conditions for bryophyte growth. Bryophytes are better able to grow under extremely low light conditions, which prevail at high latitudes (e.g., Alaback, 1982). Much more detailed information is needed on these poorly understood taxa, especially upper-canopy epiphytes, which are the most poorly collected.

Landscape-Level Diversity Patterns

Biodiversity patterns at the landscape scale have important implications to wildlife habitat and to the development of conservation strategies in temperate rainforest regions. Just as habitat generalists appear to dominate the high-latitude regions of temperate rainforest, habitat generalists also appear to dominate poor growing sites in the region, as discussed earlier. Soils differences are critical in the temperate rainforest region, because of the limited importance of moisture stress as a physiological constraint and the strong tendency toward rapid podzolization and paludification (Ugolini & Mann, 1979). Highly podzolized or paludified soils in this region have thick organic layers, are highly acidified, and although they contain significant nutrient pools, they usually have low nutrient availability for vascular plants. Most plants depend on soil mycorrhizae to extract nutrients from these organic soils. In peatlands, soils also provide poor rooting substrates and quickly dry out during rainless days during the summer. As a consequence, drought-tolerant, thick-leafed circumpolar evergreen shrubs dominate these habitats (Neiland, 1971).

Across a gradient of soil moisture and nutrient regime in the perhumid and subpolar zone, a distinct pattern in biodiversity develops. The highest diversity occurs in meadows along the ocean, lakes, and recent glacial areas. This is also the zone most likely to be affected by human disturbance. In terms of absolute number of species, forests, peatlands, and alpine meadows have the greatest richness. If species richness per unit land area is considered, then riparian forests have far and away the highest species richness, especially for nonvascular plant species. Peatland and other forest types are then a distant second, with alpine having the lowest diversity. Peatland and alpine habitats have the highest proportion of circumpolar species, whereas riparian forests have the highest endemism. A nearly opposite ranking of plant communities occurs if species packing in small survey plots (<0.25 ha) is measured, as discussed previously.

Disturbance regime also plays a key role in regulating landscape diversity. Recently disturbed sites, especially in the southern portion of the coastal forest zone, and old-growth forests have the highest level of diversity. Immediately following disturbance, graminoid and herbaceous species often colonize sites, significantly elevating diversity. Within 5 to 10 years, these species are shaded out by woody shrubs and tree seedlings (Alaback, 1982). In the perhumid and subpolar zones, these increases in species richness are less significant than in the seasonal or warm temperate zones. Following the development of a dense conifer forest canopy, most understory plants are shaded out, especially on productive, well-drained sites, leading to a substantial deterioration in wildlife habitat values (Alaback, 1982; Wallmo & Schoen, 1980).

In the perhumid zone, catastrophically disturbed upland sites may take as long as 300 to 400 years to attain the same level of diversity as old-growth forests. Riparian forests may take 500 years or more to attain maximal diversity of vascular and particularly nonvascular plant species following catastrophic disturbance (Alaback, in press). Because of the tight coupling between hydrology, geomorphology, and vegetation processes, disturbance may also destabilize these ecosystems, leading to shrub- or hardwood-dominated forest (Schrader, 1991). Some of the most diverse landscapes have a long history of disturbance, including both catastrophic and small-scale maintenance disturbance (Veblen & Alaback, this volume; Veblen, Schlegel, & Oltremari, 1983).

Across the gradient from warm temperate to subpolar rainforest, landscape patchiness and the grain of the mosaic also vary substantially. In the subpolar zone, forest occupies less than 14% of the land surface and includes a relatively depauperate forest, usually on steep slopes. Gentle terrain is dominated by wetland types such as peatland, bogs, fens, and alpine meadows. This results in an extremely fine-grained mosaic with relatively stable natural edges, particularly between forest and bog. Forest cover is a clear limiting factor for overwinter survival for a variety of animal species (Eck, 1984; Parker, 1988). In the perhumid zone, forest occupies approximately 30% of the land area but is mostly composed of open unproductive forest. Only 4%–8% of the land area is occupied by productive close-canopy forest, with high wildlife cover values. In the seasonal and warm temperate forest zones, the great majority of land area is capable of supporting forest, often in a highly productive close-canopied ecosystem (Alaback & McClellan, 1992). The mosaic in these regions is a function more of fire history (natural or human caused) rather than intrusion of wetlands or other edaphic types (Agee, 1991). In these forest types, cover is not considered as serious a constraint for wildlife as it is in more northern forests.

Structural Diversity

Old-growth forests are a distinctive feature of the natural vegetation of the entire Pacific forest region (Alaback & Juday, 1989; Franklin et al., 1981; Spies & Franklin, 1991). These forests typically develop for 2 to 5 centuries after a catastrophic disturbance, or they may be essentially ageless and are maintained by periodic windthrow or other forms of small-scale mortality or disturbance. Although old-growth forests do not necessarily contain the highest level of species diversity in any given region, they always contain the highest level of structural diversity and are a key component of regional diversity (Spies, 1991). This diversity comes from the development of multiple canopy layers, the presence of a wide range of sizes and decay classes of snags and logs, and a complex and highly structured understory layer. Overstory canopies tend to be patchy and uneven in height, allowing for the development of a variety of understory species, each responding to small-scale disturbances of different intensity, frequency, or phenology and to local microsite conditions. Small clonal herbs and many shrubs seem to thrive particularly in old-growth forests where canopy gaps are frequently created by disease or wind (Tappeiner & Alaback, 1989).

Although it is often implied that secondary forests are inherently less diverse than old-growth forests because of the age of dominant trees, secondary forests with high structural diversity are usually able to provide habitat for a wide range of forest species. Naturally created secondary forests may maintain a high level of diversity, depending on the nature of the disturbance. Fire-origin forests, for example, often display a large carryover from the previous stand, including residual dominant trees, snags, and logs (Juday, 1976; Spies & Franklin, 1991; Spies, Franklin, & Thomas, 1988). Understory propagules are often associated with residual trees as well. Multiple canopies can also develop in these younger forests if the original disturbance was patchy. Managed forests or anthropogenic forests in general are much simpler structurally (Alaback & Herman, 1988; Alaback & McClellan, 1992). Snags and logs are much less common and the overstory canopy is more uniform, leading to a more uniform depauperate understory (Alaback, 1982; Spies, 1991). What limited data are available also suggest that insects are much less diverse in these simplified managed ecosystems (Schowalter, 1989). There have not been any studies in either the mesophytic or temperate rainforest regions in which the ecological characteristics of managed forests have been directly contrasted with those of natural secondary forest. These comparisons are needed to identify the linkage between structure and function in the maintenance of biological diversity in these forests and to provide guidance for land managers on how to enhance the functional characteristics of managed forests.

Comparisons with Ecological Analogs in the Southern Hemisphere

The temperate rainforests of North America would be expected to be more diverse than their ecological equivalents in the Southern Hemisphere, because of their recent connection to a large and diverse temperate forest biome and the diversity of climates that occur in the region. Several other complicating factors both in history and current climate appear to counteract this influence, resulting in roughly equivalent species diversity in the principal rainforest regions of the world.

As shown in Figure 7.1, Chile has a significantly more equitable climate than ecologically equivalent ecosystems in North America, both in terms of annual variation in temperature and in distribution of rainfall. Chile also has a significantly more exposed coastline, with higher storm frequencies and higher wind velocities, resulting in a much more extensive subpolar rainforest than in North America. Seasonal rainforest, arising primarily as a transitional type between more northerly drier climates and more continental climates, is much more extensive in North America, perhaps again as a result of the greater continental mass in temperate North America as compared to South America. It appears that the Chilean flora has a close association with tropical and subtropical regions to the north, from which many of the species probably evolved. Although regional vascular plant diversity in Chile appears to be substantially less than that in North America, forest species richness is equal or greater in Chile than in North America (Arroyo et al., this volume). See Table 7.1. In the perhumid zone of both hemispheres, approximately 250 species occur in forest habitats (Arroyo et al., this volume; Martin, 1989). One of the principal contrasts in the forest flora of the two regions is much higher regional endemism (or smaller average geographic distributions) in Chile, whereas high regional endemism occurs only in the warm temperate zone in the Pacific coastal forest of North America.

The rainforests of Tasmania provide an interesting contrast with those of North America and Chile with respect to the influence of both evolutionary and human factors on biodiversity patterns (Brown & Read, this volume; Jarman et al., 1991; Read et al., 1990). Tasmania has a relatively isolated rainforest flora, with large extinctions associated with the development of cooler climates since the Tertiary. Poor migration cor-

Table 7.1. Estimates of species richness for selected temperate rainforest regions of the Northern and Southern Hemispheres. Land-area estimates taken from GIS-derived data presented in Kellogg (1992)

Zone	Species Richness (species/in [km^2])				
	Land Area (km^2)[1]	Number of species[1]	North America[1]	Chile[2]	Tasmania[3]
Total rainforest	207,000	450	84.65	87.42	21.26
Seasonal	37,440	450	98.40	96.80	—
Perhumid	140,000	233	45.28	49.82	—
Subpolar	28,000	205	46.10	—	—

[1] Data for North America only; species estimates from Martin (1987), Borchers and Wertheim (1992), and Henderson et al. (1989)
[2] Data from Arroyo et al. (this volume).
[3] Data from Brown and Reed (this volume).

ridors and local isolation of the flora have accentuated these losses. The flora has been further affected by a long occupation by aborigines (20,000–30,000 years), resulting in widespread burning and forest clearing. The Tasmanian rainforest has the lowest total species richness of vascular plants of any major temperate rainforest region (70 angiosperms, 6 gymnosperms, and 49 ferns; refer to Table 7.1). The bryophyte and lichen flora is also smaller than in the other rainforest regions, but to a lesser degree than vascular plant species (168 species mosses and liverworts, 217 lichen species). Thus, although the vascular flora is only 35% as rich as for rainforests in North America, bryophytes are approximately 18% richer in species than in the perhumid zone of North America (Worley, 1970), but approximately 30% fewer lichen species occur than in the perhumid region of North America (Geiser, USDA Forest Service, Petersburg, Alaska, personal communication). The Tasmanian rainforest is mostly in the seasonal and perhumid zones and has a mild oceanic climates, much like southern Chile. The Tasmanian flora is most closely associated with drier and warmer ecosystems in Australia and, to a lesser extent, New Zealand and southern South America (Brown & Read, this volume).

It appears that despite important differences in climate, history, and biogeography, significant covergence in biodiversity patterns have developed with respect to the principal temperate rainforest regions of the world. Similar levels of species richness occur within forest habitats of the two principal rainforest regions, and similar species richness of nonvascular plants has developed between relatively isolated and small temperate rainforest regions and the larger ones. This present-day similarity may not necessarily imply that these systems will respond similarly to climate change or human disturbance. If the preliminary data on low levels of genetic diversity of trees from the Southern Hemisphere are generally representative of these forests, then the Southern Hemisphere would be expected to have a higher sensitivity to climate change than the Northern Hemisphere. High levels of endemism and widespread use of fire following forest clearing (but a low natural frequency of fire) suggest a high level of impact to Southern Hemisphere forests both from climate change and human impact. In the Northern Hemisphere, by contrast, regional endemism is low, but sensitivity to changes in disturbance regime, climatic constraints, and landscape structure is more complex, leading to significant landscape-level changes in biodiversity patterns with both climate change and human disturbance.

Implications for Resource Management

Basic patterns of biodiversity of forest vegetation in the rainforest zone of North America suggest that considerable opportunity should exist for resource development activity without substantially compromising regional plant species richness. In the temperate rainforest region, the primary source of stress and change in forest ecosystems in the future should come from a combination of climate change and human disturbance or management (Alaback & McClellan, 1992). The greatest challenges for management of biodiversity will be maintaining an adequate area and distribution of specialized habitats. Historical cutting patterns suggest a strong impact on highly productive, low-elevation old-growth forests initially and rapid landscape fragmentation (e.g., Harris, 1984). Since the proportion of land in highly productive soil types rapidly declines with latitude, this implies greater sensitivity of high-latitude sites to landscape-level modification by human disturbance. Species that are specially adapted to these forest types will therefore require special attention to avoid significant population declines. Of particular concern are nonvascular plant species, which tend to be habitat specialists and are closely associated with large organic debris (logs and snags, or standing dead trees) (During, 1992; Harmon et al., 1986). Forest management for wood fiber production generally results in an eventual elimination of coarse woody debris, since trees do not grow for as long a period as they would under natural-disturbance regimes, and mortality is often minimized through thinning and by maintaining high levels of stand vigor. If current management practices are continued into the future, reduction of this key structural component to forest diversity may have direct constraints on nonvascular plant diversity, as has already been demonstrated in boreal forests (e.g., Søderström, 1988). Structural diversity also appears to play a key role in maintaining habitat for many animal populations (e.g., Hansen, Spies, Swanson, & Ohmann, 1991).

Ecosystem recovery following disturbance, especially that from forest clearing, appears to take longer in northern temperate rainforests than in other forest types (Veblen & Alaback, this volume). Old-growth forests may develop following catastrophic disturbance in as little as 200 years in the mesophytic (*Pseudotsuga*) forest type in the Oregon Cascades (Franklin et al., 1981; Spies & Franklin, 1991). In northern temperate rainforests, by contrast, many understory species appear to require the disturbance or microclimate associated with canopy gaps in mature forest to maintain viable populations (Alaback, 1982; Tappeiner & Alaback, 1989). It may take 3 or 4 centuries for upland forests, or even longer in riparian ecosystems, to develop the species richness of primary rainforest, although at some younger successional stages, high levels of diversity may be achieved for short periods of time. Riparian ecosystems provide the highest level of diversity and best representation of specialized habitats, so these should be key habitats for development of conservation strategies in temperate rainforest regions. They provide critical habitat for aquatic species, including many commercial fish species. Riparian zones also have the highest economic values for trees, are used for transportation corridors, and have high value for recreation; thus, they are a focal point for land-use decisions in general as well as for biodiversity. Relatively few pristine riparian forests exist on the Pacific coast, most of them in Alaska and British Columbia. Fewer still exist on larger stream systems. In most cases, a focus on the maintenance of biodiversity in riparian systems will require considerable attention on ecological restoration.

Careful monitoring of biodiversity changes with climate change and human land-use changes should provide a useful barometer of ecosystem- and landscape-level functioning, especially for rare or specialized species. Development of conservation strategies for maintaining biodiversity should complement management strategies that might be developed for promoting greater ecosystem sustainability and for promoting wildlife values of forests. Because of the overall similarity of climate and biodiversity patterns of the principal temperate rainforest regions of the world, these bioregions would seem to be particularly valuable places to focus future effort on comparing the cumulative effects of land-use change and climate change across the contrasting gradients of North America and the Southern Hemisphere.

References

Adams, W.T. 1992. Gene dispersal within forest tree populations. *New For*, 6, 217–240.

Adams, W.T., and Birkes, D.S. 1990. Estimating mating patterns in forest tree populations. *Proceedings of the International Workshop on Plant Biology, Biochemical Markers in Population Genetics of Forest Trees*. Institute for Agroforestry of the National Research Council of Italy, Porano-Orvieto, Italy.

Adams, J.M., and Woodward, F.I. 1989. Patterns in tree species richness as a test of the glacial extinction hypothesis. *Nature*, 339, 699–700.

Agee, J.K. 1991. Fire history of Douglas-fir forests in the Pacific Northwest. In L.F. Ruggiero, K.B. Aubry, A.B. Carey, and M.H. Huff (eds.), *Wildlife and Vegetation of Unmanaged Douglas-Fir Forests* (pp. 25–33). Portland, OR: U.S.D.A. Forest Service, Pacific Northwest Research Station, General Technical Report PNW-GTR-285.

Akasi, T., and Kawamura, T. 1977. Partition into genetic variance and dominance variance in height and butt diameter of *Chamaecyparis obtusa* seedlings. In Japanese Government Forest Experiment Station Bulletin No. 296 (pp. 11–18).

Alaback, P.B. 1982. Dynamics of understory biomass in Sitka spruce-western hemlock forests of Southeast Alaska. *Ecology*, 63, 1932–1948.

Alaback, P.B. 1991. Comparative ecology of temperate rainforests of the Americas along analogous climatic gradients. *Rev Chil Hist Nat*, 64, 399–412.

Alaback, P.B., and Herman, F.J. 1988. Long-term response of understory vegetation to stand density in *Picea–Tsuga* forests. *Can J For Res*, 18, 1522–1530.

Alaback, P.B., and Juday, G.P. 1989. Structure and composition of low elevation old-growth forests in research natural areas of Southeast Alaska. *Nat Areas J*, 9, 27–39.

Alaback, P.B., and McClellan, M. 1992. Effects of global warming on managed coastal ecosystems of western North America. In H.A. Mooney, E. Fuentes, and B.I. Kronberg (eds.), *Contrasts in Global Change Responses* Between North and South America. New York: Academic Press.

Alaback, P.B., and Tappeiner II, J.C. 1991. Response of western hemlock (*Tsuga heterophllya*) and early huckleberry (*Vaccinium ovalifolium*) seedings following forest windthrow. *Can J For Res*, 21, 534–539.

Axelrod, D.I. 1976. *History of Coniferous Forests*. Berkeley: University of California Press. Berkeley.

Axelrod, D.I., Arroyo, M.T.K., and Raven, P.H. 1991. Historical development of temperate vegetation in the Americas. *Rev Chil Hist Nat*, 64, 413–446.

Beadle, C.L., and Jarvis, P.G. 1977. The effects of shoot water status on some photosynthetic partial processes in Sitka spruce. *Physiol Plant*, 41, 7–13.

Bierzychudek, P. 1982. Life histories and demography of shade-tolerant temperate forest herbs: A review. *New Phytol*, 90, 757–776.

Borchers, S.L., Wattenbarger, J., and Ament, R. 1989. Forest plant associations of Montague Island, Chugach National Forest: Preliminary results. In E.B. Alexander (ed.), *Watershed 89, Proceedings of the Conference* (pp. 29–46). Juneau, AK: U.S.D.A. Forest Service, Region 10.

Borchers, S.L., and Wertheim, C. 1992. *Forest Plant Identification Guide: Chugach National Forest*. Juneau, AK: U.S.D.A. Forest Service, Alaska Region, R10-MB-189.

Brubaker, L.B. 1988. Vegetation history and anticipating future vegetation change. In J.K. Agee and D.R. Johnson (eds.), *Ecosystem Management for Parks and Wilderness* (pp. 41–61). Seattle: University of Washington Press.

Burley, J. 1965a. Genetic variation in *Picea sitchensis* (Bong), Carr. Unpublished doctoral dissectation, Yale University, New Haven, CT.

Burley, J. 1965b. Karyotype analysis of Sitka spruce, *Picea sitchensis* (Bong) Carr. *Silv Genet*, 14, 127–132.

Burley, J. 1966. Genetic variation in seedling development of Sitka spruve, *Picea sitchenis* (Bong) Carr. *Forestry*, 39, 68–94.

Campbell, R.K. 1979. Genecology of Douglas-fir in a watershed in the Oregon Cascades. *Ecology*, 60,1036–1050.

Cannell, M.G.R. 1984. Exploted plants—Sitka spruce. *Biologist*, 31, 255–261.
Chapin III, F.S., Walker, L.R., Fastie, C.L., and Sharman, L.C., 1994. Mechanisms of primary succession following deglaciation at Glacier Bay, Alaska. *Ecol Monograph*. 64, 2, 149–175.
Colangeli, A.M. 1991. Reproductive biology: Pollen/seed/seed orchards. In J. Daniel Lousier (ed.), Yellow Cypress: Can We Grow It? Can We Sell It? (pp. 29–30). British Columbia Ministry of Forests, FRDA Report 171.
Conkle, M.T. 1981. Isozyme variation and linkage in six conifer species. In M.T. Conkle (ed.), *Proceedings of the Sympesium on Isozymes of Narth American Forest Trees and Forest Insects*. San Francisco, CA: U.S.D.A., Forest Service, Pacific Southwest Research Station, General Technical Report PSW 48.
Copes, D.L. 1981. Isoenzyme uniformity in western red cedar seedlings from Oregon and Washington. *Can J For Res*, 11, 451–453.
Copes, D.L., and Beckwith, R.C. 1977. Isoenzyme identification of *Picea glauca*, *P. sitchensis*, and *P. lutzii* populations. *Bot Gaz*, 138, 512–521.
Cordes, L.D. 1972. *An autecological study of the Sitka spruce forest on the west coast of Vancouver Island*. Unpublished doctoral dissentation, University of British Columbia, Vancouver, British Columbia, Canada.
Daubenmire, R. 1968. Some geographic variations in *Picea sitchensis* and their ecologic interpretations. *Can J Bot*, 46, 787–798.
DeMeo, T., Martin, J.R., and West, R. 1992. *Forest Plant Association Guide, Ketchikan Area, Tongass National Forest*. Ketchikan, AK: U.S.D.A., Forest Service, Tongass National Forest, Ketchikan Area.
During, H.J. 1992. Endangered bryophytes in Europe. *Tr Ecol Evol*, 7, 253–254.
Eck, K.C. 1984. Forest characteristics and associated deer habitat values, Prince William Sound Islands In W.R. Meehan, T.R. Merrell, and T.A. Hanley (eds.), *Proceedings of the Sympesium on Fish and Wildlife Relationships in Old-Growth Forests, April 12–15, 1982* (pp. 235–246). Putland, OR: American Institute of Fishery Research Biology and U.S.D.A., Forest Service, Pacific Northwest Research Station.
Falkenhagen, E.R. 1977. Genetic variation in 38 provenances of Sitka spruce *Silv Genet*, 26, 67–75.
Falkenhagen, E.R. 1978. Parent tree variation in Sitka spruce provenances, an example of fine geographic variation. *Silv Genet*, 27, 24–28.
Farr, W.A., and Hard, J.S. 1987. *Multivariate analysis of Climate Along the Southern Coast of Alaska: Some Forestry Implications*. Portland, OR: U.S.D.A., Forest Service, Pacific Northwest Research Station, General Technical Report PNW-RP-372.
Farr, W.A., and Harris, A.S. 1979. Site index of Sitka spruce along the Pacific coast related to latitude and temperature. *For Sci*, 25, 145–153.
Fastie, C. 1990. Inference and verification in chronosequence studies at Glacier Bay. In A.M. Milner and J.D. Wood, Jr. (eds.), *Proceedings of the Second Glacier Bay Sci Symposium, September 19–22, 1988* (pp. 147–149). Glacier Bay Lodge, AK, National Park Service, Anchorage, AK.
Foster, G.S., Campbell, R.K., and Adams, W.T. 1985. Clonal selection prospects in western hemlock combining rooting traits with juvenile height growth. *Can J For Res*, 15, 488–493.
Franklin, J.F., Cromack, K., Denson, W., McKee, A.W., Maser, C., Sedell, J., Swanson, F., and Juday, G.P. 1981. *Ecological characteristics of old-growth Douglas-fir forests*. Portland, OR: U.S.D.A., Forest Service, Pacific Northwest Research Stations General Technical Report PNW-118.
Franklin, J.F., and Dyriess, C.T. 1973. *The Natural Vegetation of Washington and Oregon*. Portand, OR: U.S.D.A., Forest Service, Pacific Northwest Forest Range Experirment Station, General Technical Report PNW-8.

Franklin, J.F., Swanson, F.J., Harmon, M.E., Perry, D.A., Spies, T.A., Dale, V.H., McKee, A., Ferrell, W.K., Means, J.E., Gregory, S.V., Lattin, J.D., Schowalter, T.D., and Larsen, D. 1991. Effects of global climatic change on forests in northwestern North America. *New Env J*, 7, 233–254.

Gagnon, D., and Bradfield, G.E. 1987. Gradient analysis of west central Vancouver Island forests. *Can J Bot*, 65, 822–833.

Gholz, H.L., Fitz, F.K., and Waring, R.H. 1976. Leaf area differences associated with old-growth forest communities in the western Oregon Cascades. *Can J For Res*, 6, 49–57.

Griggs, R.F. 1934. The edge of the forest in Alaska and the reason for its position. *Ecology*, 15, 80–96.

Grossnickle, S.C. 1991. Physiological and environmental tolerances of Yellow Cypress. In J. Daniel Lousier (ed.), *Yellow Cypress: Can We Grow It? Can We Sell It?* (pp. 31–33). British Columbia Ministry of Forests, FRDA Report 171.

Grossnickle, S.C., and Russell, J.H. 1991. Gas exchange processes of yellow-cedar (*Chamaecyparis nootkatensis* (D. Don) Spach) in response to environmental variables. *Can J Bot*, 69, 2684–2691.

Hamrick, J.K., Godt, M.W., and Sherman-Broyles, S.L. 1992. Factors influencing levels of genetic diversity in woody plant species. *New For*, 6, 95–124.

Hansen, A.J., Spies, T.A., Swanson, F.J., and Ohmann, J.L. 1991. Conserving biodiversity in managed forests. *BioScience*, 41, 382–392.

Harmon, M.E., Franklin, J.F., Swanson, F.J., Sollins, P., Gregory, S.V., Lattin, J.D., Cummins, K.W., Anderson, N.H., Cline, S.P., Aumen, N.G., Sedell, J.R., Lienkaemper, G.W., and Cromack, Jr., K. 1986. Ecology of coarse woody debris in temperate ecosystems. *Adv Ecol Res*, 15, 133–302.

Harr, R.D. 1982. Fog drip in the Bull Run municipal watershed, Oregon. *Wat Res Bull*, 18, 785–789.

Harris, L.D. 1984. *The Fragmented Forest: Island Biogeographic Theory and the Preservation of Biotic Diversity*. Chicago: University of Chicago Press.

Hawk, G.M. 1977. *A comparative study of temperate* Chamaecyparis *forests*. Unpublished doctoral dissertation, Oregon State University, Corvallis, OR.

Hawkins, B.J., and Sweet, G.B. 1989a. Evolutionary interpretation of a high temperature growth response in five New Zealand forest tree species. *New Z J Bot*, 27, 101–107.

Hawkins, B.J., and Sweet, G.B. 1989b. Genetic variation in the cold tolerance of Totara. from symposium.

Hawkins, B.J., and Sweet, G.B. 1989c. Genetic variation in rimu—an investigation using isoenzyme analysis. *New Z J Bot*, 27, 83–90.

Hennon, P.E., Hansen, E.M., Shaw III, C.G. 1990. Dynamics of decline and mortality of *Chamaecyparis nootkatensis* in southeast Alaska. *Can J Bot*, 68, 651–662.

Hennon, P.E., and Loopstra, E.M. 1991. *Persistence of Western Hemolock and Western Red Cedar Trees 38 Years after Girdling at Cat Island in Southeast Alaska*. Portland, OR: U.S.D.A., Forest Service, Pacific Northwest Research Station, Research Note PNW-RN-507.

Heusser, C.J. 1960. *Lete-Pleistocene Environments of North Pacific North America. American Geograghic Society Special Publication No. 35*.

Hill, R.S., Read, J., and Busby, J.R. 1988. The temperature-dependence of photosynthesis of some Australian temperate rainforest tress and its biogeographical significance. *J Biogeog*, 15, 431–449.

Hitchcock, C.L., and Cronquist, A. 1973. Flora of the Pacific Northwest Seattle: University Washington Press.

Hulten, E. 1937. *Outline of the History of Arctic and Boreal Biota during the Quaternary Period*. Thule, Stockholm. Bokfoer. Aktieb.

Hulten, E. 1968. *Flora of Alaska and Neighboring Territories*. Stanford, CA: Stanford University Press.
Jarman, S.J., Kantvilas, G., and Brown, M.J. 1991. *Floristic and Ecological Studies in Tasmanian Rainforest*. Tasmania: Focestry Commission, Tasmanian National Rainforest Conservation Program, Techical Report No. 3.
Juday, G.P. 1976. *The location, composition, and structure of old-growth forests of the Oregon Coast Range*. Unpublished doctoral dissertation, Oregon State University, Corvallis, OR.
Kellison, R.C., and Weir, R.J. 1987. Breeding strategies in forest tree populations to buffer against elevated atmospheric carbon dioxide levels. In W.E. Shands and J.S. Hoffman (eds.), *The Greenhouse Effect, Climate Change and U.S. Forests* (pp. 285–294). Washington, D.C.: The Conservation Foundation.
Kellogg, E. (ed.) 1992. Coastal temperate rain forests: Ecological *Characteristics, status and distribution worldwide*. Portland, OR: Ecotrust, Occasional Paper Series No. 1.
Kojima, S., and Krajina, V.J. 1975. Vegetation and environment of the coastal western hemlock zone in Strathcona Provincial Park, British Columbia, Canada. *Syesis*, 8, supp 1, 1–123.
Kratz, A.M. 1975. *Vegetational analysis of the coastal* Picea sitchensis *forest zone in Olympic National Park, Washington*. Unpublished masters thesis, Western Washington University, Bellingham, WA.
Kuser, J.E. 1980. *Provenance variation in western hemlock* (Tsuga heterophylla (Raf.) Sarg.) *seedlings*. Unpublished doctoral dissertation, Oregon State University, Corvallis, OR.
Kuser, J.E., and Ching, K.K. 1980. Provenance variation in phenology and cold hardiness of western hemlock seedlings. *For Sci*, 26, 463–470.
Kuuluvainen, T. 1992. Tree architectures adapted to efficient light utilization: Is there a basis for latitudinal gradients? *Oikos*, 65, 275–284.
Little, Jr., E.L. 1971. *Atlas of United States Trees*. Vol. 1: *Confers and Important Hardwoods*. Washington, D.C.: U.S.D.A., Miscellaneous Publication 1146.
Little, Jr., E.L. 1976. *Atlas of United States Trees*. Vol. 3: *Minor Western Hardwoods*. Washington, D.C.: U.S.D.A., Miscellaneous Publications 1314.
Martin, J.R. 1989. Vegetation and environment in old growth forests of northern southeast Alaska: A plant association classification. Unpublished masters thesis, Arizona State University. Tempe, AZ.
Meager, M.D. 1976. Studies of variation in hemlock (*Tsuga*) populations and individuals from southern British Columbia. Unpublished doctoral dissertation, University of British Columbia, Vancouver, British Columbia.
McMullen, L.H. 1976. Effect of temperature on oviposition and brood development of *Pissodes strobi* (Coleoptera: Curculionidae). *Can Ent*, 108, 1167–1172.
Millar, C.I., and Marshall, K.A. 1991. Allozyme variation of Port-Orford cedar (*Chamaecyparis lawsoniana*): Implications for genetic conservations. *For Sci*, 37, 1060–1077.
Muller, M.C. 1977. A preliminary checklist of the vascular plants in Southeastern Alaska. Juneau, AK: U.S.D.A., Forest Service, Region 10, Administrative Document 112, Wildlife and Fisheries Habitat Management Notes.
Munz, P.A., and Keck, D.D. 1959. *A California Flora*. Berkeley: University of California Press.
Neiland, B.J. 1971. The forest bog complex of southeast Alaska. *Vegetatio*, 26, 3–62.
Owens, J.N., and Molder, M. 1975. Pollination, female gametophyte, and embryo and seed development in yellow cedar (*Chamaecyparis nootkatensis*). *Can J Bot*, 53, 186–199.

Parker, K.L. 1988. Effects of heat, cold, and rain on coastal black-tailed deer. *Can J Zool*, 66, 2475–2483.

Peters, R.L. 1990. Effects of global warming on forests. *For Ecol Mgt*, 35, 13–33.

Pojar, J., Klinka, K., and Meidinger, D.V. 1987. Biogeoclimatic ecosystem classification in British Columbia. *For Ecol Mgt*, 22, 119–154.

Quay, E. 1982. *Structure and dynamics of old growth Sitka spruce* Picea sitchensis Parry. Forests of the Oregon Coastal Range. Unpublish doctoral dissertation, Oregon State University, Corvallis, OR.

Rapoport, E.H. 1982. (B. Drausal, Trans.) Areography: Geographical strategies of species. Vol. 1. New York: Pergamon Press.

Read, J., and Hill, R.S. 1985. Photosynthetic responses to light of Australian and Chilean species of *Nothofagus* and their relevance to the rainforest dynamics. *New Phytol*, 101, 731–742.

Roche, L. 1969. A genecological study of the genus *Picea* in British Columbia. *New Phytol*, 63, 505–550.

Russell, J., and Cartwright, C. 1991. The genetics of yellow cypress. In J.D. Lousier (ed.), *Yellow Cypress: Can We Grow It? Can We Sell It?* (pp. 34–35). British Columia Ministry of Forests, FRDA Report 171.

Ruth, R.H., and Harris, A.S. 1979. Management of western hemlock-Sitka spruce forest for timber production. Poutland, OR: U.S.D.A., Forest Service, General Technical Report PNW-88.

Schowalter, T.D. 1989. Canopy arthropod community structure and herbivory in old-growth and regenerating forests in westerm Oregon. *Can J For Res*, 19, 318–322.

Schrader, B. 1991. *Vegetation and conifer seedling response to clearcutting of alluvial spruce sites in Southeast Alaska*. Unpublished masters thesis, University of Washington, Seattle, WA.

Shiraishi, S., Kaminaka, H., and Ohyama, N. 1987. Genetic variation and differentiation recognized at two allozyme loci in Hinoki (*Chamaecyparis obtusa*). *J Japan For Soc*, 69, 88–93.

Søderström, L. 1988. The occurrence of epixylic bryophyte and lichen species in an old natural and a managed forest stand in northeast Sweden. *Biol Cons*, 45, 169–178.

Søegaard, B. 1966. Variation and inheritance of resistance to attack by *Didymascella thujina* in western red cedar and related species. In H.D. Gerhold et al. (eds.), *Breeding Pest-Resistent Trees* (pp. 83–87). New York: Pergamon Press.

Spies, T.A. 1991. Plant species diversity and occurrence in young, mature, and old growth Douglas-fir stands in western Oregon and Washington. In L.F. Ruggiero, K.B. Aubry, A.B. Carey, and M.H. Huff (eds.), *Wildlife and Vegetation of Unmanaged Douglas-Fir Forests* (pp. 111–121). Portland, OR: U.S.D.A., Forest Service, General Technical Report PNW-6TR-285.

Spies, T.A., and Franklin, J.F. 1991. The structure of natural young, mature, and old-growth Douglas-fir forests Oregon and Washington. In. L.F. Ruggiero, K.B. Aubry, A.B. Carey, and M.H. Huff (eds.), *Wildlife and Vegetation of Unmanaged Douglas-Fir Forests* (pp. 91–109). Portland, OR: U.S.D.A., Forest Service, General Technical Report PNW-6TR-285.

Spies, T.A. Franklin, J.F. and Thomas, T.B. 1988. Coarse woody debris in Douglas-fir forests of western Oregon and Washington. *Ecology*, 69, 1689–1702.

Sprugel, D. 1989. The relationship of evergreenness, crown architecture, and leaf size. *Am Nat*, 133, 465–479.

Stephenson, N.L. 1990. Climatic control of vegetation distribution: The role of the water balance. *Am Nat*, 135(5), 649–670.

Tappeiner II, J.C., and Alaback, P.B. 1989. Early establishment and vegetative growth of understory species in the western hemlock—Sitka spruce forests of Southeast Alaska. *Can J For Res*, 67, 318–326.
Taylor, R.L., and MacBryde, B. 1977. *Vascular Plants of British Columbia: A Descriptive Resource Inventory*. Vancouver: University of British Columbia Press, The Botanic Garden, University of British Columbia, Technical Bulletion 4.
Ugolini, F.C., and Mann, D.H. 1979. Biopedological origin of peatlands in southeast Alaska. *Nature*, 281, 366–368.
Veblen, T.T., Schlegel, F.M., and Oltremari, J.V. 1983. Temperate broadleaved evergreen forests of South America. In J.D. Ovington (ed.), *Ecosystems of the World*. Vol. 10: *Temperate Broad Leaved Evergreen Forests* (pp. 5–31). Amsterdan: E lsevier.
Ver Hoef, J. 1985. *Vegetation patterms of Two Oceas in Southeast Alaska's Forests*. Unpublished MS thesis, University of Alaska, Fairbanks, AK.
Ver Hoef, J.M., Neiland, B.J. and Glenn-Lewin, D.C. 1988. Vegetation gradient analysis of two sites in Southeast Alaska. *NW Sci*, 62, 171–180.
Von Rudloff, E., and Lapp, M.S. 1979. Population variation in leaf oil terpene composition of western red cedar, *Thuja plicata*. *Can J Bot*, 57, 476–479.
Wallmo, O.C., and Schoen, J.W. 1980. Response of deer to secondary forest succession in southeast Alaska. *For Sci*, 26, 448–462.
Waring, R.H., and Franklin, J.F. 1979. Evergreen coniferous forests of the Pacific Northwest. *Science*, 204, 1380–1386.
Waring, R.H., and Major, J. 1964. Some vegetation of the California coastal redwood region in relation to gradients of moisture, nutrients, light, and temperature. *Ecol Monog*, 34, 167–215.
Watt, W.R., Nelson, R.E., and Jarvis, P.J. 1976. Photosynthesis in Sitka spruce. *J App Ecol*, 13, 633–638.
Whittaker, R.H. 1960. Vegetation of the Siskiyou Mountains, Oregon and California. *Ecol Monog*, 30, 279–338.
Willson, M.F. 1991. Seed dispersal by frugivorous animals in temperate forests. *Rev Chil Hist Nat*, 64, 537–554.
Worley, I. 1970. A checklist of mosses of Alaska. *Bryologist*, 73, 59–71.
Wright, J.W. 1955. Species crossability in spruce in relation to distribution and taxonomy. *For Sci*, 1, 319–349.
Zobel, D.B., McKee, A., and Hawk, G.M. 1976. Relationships of environment to composition, structure, and diversity of forest communities of the central western Cascades of Oregon. *Ecol Monog*, 46, 135–156.

8. Phytogeographic Relationships and Regional Richness Patterns of the Cool Temperate Rainforest Flora of Southern South America

Mary T. Kalin Arroyo, Magaly Riveros, Alejandro Peñaloza, Lohengrin Cavieres, and Ana Maria Faggi

The western margin of southern South America bears cool temperate rainforest over a narrow landmass of some 17.5° of latitude, encompassing great physiographic diversity. The South American rainforest zone constitutes a remarkable forest island, widely separated from the closest wet, closed forest types on the South American continent by 1500 km to 2000 km. To the north, it is bound by seasonal, mediterranean-type climate scrublands and one of the driest deserts on earth, the Atacama. To the east, the rainforest intergrades with a narrow band of deciduous *Nothofagus* or *Austrocedrus chilensis* coniferous forest, followed by dry Patagonian steppe. Southward is the Antarctica continent, and westward the Pacific Ocean. In this forest island are found high productivity levels (Schmidt & Lara, 1985), the second-longest-living tree in the world—*Fitzroya cupressoides*, attaining more than 3600 years (Lara, 1993), and many obligately outbred woody species (Riveros, 1991).

Cool temperate rainforest also occurs along the Pacific coast of North America along an analagous climatic gradient (Alaback, 1991; Franklin, 1988). There, however, it commences at a higher latitude on account of the shallower temperature gradient in the Northern Hemisphere (Alaback, 1991). As in South America, rainforest is bordered by mediterranean-type climate vegetation type with a summer drought to the south. However, in contrast with South America, the rainforest region shows good eastward continuity, with extensive areas of coniferous forest comprising the vast boreal zone and forests of the Rocky Mountains (Elliot-Fisk, 1988; Peet, 1988).

Although much phytosociological work has been carried out in the cool temperate rainforests of southern South America (e.g., Dimitri & Correa, 1966–1967; Dollenz, 1980; Oberdörfer, 1960; Pisano, 1977, 1989; Ramírez & Figueroa, 1985, 1987; Schmithüsen, 1958; Villagrán, Soto, & Serey, 1974) and forest dynamics has been a strong focus (e.g., Armesto & Figueroa, 1987; Donoso, 1984a, 1993; Donoso, Escobar, & Urrutia, 1985; Veblen, 1982, 1985; Veblen et al., 1991), the overall regional diversity patterns have only been touched upon (e.g., Alaback, 1991; Dimitri, 1972). The insular nature of the South America rainforest zone, a condition in place since the late Tertiary, and therefore prior to the Pleistocene, could lead to the prediction of lower diversity than in western North American rainforests. This paper will show, however, that complex historical and climatic factors seem to have led to an opposite effect, insofar as vascular plant diversity is concerned. While the Tertiary and Quaternary left North American rainforests gymnosperm-dominated, with angiosperms becoming prominent only in specialized habitats and at the early stages of succession (Franklin, 1988), those of South America comprise an intergrading series of angiosperm-dominated, gymnosperm-dominated, and mixed gymnosperm–angiosperm forests (Armesto & Figueroa, 1987; Armesto, Mitchell, & Villagrán, 1985; Donoso et al., 1985; Veblen, Schlegel, & Oltremari, 1983) of strikingly different aspect and diversity.

After outlining the major climatic trends, in order to lay a background to present diversity patterns, we first show that the temperate rainforest constitutes an extraordinarily rich floristic assemblage, linking ancient continents, and as such is quite distinct from western North America temperate rainforest. Second, we address the life-form composition and the overall richness of the rainforest flora. Finally, the latitudinal gradient in species richness is focused on. In the northern hemisphere, mean latitudinal range increases significantly with latitude in several groups of organisms (Rapoport, 1975; Stevens, 1989). Stevens (1989) related this trend to adaptation to greater annual temperature extremes at higher latitudes. In this paper we show that there is no latitudinal increase in annual temperature range throughout the entire cool temperate rainforest zone in southern South America. It is of theoretical interest, thus to determine whether Rapoport's rule is a feature of this southern hemisphere vegetation type.

The Broad Climatic Patterns

Conditions for existence of rainforest in southern South America derive from a maritime climate with strong westerly flow along the Pacific coast during the summer and winter months, an intense rainshadow effect of the Andean mountains, and the cool equatorward-running Humboldt current. South of 30° S, the South Pacific High is displaced to around 39° to 40° S over summer months, determining a mediterranean-type climate with prolonged summer drought north of around 39° to 40° S. The Humboldt current maintains cool conditions along the entire Pacific coast, leading to humid and foggy conditions especially in coastal locations. Along this gradient, total precipitation increases from around 1500 mm at latitude 39° S to close to 4500 mm in coastal areas at latitude 47° S. Inland from 45° to 55° S, precipitation is much lower (Figure 8.1a). In the extreme southwest, precipitation can be around 1300 mm to 1400 mm (Figure 8.1a). Mean summer and winter temperature conditions along the southern Pacific have been well documented by Alaback (1991), who showed that greater summer warmth is seen in relation to the Pacific Northwest coast. Mean annual temperature trends, are shown Figure 8.1b.

A striking feature of the general rainforest region that Alaback (1991) failed to mention in his comparison of rainforest climates is high climatic equability. Figure 8.2a shows that the annual temperature range for coastal stations in South America rarely exceeds 7°C. Using the mean maximum of the warmest month and the mean minimum of the coldest month (Figure 8.2c), the temperature range is still low—not exceeding 20°C. More interestingly, there is no increase in annual temperature range along the entire Pacific coast in southern South America. There are few data for inland locations in the general rainforest area. Lonquimay (38°26′ S, 900 m) has an annual temperature range of 13.2°C. The corresponding value to Balmaceda (45°54′ S, 520 m), which is farther to the east than the rainforest area, is 12.1°C, suggesting that temperature moderation is characteristic of the rainforest zone in general. This tendency derives from the tapering southern South America landmass, leading to a decreased land/ocean ratio with latitude. A fairly low annual temperature range also characterizes mediterranean-type climate latitudes and the seasonal rainforest area in western North America (Figure 8.2b,d). However, as the perhumid and boreal rainforest regions are approached, equability diminishes, on account of an expanding land/ocean ratio. Bailey (1964) developed a nomogram that expresses equability in relation to a mean annual earth temperature of 14°C. The equability rating of all South America cool temperate rainforest types falls in the 50% to 70% range, which comes very close to areas in southeastern Brazil and northern Australia as illustrated by Axelrod (1992) (Figure 8.3). The corresponding equability ratings for the seasonal and perhumid rainforest areas of western North America overlap with those of southern South America. However, that of the boreal rainforest is considerably lower (40%). The South American temperate rainforest zone thus exhibits not only a warmer winter temperature, but greater overall regional uniformity in terms of temperature extremes than its North American counterpart.

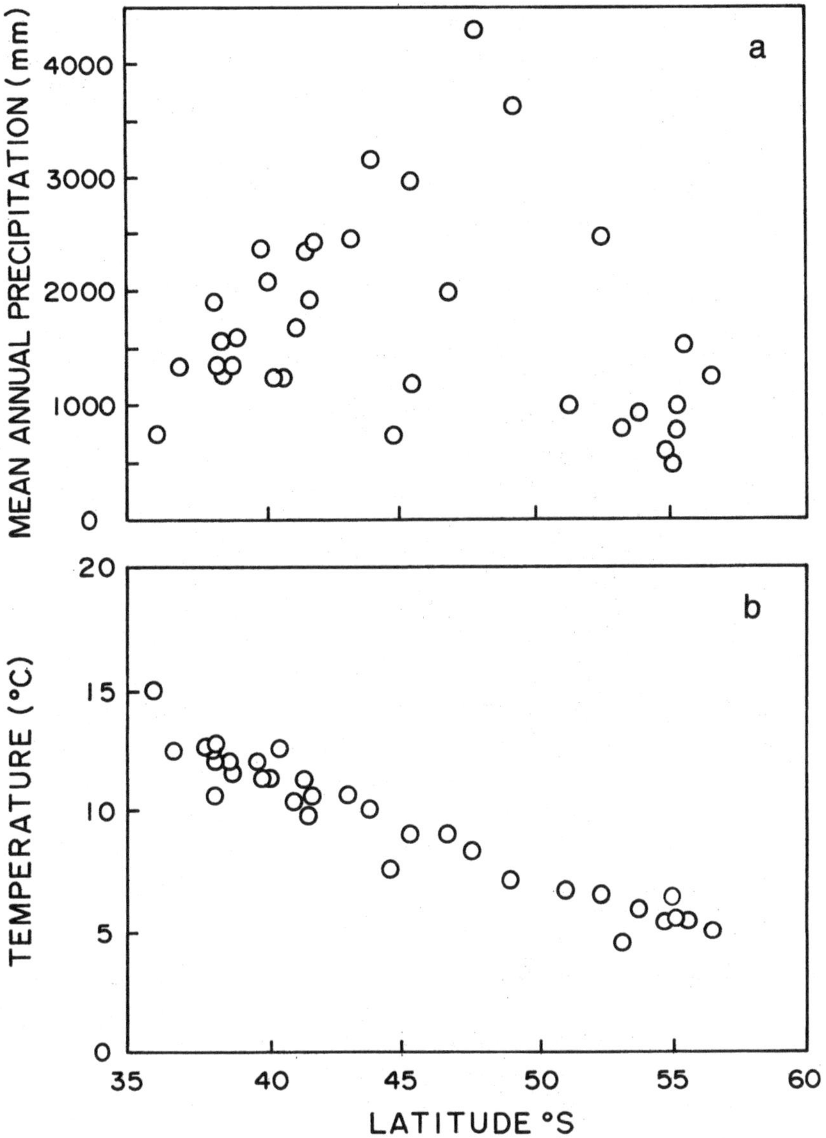

Figure 8.1. Annual precipitation (**a**) and mean temperature (**b**) tendencies in southern Chile.

For purposes of regional intercontinental comparisons in diversity it becomes important to define the rainforest zone with greater precision. Alaback (1991) developed useful criteria for delimiting climatically analagous areas of cool temperate rainforest in North and South America. His criteria include more than 1400 mm precipitation, of which 10% occurs in the summer months, a July (northern) or January (southern)

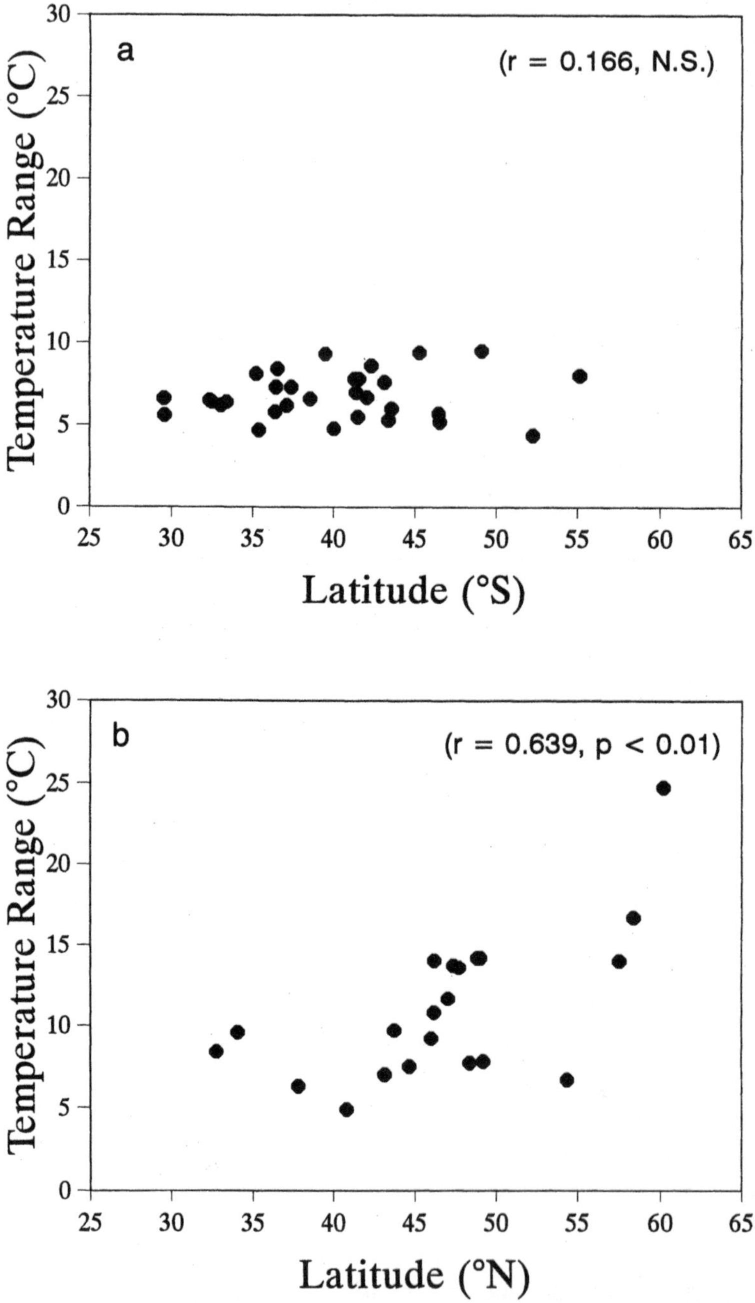

Figure 8.2. Relationship between latitude and annual temperature range for coastal stations in Chile and the Pacific Northwest. (**a**) Chile, temperature range based on means of coldest and warmest months. (**b**) Pacific Northwest, temperature range based on means of coldest and warmest months. (**c**) Chile: tem-

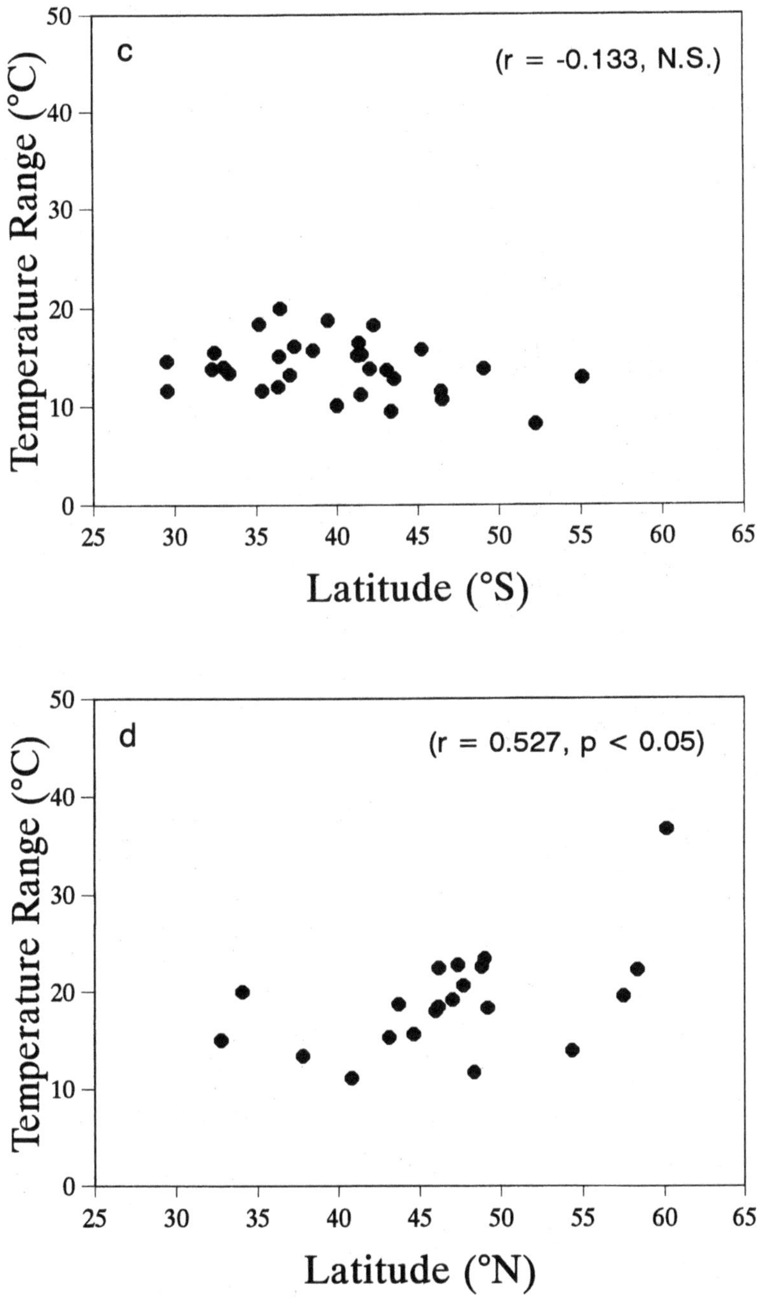

perate range based on highest mean monthly maximum and lowest mean monthly minimum. (**d**) Pacific Northwest: temperature range based on highest mean monthly maximum and lowest mean monthly minimum.

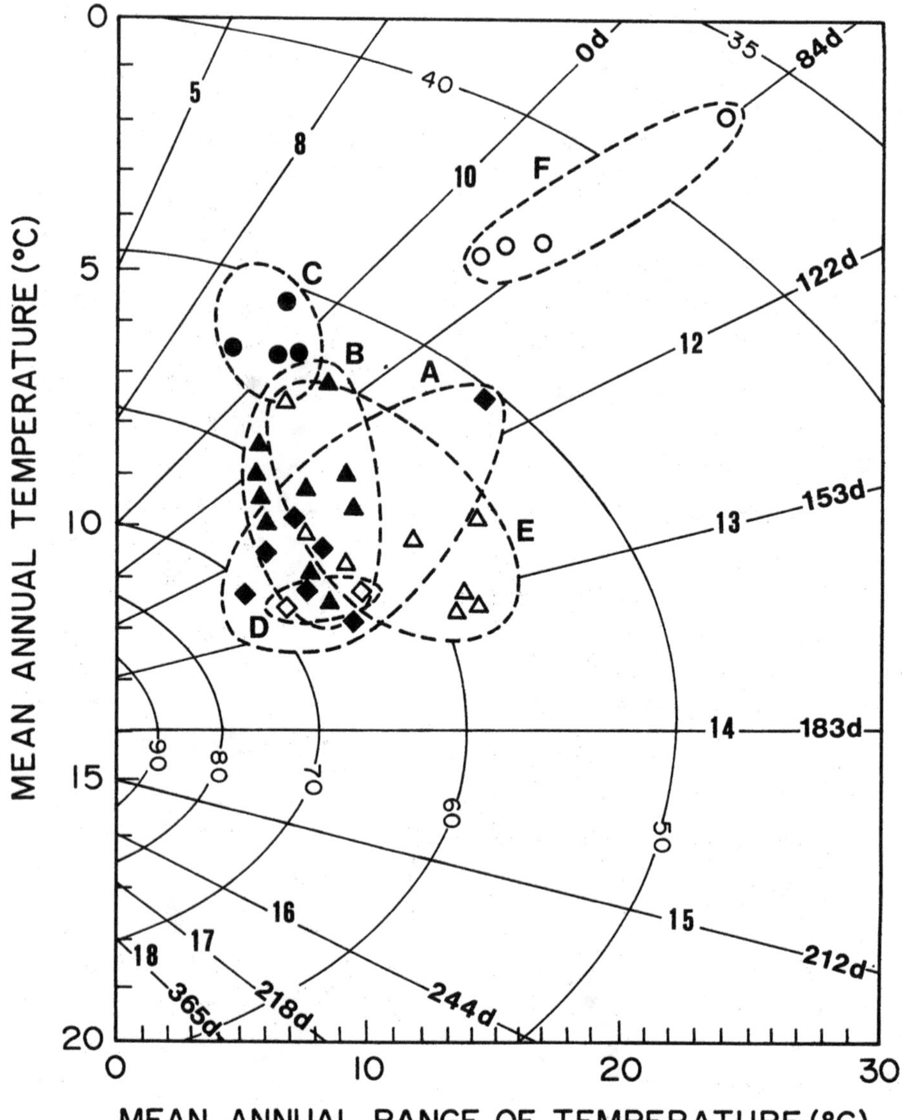

Figure 8.3. Position of various rainforest types in southern South America and western North America on a Bailey (1964) nomogram. The arcs describe degree of equability. A theoretical mean temperature of 14°C and no annual temperature range has an equability rating of 100. Data are the coastal stations in Figure 8.2, plus additional inland stations. A: ◆ Seasonal rainforest, South America. B: ▲ Perhumid rainforest, South America. ● C: Boreal rainforest, South America. D: ◇ Seasonal rainforest, North America. E: △ Perhumid rainforest, North America. F: ○ Boreal rainforest, North America. Rainforest zones as in Weigland (1992).

isotherm equal or inferior to 16°C, cloudy summers, and absence of natural fire as useful criteria. We have relaxed the 10% summer rainfall requirement to allow for 9% rainfall, given the existence of areas with over 150 mm on the Pacific coast of Chile that still receive as much as 150 mm rainfall over the summer months, yet under 10% of the annual total.

The rainforest zone delimited here is depicted in Figure 8.4. The northern limit is set at around 38°30′ S in the Andes to include areas

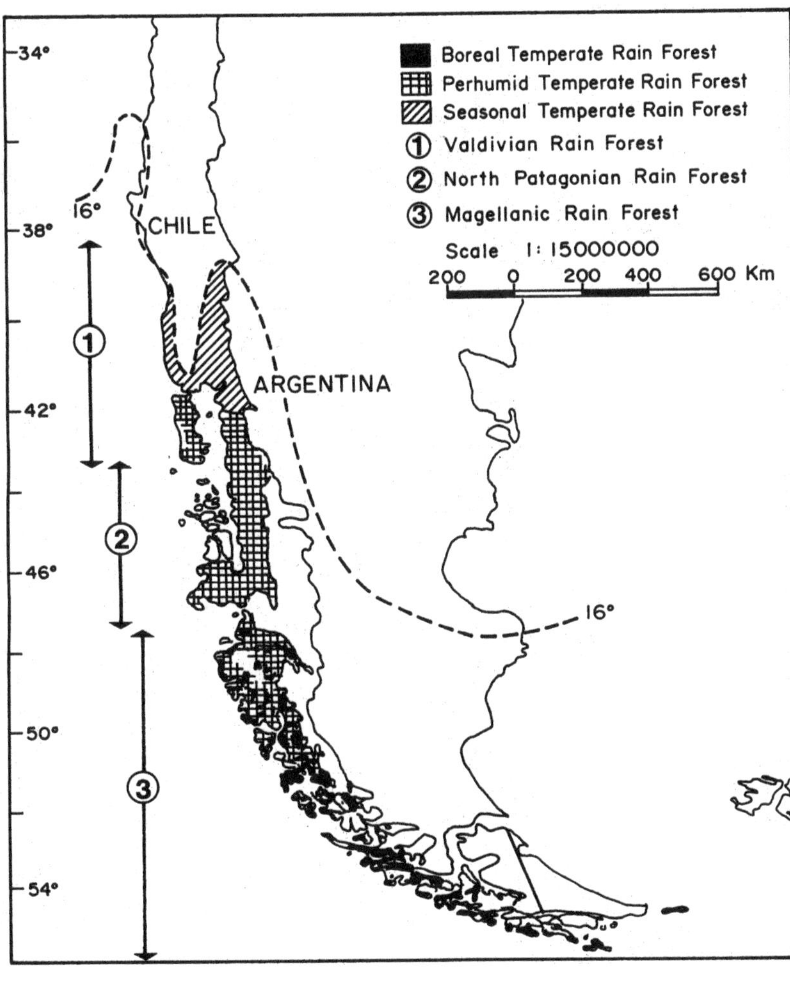

Figure 8.4. Distribution of cool temperate rainforest in southern South America, based on criteria outlined in text. The major rainforest zones according to Veblen et al. (1983) (vertical lines), and divisions according to Weigand (1992) (shaded areas) are shown.

above 1000 m on Volcán Llaima and Parque Nacional Conguillio, and at around Puerto Dominguez (38°54′ S; 1580 mm annual precipitation, 9.6% in summer) north of Valdivia on the coast. The central depression (mainly summer-deciduous forest) is excluded to 41° S because of a mean January temperature exceeding 16°C. The northeastern limit crosses into Argentina to include the arc west of Lago Espejo, Isla Victoria, and the western part of Lago Menéndez. Isla Victoria receives more than 1600 mm precipitation (Dimitri, 1972) of which an estimated 9.3% is received in the summer interval (defined here as 20 December to 20 March). The more southerly eastern limit of the rainforest is seen as running west of Aysen to meet the western borders of the North Patagonian and South Patagonian ice fields.

The southern limit of rainforest, however, is still problematical. The Cape Horn Islands are tentatively included. Recorded precipitation at Bahía Orange (55°27′ S, 68°06′ W), just north of the islands over the years 1982 to 1983 was 1483 mm (Pisano, 1980a), while the Diego Ramírez Islands to the southwest receive an average of 1218 mm precipitation annually. The Cape Horn Islands might receive less than 1400 mm precipitation, but given that they are limiting and provide the only good floristic data (Dollenz, 1980, 1981, 1982a,b; Pisano, 1980a, 1980b) for the southern extreme of the rainforest, their inclusion seems justified.

The rainforest can be further divided into seasonal and nonseasonal types, and the latter into perhumid and boreal sections (Weigand, Mitchell, & Morgan 1992). The first limit sits around 42° S, but with all of the island of Chiloé included in nonseasonal rainforest (Figure 8.4). The limit of the boreal rainforest is around 51°30′ S. The three major rainforest zones as defined by Veblen et al. (1983) are also useful for regional comparisons (Figure 8.4). These comprise the Valdivian rainforest zone, to 43°20′ S, the North Patagonian rainforest zone, to 47°30′ S, and the Magellanic rainforest zone to 55°30′ S.

It should be noted that climatically defined rainforest partially includes high-elevation deciduous *Nothofagus pumilio* and *N. antarctica* forest and *Araucaria araucana* forest, generally not considered rainforest on floristic grounds (c.f., Oberdörfer 1960; Schmithüsen 1958). On the other hand, excluded are the northern extremes of the Valdivian rainforest and southeastern extreme of the Magellanic rainforest as defined by Oberdörfer (1960).

Origins of the Rainforest Flora

Knowledge of the origins of the present-day rainforest flora can provide important clues to understanding ecological dynamics. Many rainforest trees are long lived and must evolve slowly. Ecophysiological constraints conditioned by past evolutionary history can be expected (Arroyo, Armesto, Squeo, & Gutiérrez, 1993). If the origins of taxa can be traced, predictions about their responses in present day communities and to global change can be made and tested (c.f., Read & Hill, 1989).

The basic floristic differences between North American and South American temperate forests, including other forest types, can be related to their Cretaceous and Tertiary histories. Each forest area has been assembled out of floristic elements that evolved independently on either side of the tropics, on landmasses differing markedly in degree of continentality (Axelrod, Arroyo, & Raven, 1991). The larger North American continent, with a more continental climate, favored deciduousness among temperate angiosperm trees, whereas the evergreen habit was maintained on the narrower southern South American landmass with greater overall climatic equability.

The origin of the rainforest in southern South America has been particularly complex due to a long-standing relationship of the South American continent with Antarctica and other southern continents throughout the Cretaceous and early Tertiary (Dalziel, 1992). A connection across Antarctica, then much warmer, facilitated migration of elements all the way across from the northern part of the Australian land mass, at that time located closer to Antarctica. Simultaneously, there was excellent continuity into high tropical latitudes in South America (Arroyo et al., 1993). This is in major contrast to current high-latitude areas of North America, which throughout the late Cretaceous and early Tertiary, although warmer, lay at the edge of a continental mass (Dalziel, 1992).

This critical period in the history of southern South American rainforest has left strong imprints on its present day floristic composition. Table 8.1 which gives the current distributions of the woody genera recorded in cool temperate rainforest of southern South America today, shows that eighteen genera, including many important rainforest trees, exhibit wide geographical disjunctions in the southern lands (e.g., *Araucaria*, *Aristotelia*, *Caldcluvia*, *Discaria*, *Eucryphia*, *Gevuina*, *Griselinia*, *Hebe*, *Laurelia*, *Lomatia*, *Luzuriaga*, *Muehlenbeckia*, *Nothofagus*, *Podocarpus*, *Pseudopanax*, *Prumnopitys*, *Sophora* section *Edwardsia*, and *Weinmannia*). This list grows when the systematic relationships of a number of rainforest genera endemic to the Chilean phytogeographic region are considered (Table 8.1). The common ancestor of *Mitraria*, *Asteranthera*, and *Sarmienta* (South America), and *Fieldia* from southeastern Australia and *Lenbrassia* of Queensland, all monotypic genera of Tribe Mitrarieae (Van Balgooy, 1984) of Gesneriaceae, must have been centered in southern Gondwanaland. *Austrocedrus*, closely related to New Zealand *Libocedrus*, South America *Lebetanthus*, an endemic genus of climbing shrubs of the Magellanic rainforest, closely related to endemic *Prionotes* of Tasmania (Arroyo, 1975); and the dominant tree *Laureliopsis* of the Valdivian rainforest, closely related to *Atherosperma*, restricted to Tasmania and southeast Australia (Schodde, 1983), also belong here. Although not fully resolved, South American *Drimys* seems to be fairly closely related to

Table 8.1. Other Distribution Areas of 82 Genera with Woody Species in South American Temperate Rainforest

Acrisione (2, 1)	ENDEMIC
Adesmia (230, 1)	E Patagonia, N Andes to Peru
Aextoxicon (1, 1)	ENDEMIC
Amomyrtus (2, 2)	ENDEMIC
Antidaphne (7, 1)	Neotrop. lowlands, Trop. Andes to Mexico (d2)
Araucaria (18, 1)	AUS, NC, NG, Norfolk Is., Loyalty Is., Isle of Pines, NE Argentina-SE Brazil (d1), (d2)
Aristotelia (5, 1)	AUS, NZ, TAS (d1)
Asteranthera (1, 1)	ENDEMIC
Austrocedrus (1, 1)	ENDEMIC
Azara (10, 3)	NE Argentina-SE Brazil (d2)
Baccharis (ca. 350, 8)	N and S America
Berberis (450, 11)	Eurasia, America, N Africa
Blepharocalyx (3, 1)	Trop. Andes-SE Caribbean, SE Brazil (d2)
Blechnum (ca. 220, 1)	Mainly Southern Hemisphere
Boquila (1, 1)	ENDEMIC
Buddleja (ca. 100, 1)	Warm, especially E Asia
Caldcluvia (11, 1)	Trop. AUS, NG, Malaysia (d1)
Campsidium (1, 1)	ENDEMIC
Chiliotrichum (2, 2)	S South America
Chusquea (ca. 96, 6)	NE Argentina-SE Brazil, Trop. Andes to C America (d2)
Cissus (350, 1)	Widespread tropical and warm
Coriaria (20, 1)	NZ, Trop. Andes (Peru)-Mexico, Old World, Pacific (d1), (d2)
Corynabutilon (1)	Segreg. of *Abutilon* (100). Tropical and Warm
Crinodendron (4, 1)	NW Argentina-C Bolivia (d2), SE Brazil (d2)
Dasyphyllum (36, 1)	S Brazil-Paraguay-Trop. Andes to Colombia-Venezuela (d2)
Desfontainia (1, 1)	Trop. Andes-Costa Rica (d2)
Desmaria (1, 1)	ENDEMIC
Discaria (8, 1)	SE AUS, TAS, NZ (d1)
Drimys (6, 2)	Trop. S America (d2)
Elytropus (1, 1)	ENDEMIC
Embothrium (1, 1)	ENDEMIC
Empetrum (2, 1)	N Temp and Arctic
Ercilla (2, 2)	ENDEMIC
Escallonia (39, 6)	JF, Andes to Costa Rica, NE Argentina, SE Brazil (d2)
Eucryphia (5, 1)	SE AUS, TAS (d1)
Fitzroya (1, 1)	ENDEMIC*,**
Fuchsia (sect. Quelusia) (9, 1)	SE Brazil (d2)
Gaultheria (ca. 170, 8)[1]	Widespread
Gevuina (3, 1)	Trop. AUS, NG (d1)
Griselinia (ca. 6, 4)	NZ, SE Brazil (d1), (d2)
Hebe (ca. 75, 2)	SE AUS, NZ, NG (d1)
Hydrangea (sect.) Cornidia (11, 1)	Trop. Andes-Mexico (d2)

Table 8.1. *Continued*

Lapageria (1, 1)	ENDEMIC
Latua (1, 1)	ENDEMIC
Laurelia (2, 1)	NZ (d1)
Laureliopsis (1, 1)	ENDEMIC**
Lardizabala (1, 1)	ENDEMIC
Lebetanthus (1, 1)	ENDEMIC**
Lepidoceras (2, 1)	Trop. Andes (Peru) (d2)
Lomatia (ca. 5, 3)	AUS, TAS, Trop. Andes to Ecuador (d1), (d2)
Luma (2, 2)	ENDEMIC
Luzuriaga (4, 3)	NZ (d1)
Maytenus (ca. 225, 4)	Widespread trop. to warm incl. AUS
Misodendrum (8, 8)	ENDEMIC
Mitraria (1, 1)	ENDEMIC
Muehlenbeckia (15, 1)	AUS, NZ (d1)
Mutisia (59, 2)	C & N Andes, SE Brazil-NE Argentina
Myoschilos (1, 1)	ENDEMIC
Myrceugenia (40, 5)	JF, NE Argentina-SE Brazil (d2)
Myrteola (3, 1)	JF, Trop. Andes, Guyana Highl. (d2)
Notanthera (1, 1)	ENDEMIC
Nothofagus (35, 7)	AUS, NC, NG, TAS (d1)
Ovidia (2, 2)	ENDEMIC
Persea (150, 1)	Amer. and Asian tropics
Philesia (1, 1)	ENDEMIC**
Pilgerodendron (1, 1)	ENDEMIC*,**
(Eu) Podocarpus (2)	NZ, NC (d1)
Prumnopitys (10, 1)	NZ, TAS, NC, Trop. Asia, N Andes-Costa Rica (d1), (d2)
Pseudopanax (6, 2)	NZ, TAS (d1)
Rhamnus (125, 1)	Widespread
Rhaphithamnus (2, 1)	JF, N Andes (Peru) (d2)
Ribes (150, 4)	Temp. N Hemisp, Andes
Sarmienta (1, 1)	ENDEMIC
Saxegothaea (1, 1)	ENDEMIC
Schinus (27, 1)	Neotropical
Senecio (ca. 1500, 1)	Widespread
Solanum (ca. 1400, 2)	Widespread
Sophora (Sect. Edwardsia) (6, 1)	NZ, JF, Easter Island (d1)
Tepualia (1, 1)	ENDEMIC
Tristerix (11, 1)	E Andes to Colombia
Ugni (5–15, 2)	JF, Trop. Amer. (d2)
Weinmannia (150, 1)	NZ, Madagascar, Asia, Trop. Andes, SE Brazil (d1) (d2)

AUS = mainland Australia; NC = New Caledonia; NG = New Guinea; NZ = New Zealand; TAS = Tasmania. ENDEMIC: endemic to the Chilean phytogeographic region (see text). d1: disjunct in southern lands; d2: disjunct within South America; * endemic to rainforest (climatic definition; see text); ** genera endemic to rainforest (floristic definition; see text). Figures in parenthesis: total species/genus, species in the southern South American temperature rainforest, respectively. [1] includes *Pernettya*.

the endemic Tasmanian genus *Tasmannia* (Ehrendorfer, Silberhauser-Gottsberger, & Gottsberg, 1979). Dawson (1972) suggests that monotypic *Tepualia* in South America has affinity to *Mearnsia*, today found in New Zealand, New Caledonia, New Guinea, and on the Philippines and Solomon Islands. *Myrteola* seems to be closely related to *Myrtastrum* of New Caledonia and *Neomyrtus* of New Zealand (Landrum, 1988). The fossil record increases this already impressive list. Fossil pollen of South American *Embothrium* has been found in Australia (M. Crisp, personal communication). Excluding suspected cases of recent long-distance dispersal (*Hebe*, *Sophora* section *Edwardsia*), and *Weinmannia*, which might have reached South America and New Zealand independently, the above shows that there are about 21 woody angiosperm genera in the cool temperate rainforest of southern South America, whose origins can be traced to a southern Gondwanaland tie. Assuming that all seven genera of South American rainforest gymnosperms evolved in southern Gondwanaland, at least one third of woody genera of the rainforest are of southern Gondwanaland origin.

To these Gondwanaland elements, not surprisingly considering the interstitial continental position of the rainforest zone in the Tertiary and Cretaceous, a substantial neotropical element can be added. Table 8.1 shows that there are 19 woody genera or sections of woody genera (about 25%) reported to occur in southern South America rainforest today that are centered in South America with species growing in forest habitats at subtropical and tropical latitudes. This substantial neotropical element includes *Antidaphne*, *Azara*, *Blepharocalyx*, *Chusquea*, *Crinodendron*, *Dasyphyllum*, *Desfontainia*, *Drimys*, *Escallonia*, *Fuchsia* section *Quelusia*, *Hydrangea* section *Cornidia*, *Lepidoceras*, *Mutisia*, *Myrceugenia*, *Myrteola*, *Rhaphithamnus*, *Schinus*, *Tristerix*, and *Ugni*, among which are many important rainforest trees and shrubs. The genus *Dasyphyllum* (Mutisiae) has recently been considered on molecular grounds to be basal to the entire Asteraceae family (Kim, Jansen, Wallace, Michaels, & Palmero, 1992). Undoubtedly endemic *Luma*, closely related to *Myrceugenia* (Landrum, 1981) of the tribe Myrteae, also belongs here.

The overall distribution patterns of these neotropical genera are surprisingly concordant (Arroyo, Cavieres, Marticorena, & Muñoz, 1995; Landrum, 1981). Absence from the Amazon basin and surrounding areas is commonplace. Typically there will be one or few species in the cool temperate rainforest, a few in the adjacent mediterranean scrublands in central Chile, and well-defined disjunctions on the eastern side of the continent in the region of northeastern Argentina–southeastern Brazil and/or in the northern tropical Andes. Some genera (e.g., *Crinodendron*, *Azara*) show an intervening disjunction in north-central Argentina–southern Bolivia. Disjunctions may be expressed at the infraspecific level (e.g., *Maytenus boaria*, *Lomatia hirsuta*, *Rhaphithamnus spinosus*, *Myrteola myrsinites*), at the infra- and interspecific levels (e.g., *Myrceu-*

genia) and at the sectional level (e.g., *Dasyphyllum*) (Cabrera, 1959). Such disjunctions appear to have been produced mostly by regional segregation resulting from development of aridity east of the Andes and in the area of the Atacama in the Oligocene–Miocene interval (Arroyo, Squeo, Armesto, & Villagrán, 1988; Galli-Olliver, 1967). The links to tropical latitudes are nevertheless complex, and there is evident relationship with the southern Gondwanaland track in some cases at an earlier stage (e.g., *Myrteola, Drimys*). Subtropical and tropical members of these genera frequently prefer cool, upland habitats, and thus could have migrated into southern South America in the Eocene cooling, occupying the higher elevations of the incipient Andes, and descending to their present lower elevations in southern South America as of the Pliocene. However that primitive members can occur in the Chilean phytogeographical area in some genera (*Antidaphne*, Kuijt, 1988; *Dasyphyllum*, Cabrera, 1959; and *Fuchsia* section *Quelusia*, Berry, 1989), might indicate northward movement along mountainous corridors.

As of the upper Tertiary, with the Andean uplift and development of the cold Humboldt current along the Pacific coast, the rainforest zone became progressively isolated with development of arid climates to the east and north. At this stage, it became open to intrusion by new floristic elements diversifying in adjacent vegetation types. Typical here are species of *Chiliotrichum* and *Adesmia*, which are better represented in drier communities and only just get into rainforest. A boreal track is also suggested by the occurrence of such genera as *Ribes*, *Empetrum*, *Baccharis*, and *Berberis*, common to the northern hemisphere, including Europe, Asia, and North American rainforest. Most of these genera, judging by their current distributions, would have reached southern South America along the Andean chain. The bipolar distribution of *Empetrum*, nevertheless, is consistent with long-distance dispersal (Moore, Harbourne, & Williams, 1970).

Relating to the late Tertiary and previous periods, the Chilean phytogeographic region also exhibits a very high level of endemism (Arroyo et al., 1993). Twenty-eight genera of the total of 82 woody genera in rainforest (34%) and one family (*Aextoxicaceae*) are regional endemics entirely restricted to the Chilean phytogeographic region, defined as the entire temperate forest zone of Chile bordering into Argentina, together with the mediterranean shrublands of central Chile (Table 8.1). *Pilgerodendron* and *Fitzroya* are practically restricted to the rainforest habitat, as defined here, and can be considered as true rainforest endemic genera. Empirical evidence suggests many of the endemic genera are of great antiquity and that both Gondwanaland and Neotropical origins have been involved. Such genera tend to be taxonomically isolated (e.g., *Aextoxicon*, *Lardizabala*, *Boquila*, *Fitzroya*) and 79% are monotypic, suggesting relictual status. The systematic affinities of the family *Aextoxicaceae*, a major rainforest tree, are highly uncertain. Endemism is especially well represented among genera of vines (50%) and hemiparasites (53%). Although there are proportionally fewer genera of trees (28%) and shrubs (20%), their numbers are substantial (Table 8.1). In that such genera comprise such a high proportion of the woody rainforest flora, they warrant more detailed study.

The Quaternary brought large-scale upheavals in the high-latitude biota in each hemisphere. The Pleistocene history of southern rainforest is now fairly well known (e.g., Ashworth, Markgraf, & Villagrán, 1991; Heusser, 1983; Markgraf, 1989; Villagrán, 1988, 1991), and will not be discussed in any detail here, except to highlight a few interesting contrasts that could bear on the patterns to be discussed later.

In South America, the lowland full glacial limit lay around 42° S, such that a large proportion of the present seasonal rainforest zone (38°30′ to 2°00′ S) was not glaciated. Palynological evidence suggests that the central west coast and much of south-central Chile supported a high precipitation regime during the last glaciation maximum (Ashworth et al., 1991; Heusser, 1983; Villagrán 1988), whereas higher latitudes became cooler and drier (e.g., Ashworth et al., 1991). With high precipitation favored at the northern extreme of the rainforest zone in the Pleistocene, rainforest species were able to descend into, or persist as, relic populations in the nonglaciated Coast Range and northern corner of the island of Chiloé, without being totally eliminated from the area (Villagrán, 1991).

In the Pacific Northwest coast, the Pleistocene history seems to have differed, in that climate was dry at the southern end of the ice during the last glacial maximum (Brubaker, 1991). Here glaciers covered the Cascade Range and Olympic Mountains from 22,000 to 29,000 b.p. with the lowland areas still ice-free at this stage. At 18,000 b.p., ice advanced from southern British Colombia into the coastal Olympic Peninsula and Puget Sound lowland (Waitt & Thorson, 1983) (c. 48° N). In the Pacific Northwest, according to Brubaker (1991), during the full glacial (18,000 b.p.), a cold-dry easterly airflow prevailed over the Pacific Northwest, particularly in winter, favoring tundra and xeric treeline species, such as Engelmann spruce south of the ice limit. According to the same author, there is no evidence that such present dominants as Douglas fir lay immediately to the south of the ice sheet in western Washington during late glacial times, as seems to have been the case for their analogs in southern South America. Thus the nonglaciated section of the present western North American rainforest, seems to have suffered greater climatic upheaval and might have been recruited from smaller or more distant refugia than its southern analog.

In summary, the South America rainforest has been drawn from a wide array of phytogeographic elements with notable survival of those elements that must have evolved under warmer Tertiary and Cretaceous climates. It is difficult to estimate how many more of these elements might have existed before the Pleistocene. What will become clear in the following paragraphs is that the net balance of Tertiary and Quaternary events has favored greater richness in the southern forests.

Diversity Patterns and Life Forms

As many authors have pointed out (e.g., Cowling, Holmes, & Rebels, 1992; Diamond, 1988) species diversity is a complex phenomenon because of difficulties in differentiating between historical and ecological processes. Evolutionary events, such as speciation and extinction, are affected both by history and ecological dynamics. Ideally, moreover, data on alpha, beta, and gamma diversity are needed. For meaningful assessments of biodiversity, as opposed to species richness *per se*, attention must also be given to the number of phylogenetic lineages contained among the species present (Arroyo, Raven, & Sarukhan, 1992). Most current methods designed to address this problem (e.g., Williams, Humphries, & Vane-Wright, 1991) depend on the existence of cladistic data and are applicable only for well-studied groups of organisms. The nearest estimate of phylogenetic diversity for regional comparisons involving large numbers of species at the moment is the number of families and genera present. The data available for southern South American cool temperate rainforest is presently too sketchy and too variable methodwise to consider the individual diversity components outlined above. Thus, basically, we will consider the broad regional product of alpha, beta, and gamma diversity, along with some limited data for equal-sized plots.

A consideration of diversity in the rainforest habitat immediately raises the question of what should be considered a rainforest species. We considered a rainforest species as that reported to grow in or under the influence of a tall-to-moderately tall forest canopy. This definition was adopted in view of the very open canopies of several rainforest types in southern South America (e.g., *Fitzroya cupressoides*, *Araucaria araucana* forest). The definition admits natural gap species, which may also be frequently found with manmade cutting of the forest canopy (e.g., *Fuchsia magellanica* and *Gaultheria* spp.), and species of flush and bog areas occurring under the forest canopy in Magellanic rainforest. However, weedy herbaceous species found occasionally on the forest edge and in disturbed secondary forest habitats have been excluded. True Magellanic moorland and krummholz are not included. We have also excluded *ñadis*, on account of their openness, and distribution mainly in the central depression (Ramírez, San Martin, Figueroa, MacDonald, & Ferrada, 1991).

Tables 8.2 and 8.3 show the total number of species found in the rainforest habitat. Ongoing work suggests that, including all habitat types, there are more than 1300 vascular plant species south of latitude 40° S in Chile (Arroyo et al., unpublished data). For the larger area south of

Table 8.2. Life Forms in the Temperate Rainforest Flora of Southern South America

Life-Form	Fam.	Gen.	Spp.	Spp. per Genus		
				Mean	S.d.	Range
ENTIRE RAINFOREST AREA (*N* = 443)						
Total woody	50	82	160	1.95	1.99	1–11
Trees	20	32	44	1.38	1.11	1–7
Shrubs	25	35	81	2.31	2.34	1–10
Vines, climbing, shrubs[1]	12	17	22	1.29	0.57	1–3
Hemiparasites	3	6	13	2.17	2.61	1–8
Total herbaceous	55	128	283	2.21	2.51	1–19
Climbers	5	5	9	1.80	1.17	1–4
Epiphytes	8	14	31	2.21	3.32	1–14
Ground herbs	50	116	243	2.10	2.11	1–13
NONSEASONAL RAINFOREST (*N* = 249)						
Total woody	45	67	110	1.64	1.28	1–6
Trees	18	27	37	1.37	0.87	1–5
Shrubs	18	24	47	1.96	1.62	1–6
Vines, climbing, shrubs[1]	11	15	19	1.27	0.57	1–3
Hemiparasites	3	5	7	1.4	0.80	1–3
Total herbaceous	41	77	139	1.81	2.24	1–18
Climbers	2	2	2	1.00	—	—
Epiphytes	6	11	27	2.46	3.70	1–14
Ground herbs	37	67	110	1.64	1.35	1–8

[1] Includes one woody epiphyte.

36° S, this number increases to more than 2000 (Arroyo, unpublished data). For the general floristic region (excluding Patagonian steppe), there are additional species in the drier forest zone on the eastern side of

Table 8.3. Taxonomic Composition of the Vascular Plant Flora of Southern South American Temperate Rainforest

Group	Fam.	Gen.	Spp.	Species per Genus		
				Mean	S.d.	Range
Angiosperms	81	175	369	2.11	2.13	1–13
Dicotyledons	68	133	275	2.07	2.06	1–12
Monocotyledons	13	42	94	2.24	2.33	1–13
Gymnosperms	3	7	8	1.14	0.35	1–2
Pteridophytes	12	23	66	2.87	4.00	1–19
Total	96	205	443	2.16	2.40	1–19

the Andes (Dimitri, 1972). There are probably some 850 to 900 vascular plant species in all temperate forest types in southern South America, (excluding low-elevation sclerophyllous forest and matorral) (Arroyo et al., unpublished data). The climatically defined cool temperate rainforest as a whole (including seasonal and nonseasonal rainforest), as revealed by the present survey, contains a total of 443 vascular plant species in 205 genera and 96 families (see Tables 8.2, 8.3 and the appendix). These data, which were assembled from herbarium records and numerous published phytosociological studies, floristic lists, and regional floristic works with good habitat data, with elimination of some species of doubtful identification, are probably fairly representative of the rainforest flora, although some additions are likely. Unless there was at least one explicit reference to existence of a species in forest, it was not considered. Such woody species as *Acrisione denticulata*, *Azara serrata*, *A. alpina*, and *Citronella mucronata* are found within the latitudinal range of rainforest, but cannot be corroborated for the rainforest habitat explicitly. *Berberidopsis corallina* might also occur in rainforest, as might the more typical Magellanic moorland endemic *Lepidothamnus fonckii*, occasionally. Thus, nothwithstanding some minor additions, approximately one third of all vascular plant diversity south of 40° S, and close to one half of all temperate forest species can be found in the rainforest habitat. Many such species, of course, also occur in adjacent habitats.

Considering seasonal and nonseasonal rainforest combined, the woody flora comprises 160 (36%) woody species derived from 82 genera and 50 families, and 283 (64%) herbaceous species in 128 genera and 55 families (Table 8.2). There are 44 species of trees from 32 genera and 20 families. Reflecting the high degree of generic endemism in southern South American rainforest, 26 (81%) of the tree genera present in rainforest are represented by a single species. However, a comparison of the appendix with Table 8.1 shows there are many nonendemic genera with single species. A low number of species per genus tends to be common to all taxonomic groups but ferns in the rainforest habitat (Table 8.4).

The largest tree genus in rainforest is *Nothofagus* (7 spp), followed by *Lomatia* (Proteaceae) (3 spp.), and *Podocarpus*, *Amomyrtus*, *Myrceugenia* and *Maytenus* (2 spp. each). Although the largest family for trees is the Fagaceae, the Myrtaceae far exceed Fagaceae in species richness when sedentary and climbing shrubs (9 spp.) are added to trees (6 spp.). Only a minority of the tree genera (seven, or 16%) are gymnosperms (*Araucaria*, *Austrocedrus*, *Fitzroya*, *Pilgerodendron*, *Podocarpus*, *Prumnopitys*, and *Saxegothaea*). In North America, gymnosperms are dominant throughout and seem to thrive in cool coastal conditions (Franklin, 1988). Some South American gymnosperms tend to occur in upland habitats and/or areas with poor soils (*Fitzroya*, *Araucaria*, and *Pilgerodendron*), while others (e.g., *Podocarpus* spp., *Saxegothaea*) form stable communities with angiosperm trees (Veblen, 1982; Veblen, Delmastro, & Schlutter, 1976). The importance of competition from faster-growing angiosperms suggested by Schmithüsen (1960) as an explanation for the distribution of some South American gymnosperms was not supported by the present-day regeneration patterns of pure and mixed stands of *Araucaria araucana* with *Nothofagus* species (Veblen, 1982). Rather *A. araucana* is seen as a stress-tolerant strategist that has always favored harsh habitats, independently of its relictual status.

Regal (1982) argued that anemophilous trees in general should be restricted to marginal sites where pollen movement between conspecifics would not be limited. This explanation seems unlikely for the ecological distribution of South American rainforest gymnosperms, which, as noted above, can be often found in mixed stands with angiosperm trees. Midgeley and Bond (1991), recently showed that isolated individuals of South African *Podocarpus falcatus* separated by 70 m from the nearest reproductive male produced mature seeds, indicating that pollen can move considerable distances without total dilution in mixed forests.

The three major rainforest zones (Table 8.4) shows progressive southward floristic impoverishment. We could only find records of 149 and 102 vascular plant species in the North Patagonian and Magellanic rainforest zones, respectively. The Magellanic zone is quite well known (e.g., Dollenz, 1982b; Pisano, 1977, 1980a, 1980b, 1989). However, modest increases in species richness in the North Patagonian zone are expected due to poor exploration. It is interesting that only 20 of all rainforest species (mostly herbaceous species of low abundance, but including a few ferns) have not been reported in the northern Valdivian rainforest zone. On the other hand, some 255 (58%) species, on present knowledge, are not found in rainforest south of this zone. This implies that practically all species in the cool temperate rainforest can be found in the region bound by its northern climatic limit and the southern extreme of Chiloé, and that a very significant proportion of the species are only found here.

Consideration of nonseasonal rainforest (Table 8.2) (this provides the best comparison for floristic data supplied by P. Alaback, this volume) shows that there are some 249 species. The number of woody species, and particularly the number of tree species, remains very high. Proportionately there are more woody species here than in the total rainforest area (44%), and only 7 tree species are absent from nonseasonal rainforest. There are also still many shrubs.

Undoubtedly, it is the life-form spectrum that most sets South American cool temperate rainforest apart from North American cool temperate rainforest. Apart from trees, shrubs, hemiparasites, and ground herbs, there are significant contributions of woody vines/climbing shrubs, herbaceous vines, and epiphytic angiosperms, virtually unknown in North American rainforest (Alaback, this volume; Franklin, 1988). The cool temperate rainforest of Tasmania also lacks vines. However, epiphytic

Table 8.4. Comparison of Species Richness in Veblen et al. (1983) Temperate Rainforest Zones in Southern South America (it should be noted that these are not idental to Oberdörfer's (1960) floristic units. Life-form diversity is measured as $H' = -\Sigma p_i \log p_i$, where p_i is the proportional representation of each woody life form. The number of fern species in each zone is also given)

Zone	Total Species	Total Woody	Trees	Shrubs	Woody Climbers	Hemiparasites	H′	Ferns
Valdivian (38°30′–43°20′ S)	423	155	44	76	22	13	0.518	63
North Patagonian (43°20′–47°30′ S)	149	73	31	25	15	2	0.501	45
Magellanic (47°30′–55°30′ S)	102	41	13	20	6	2	0.496	20

ferns are abundant (Jarman, Brown, & Kantvilas, 1984). In the New Zealand cool temperate rainforest, nevertheless, vines and vascular epiphytes can be numerous (Dickinson, Mark, & Dawkins, 1993).

South American forests also differ from their northern counterparts in the proportional representation of shrubs. Although the number of shrubs in the nonseasonal rainforest is similar to that reported by Alaback (this volume), they are in lower proportion relative to trees and all woody species. This lower representation of shrubs might reflect an often dense understory of shade-tolerant angiosperm saplings. South American shrubs tend to be more common in gaps and on the forest edge, failing to form a continuous understory. Comparison of 10 m × 10 m plots for three areas of rainforest in which all life forms were considered (Figure 8.5), shows that shrubs tend to be more numerous in the southernmost rainforests, where there is strong dominance concentration of *Nothofagus betuloides*

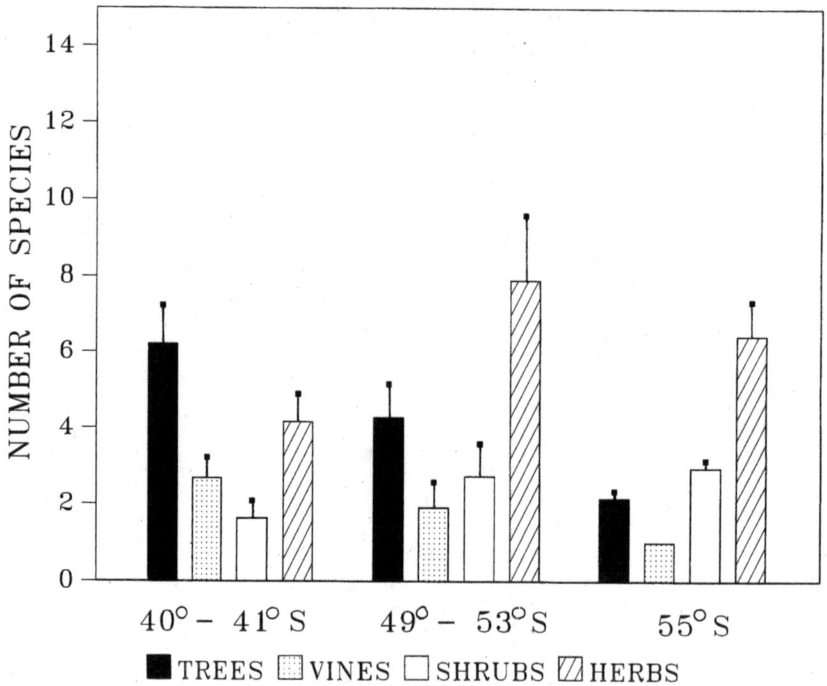

Figure 8.5. Number of tree, vine, shrub, and herb species in 10 m × 10 m plots from three areas of rainforest in southern South America. For trees, a species was considered present, independently of whether it was found in the adult or sapling stage. Sources of data: 40° to 41° S: Villagrán et al. (1974) and additional plots in Parque Nacional Puyehue (N = 19); 49° to 53° S: Dollenz (1982b, 1985–1986) (N = 11); 55° S: Dollenz (1980a, 1981, 1982) (N = 25). Vertical lines are confidence intervals of mean for all plots.

and there are fewer shade-tolerant angiosperms in the understory. In the southern forests, absence of the woody bamboo *Chusquea*, a very abundant plant of the understory north of about 48° S, is undoubtedly another factor permitting higher shrub diversity.

For woody life forms, the cool temperate rainforest of southern South America seems to resemble warm temperate rainforest more closely than it does other cool temperate rainforest. Fisher (1985) gave figures for trees and shrubs combined, woody lianes and slender climbers combined, and ferns for several sites in New South Wales classified as transitional between warm temperate and subtropical rainforest. These are reproduced here and compared to the data for the three southern South American rainforest zones (Table 8.5). The South American rainforest life forms are very similar, with only moderately fewer climbers in the Valdivian zone in relation to the warm temperate forest compared. Six species of vines/climbing shrubs may still be found in the species-poor high-latitude Magellanic rainforest (Table 8.4) (*Campsidium valdivicum*, *Griselinia ruscifolia*, *Lebetanthus myrsinites*, *Luzuriaga polyphylla*, *Mitraria coccinea*, and *Philesia magellanica*), with four extending as far south as 53° S. These records must constitute some of the highest latitudinal records for vines in the world. On the other hand, they are consistent with the very equable climate at high latitudes in South American rainforest.

Both the number and proportion of herbaceous vascular plant species in South American rainforest are, at face value, very high (Table 8.2). Slightly more herbaceous species are seen in South American nonseasonal

Table 8.5. Comparison of Life-Forms Spectra in Cool Temperate Rainforest in Southern South America and Warm Temperate Rainforest in Australia

	Percent species		
	Trees and Shrubs	Climbers	Ferns
Warm temperate rainforest (Australia)			
Liverpool Range, 32° S	52	20	28
Gibraltar Range, 29° S	57	21	22
Cool temperate rainforest[1] (South America)			
Valdivian zone	56	14	30
North Patagonian zone	48	13	39
Magellanic zone	58	11	31

[1] See Table 8.5 for latitudinal limits of South America rainforest zones.
[2] Data for Australia are from Fisher (1985).
[3] South American data adapted to Fisher's life-form categories. One South American fern, considered a shrub in Table 8.2, has been included only under the fern category for this comparison.

rainforest in comparison to its to western North America analog (c.f., Alaback, this volume), with approximately equal numbers of ground herbs. However, in South America, many ground herbs can be locally distributed species, reflecting ecotonal effects. Alpine species commonly invade the open canopy of high elevation *Araucaria araucana* forest and well-lighted areas of deciduous *Nothofagus antarctica* and *N. pumilio* forests (e.g., *Adesmia exilis*, *A. retusa*, *Perezia pedicularidifolia*, *Ourisia breviflora*, *Schizeilema ranunculus*, and *Calceolaria biflora*). Deciduous *Nothofagus* forest also tends to contain more of the core rainforest herbs than lower elevation forests. Veblen, Ashton, Schlegel, and Veblen (1978) have related this last tendency to greater dominance of *Chusquea* in the understory of low elevations forest. Other ecotonal effects are seen in the sporadic appearence of Magellanic moorland species, such as *Astelia pumila*, *Oreobolus obtusangulus*, *Pinguicula antarctica*, *Gaimardia australis*, *Donatia fasicularis*, *Drosera uniflora*, *Tapeinia pumila*, *Rostkovia magellanica*, and *Tetroncium magellanicum* in open-canopy *Pilgerodendron uviferum* and *Fitzroya cupressoides* forests growing on poorly drained, acid soils. If some of these species were disregarded, the nonfern, herbaceous flora would become considerably reduced in size.

Considering equal-size plots again (Figure 8.5) in Valdivian rainforest, the number of herbaceous species in the understory can be very low. However, it increases noticeably in far southern sites (Figure 8.5) again, probably reflecting the smaller number of angiosperm trees forming a shrubby stratum in the understory and the absence there of *Chusquea*. In that rare herbaceous species occur under or within influence of the canopy, however, they must be considered rainforest species. Some common herbaceous species in rainforest are *Acaena ovalifolia*, *Adenocaulon chilense*, *Calceolaria filicaulis*, *Chrysosplenium valdivicum*, *Codonorchis lessoni*, *Dysopsis glechomoides*, *Fragaria chiloensis*, *Gunnera tinctoria*, *Iocenes acanthifolius*, *Lagenifera* spp., *Leucheria* spp., *Nertera granadensis*, *Pilea* spp., *Rubus* spp., *Senecio otites*, *Uncinia* spp., *Viola* spp.

Finally, considering all life forms for an area about one half the size (Weigand et al., 1992), overall richness seems to be considerably higher in the cool temperate rainforest of southern South America as compared with North America (comparing our nonseasonal rainforest to data supplied by P. Alaback, this volume). Perhaps more importantly, there are striking differences in generic and familial diversity among the woody and herbaceous members. From generic data supplied by P. Alaback, the South American forest houses close to twice as many woody and herbaceous genera, using again the comparison for nonseasonal rainforest. South American cool temperate rainforest could also turn out to be richer than Tasmanian rainforest, which has around 125 species of vascular plants (Jarman et al., 1991). However, given that the rainforest area in Tasmania is about one tenth the size of that in southern South America, a final conclusion in this connection must await availablity of data for areas more comparable in size.

Latitudinal Trends in Species Richness

Along the 17.5° of latitude covered by the rainforest in southern South America, climate becomes progressively cooler, and the land mass with sufficient rainfall for rainforest narrows (Figure 8.4). Very strong glacial effects are evident on the landscape in the far southwest. A more detailed examination of the latitudinal species gradient shows that there are some striking differences for tree species in relation to western North America (Figure 8.6). Some of these differences were alluded to earlier by Alaback (1991).

Alaback (1991) showed tree-species richness increases monotonically from only 5 species per degree of latitude in the Alaskan rainforest to around 17 species at latitude 40° N. Based on preliminary data for southern South America, he detected an increase in richness north of the principal ice fields. With access to more complete data, the trend for South American trees can now be refined considerably, with evidence of an even a greater difference between the two continents than originally seen (Figure 8.6a). There is indeed a suggestion of some reduction in species richness around the ice field region (47° S) to a final 7 species at 55° S. Progressing northward, a major increase in trees is now perceived around latitude 43° S to a total of 43 species at latitude 40° S (Figure 8.6a). This is unlikely to be a collecting artifact, as the level of the island of Chiloé, where the change is expressed, has been very well studied. These 43 tree species must be compared with about 17 trees species at latitude 40° N in western North America.

Three factors seem relevant in relation to these contrasting patterns for trees. The dramatic increase in tree species in South America to some extent parallels the transition to seasonal rainforest with a high angiosperm component. It could reflect the presence of many angiosperm trees (weighing heavily in the species count), which are less tolerant of cold summer temperatures farther poleward than gymnosperms in western North America. Some support for this is seen in the rather contrasting monotonic drop off in species richness for *Nothofagus* and gymnosperms (Figure 8.6a). *Nothofagus* is a cold-hardy genus (Alberdi, Romero, Rios, & Wenzel, 1985). Arguing against this interpretation, however, is the very strong latitudinal reduction in species richness in ferns, a group of plants well adapted to cold conditions, which practically parallels that seen in tree species in general (Figure 8.6c).

Contrasting Pleistocene histories, already referred to, are perhaps another factor contributing to this striking interhemispheric difference.

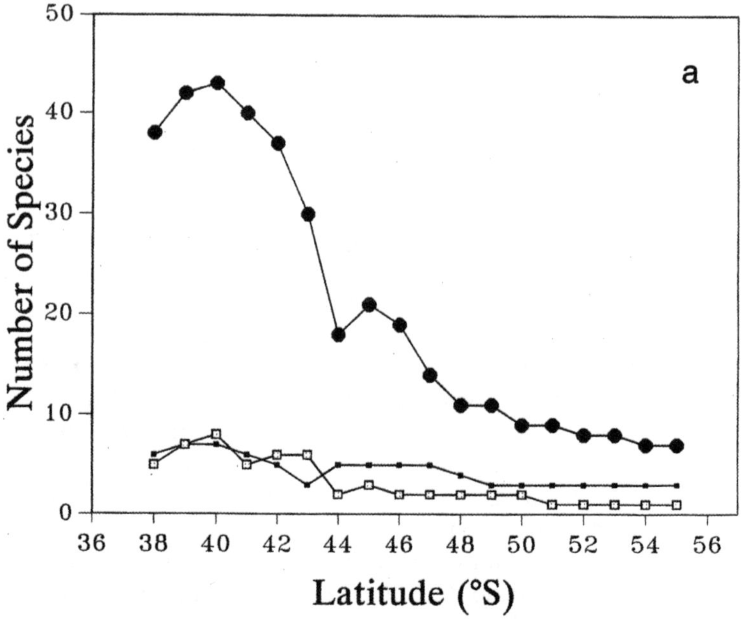

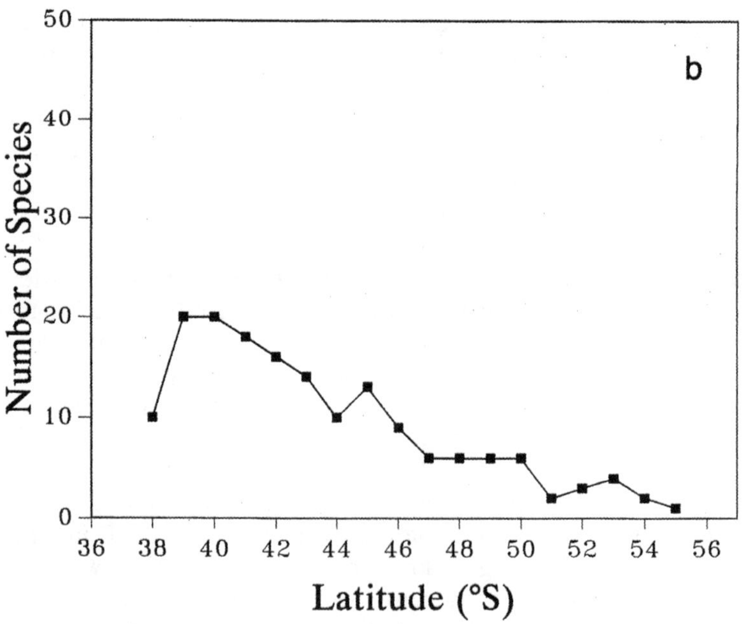

Figure 8.6. Number of species per degree of latitude in the rainforest area. (**a**) Tree species: all tree species (circles); *Nothofagus* spp. (small squares); gymnosperms (open squares). (**b**) Woody wines, woody epiphytes and climbing shrubs. (**c**) Ferns: all ferns (circles); Hymenophyllaceae (squares).

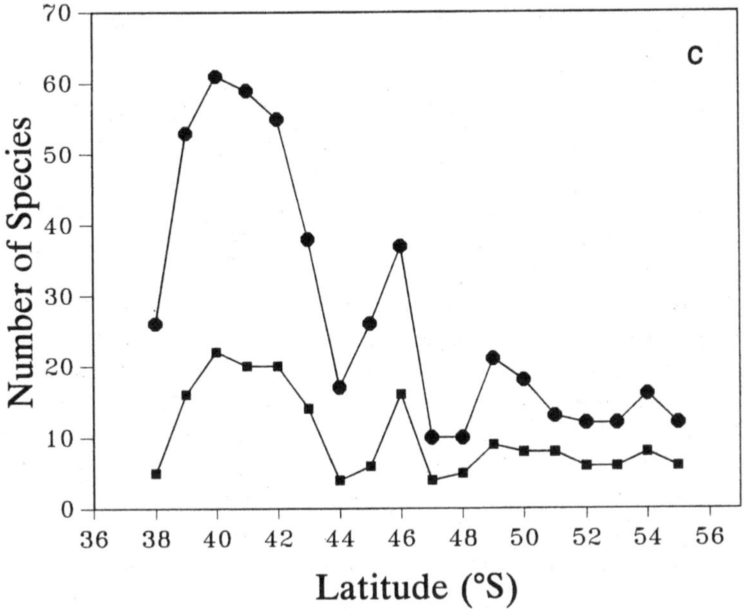

Figure 8.6. *Continued*

With high precipitation favored in the cold periods in midlatitudes of the present rainforest area in southern South America, species in the northern part of the rainforest zone were able to descend into, or persist as, relic populations in the nonglaciated Coast Range and northern corner of the island of Chiloé, without being eliminated totally from the general area (Villagrán, 1991). *Fitzroya cupressoides*, whose Andean populations were only recently reestablished from refuges, is a typical example of this pattern. Effectively, the coastal areas of the seasonal rainforest zone would have acted as an important in situ refugium. In contrast, in North America, refugia were possibly located away from the rainforest border, which supported a cold, xeric climate (Brubaker, 1991). Perhaps the pre-Pleistocene rainforest in South America was much more uniform in tree species richness latitudinally than is seen today, and many trees are only just moving back. Some support for this is seen in maintenance of high life-form diversity in the Magellanic rainforest (Table 8.4). The latter might suggest that the present southermost rainforest flora is an impoverished version of a potentially richer one.

The entire North American rainforest zone, including its southern portion, in contrast, seems to have changed climatic signals in the Pleistocene. A smoother latitudinal gradient in tree species richness would expected as a product of more gradual reestablishment of the rainforest flora. This contrasting hypothesis for the South American rainforest could be easily tested with simple transplant experiments and the determination of ecophysiological tolerances.

A third effect worthy of note in relation to the latitudinal species-richness pattern concerns the nature of the ecotone between rainforest and mediterranean-type scrublands on the west coast of Chile. Numerous species of vines and trees considered part of the rainforest flora in southern South America persist well north of the rainforest zone in mediterranean-type climate latitudes (Arroyo et al., 1995; San Martin, Troncoso, & Ramírez, 1986; Troncoso & San Martin, 1988). These species tend to segregate out from the more xeric elements in mediterranean-type climate scrub over local humidity gradients (San Martin, Figueroa, & Ramírez, 1984). Included here are low-elevation dominant rainforest trees, as *Aextoxicon punctatum*, existing under 2000 mm rainfall in southern Chile, which can be still found in cool coastal ravines with around 500 mm rainfall at 35° S and a long period of summer drought, and even further north (30° S), where there is coastal fog. *Hydrangea serratifolia* and *Elytropus chilensis*, two important rainforest vines, extend well up the coast in discontinuous populations. On the Chilean coast, there is very gradual turnover along the latitudinal gradient. Such a gradual ecotone should have facilitated floristic migration in both latitudinal directions, with many "mediterranean-type" trees occurring in the seasonal rainforest.

Finally, it is perhaps surprising that only two of the regionally endemic genera occurring in rainforest are strictly confined to rainforest. This situation also has to do with the very gradual ecotone between rainforest and mediterranean scrublands just referred to in central Chile. Additionally noncorrespondence between floristically and climatically defined rainforest comes into play. Most notably, floristically defined Magellanic rainforest extends far eastward of the climatic boundary to include areas with as little as 500 mm rainfall on the southern coast of Tierra del Fuego (Pisano, 1977, 1989). The floristic limit of theValdivian rainforest lies at 37°45′ S, to the north of the rainforest limit. Under a floristic definition *Philesia*, *Laureliopsis*, and *Lebetanthus* would correspond to rainforest endemic genera. However an untold number of "typical" monotypic rainforest elements, such as *Asteranthera*, *Elytropus*, *Mitraria*, extend well north of both rainforest limits.

Rapoport's Rule

Rapoport (1975) drew attention to a trend for the range size of organisms to increase with latitude. This trend has now been corroborated for the Northern Hemisphere for tree species, marine mollusks, fresh water and coastal fishes, reptiles, and amphibians by Stevens (1989). An analagous trend occurs along some elevational gradients (Stevens, 1992). The same author illustrated that the latitudinal trend in the Northern Hemisphere is paralleled by an increase in the temperature extremes an organism must experience throughout its lifetime. Stevens (1989) sought a causal explanation of Rapoport's rule from the latter correlation, whereby high-latitude organisms, in being adapted to a wider range of temperature conditions, should be able to occupy larger latitudinal ranges and will be less likely to be habitat specialists. The reduced range of temperature conditions experienced by organisms in low-latitude environments promotes the evolution of habitat specialization. Stevens (1989) further argued that Rapoport's rule could be a causal factor of the typical latitudinal gradient in species richness. As a particular species range increases, space for evolution of other species is preempted. We have looked for these correlations in the trees species of the cool temperate rainforest of southern South America, in view of the absence of an increase in annual temperate range along the entire latitudinal extent of rainforest, discussed earlier.

Stevens (1989) demonstrated latitudinal increase in temperature extremes by comparing the absolute maximum and minimum temperatures recorded. This kind of data was unavailable to us for South America. Moreover, use of these values seems dubious, since a brief temperature pulse is unlikely to have any long-standing biological impact. In any case, to satisfy ourselves that the annual temperature ranges as used here show the same tendency, we compared differences in absolute extremes with the two measures used here for the Northern Hemisphere stations shown in Figure 8.3, obtaining highly positively correlations ($r = 0.811$; $p < 0.001$ for the range based on means; $r = 0.829$; $p < 0.001$ for range based on highest mean maximum and lowest mean minimum). The slope of the curve was 1.09 in the second case and 1.3 in the first, indicating very faithful representation in the second case.

Figure 8.7 shows the mean latitudinal range of tree species occurring at two degree latitudinal intervals from 38° to 55° S in southern South America. At 38° S, the mean latitudinal range was the equivalent to around 14°, whereas by 55° S, it has increased to around 21°, indicating a substantial increase in range size for trees over the latitudinal gradient studied. It can be seen that this increase persists, in spite of no evident increase in annual temperature range as we have measured it.

Discussion and Conclusions

This paper has shown that in spite of the evident isolation of southern South American rainforest from other wet forest habitats in South America today, there seem to be considerably more genera and species,

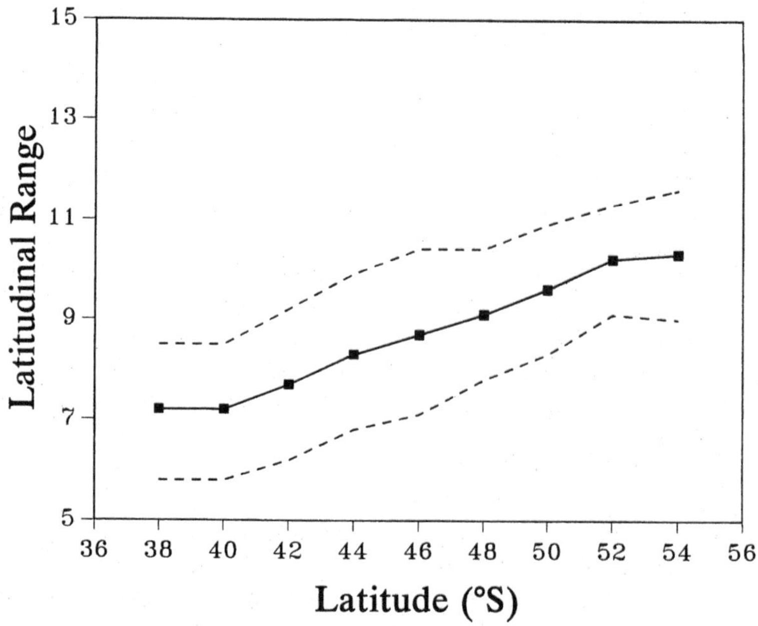

Figure 8.7. Relationship between mean range size and latitude for tree species in the cool temperate rainforest of southern South America. Ranges are total ranges, including extensions outside of the rainforest area in southern South America and continental and intercontinental disjunctions. Range size is given in units of 2°. Thus, to find the mean latitudinal range of species at any one latitude, the value on the *y*-axis should be doubled. Open symbols: North America, closed symbols: South America. The center line traces the means. The lateral lines trace the 95% interval of the means.

especially more woody species—in the south than in its Pacific Northwest analog. The relatively high richness for high-latitude temperate forest is accompanied by an usually diverse life-form spectrum, with an abundance of life forms more commonly represented in warmer forest types.

Our historical analysis indicated a marked tendency for relictualism and retention of many angiosperm genera that probably evolved in southern South America when warmer Tertiary and Cretaceous climates were in place. The physiographic placement of southern South America at the crossroads between two major sources of biota, and retention of many relic genera, may be considered the principle factors leading to the unsually high generic diversity and high number of woody species in cool temperate rainforest today.

Low dominance concentration in cool temperate rainforest of southern South America has frequently provoked comparison with subtropical forest (e.g., Alaback, 1991). The annual temperature range along the entire Pacific coast in southern South America was seen to be very low and equability high. This last feature has favored the retention of Cretaceous and Tertiary tropical and subtropical elements at temperate latitudes in many parts of the world (Axelrod, 1992). Relic genera and other ancient angiosperm elements owe their presence in the South American rainforest today, without much doubt, to this equable climate. What is remarkable about the rainforest, considering that it is a residue of an initially much larger forest province, is the comparatively little intrusion by more recent floristic elements. In this sense, it has shown remarkable stability and resilience as a floristic assemblage.

The arguments outlined above have stressed the importance of historical factors in the diversity of the southern temperate rainforests. It is more appropriate to rephrase the above, with equable climate serving as the baseline condition for the maintenance of the many ancient floristic elements that converged on southern South America. However, ecological factors must be operating to maintain this diversity. The role of large-scale disturbance in South American rainforest needs to be examined critically. For example, has it increased or reduced diversity over evolutionary time by favoring certain taxonomic groups? A great variety of temperature environments cutting east–west across the two major north–south gradients for temperature and precipitation has probably also been important, in that species have wide ranges of conditions to chose from and have evolved in close contact with many other species.

Our demonstration of Rapoport's rule for a gradient with no evident increase in annual temperature range was unexpected in the context of Steven's (1989) hypothesis. What explanation then can be offered for Rapoport's rule in our case? Rickles (cited as personal communication in Stevens, 1992) suggests that Rapoport's rule might reflect how species at different latitudes are distributed in relation to elevation. The species occurring in high latitudes can undergo strong altitudinal displacement northward along north–south trending ranges. In contrast, many more northerly, low-elevation species cannot find equivalent environmental conditions south-ward. As a consequence, in the northern part of a latitudinal gradient, a combination of species with short and long ranges would be found; whereas at higher latitudes, only species with long latitudinal ranges would be seen, leading to lower average range size at the lower latitudes. A detailed analysis of altitudinal ranges might show this to be the case in our gradient. Moreover, in our case, the more northerly, low-elevation taxa should be constrained from expanding their potential ranges northward because of aridity. It is worthwhile pointing out that three tree species in cool temperate rainforest (*Lomatia hirsuta*, *Maytenus magellanica*, and *Rhaphithamnus spinosus*) can also be found at high elevations in the northern tropical Andes and in the southern Brazilian uplands with total latitudinal ranges of 39 to 42°. A more appropriate test for Steven's (1989) explanation of Rapoport's rule would be on Australia, where well-developed altitudinal temperature corridors are less conspicuous. Of course, verification of the existence of Rapoport's rule under these conditions becomes a necessary first step.

The tendency for few species per genus seen in southern rainforest also seems to characterize North American and Tasmanian rainforest (Tasmania, Jarman et al., 1991; Alaback, this volume). These are interesting findings. Documented cases of hybridization are rare in southern rainforest. For the latter, this is counterintuitive, considering the importance of large-scale disturbance due to landslides and volcanic activity (Veblen & Ashton, 1978; Veblen, Donoso, Schlegel, & Escobar, 1981). If anything, the South American environment should provide an ideal setting for hybridization and reticulate evolution. We hypothesize that lack of hybridization is a direct consequence of a generically rich flora, with low number of species per genus. Together, these ensure little opportunities for cogeneric encounters, thus minimising hybridization. Interestingly, the best-documented cases of hybridization in forest species in southern South America are in the genus *Nothofagus* (Donoso, 1984b, 1987; Donoso & Atienza, 1983), which is also the largest tree genus in the southern South American forest flora.

As a general theme in rainforest research, it would be useful to develop testable hypotheses for the causes of low numbers of species per genus. Steven's (1989) tries to explain low species numbers in general at high latitudes in a biogeographic framework. It seems particularly relevant in this context that few of the "taxonomic relics" in rainforest are relic in the ecological sense. Some in fact are dominant (e.g., *Aextoxicon*, *Fitzroya*, *Pilgerodendron*, *Laureliopsis*) and many are fairly common (e.g., *Embothrium*, *Saxegothaea*). Such wide geographic distributions in monotypic genera would seem to indicate that the forcing factor for low number of species per genus is not that such genera are out of phase with their present environment, but that they are prevented in some way from speciating.

Genetic constraints could have developed in the Pleistocene in some long-lived species like *Fitzroya cupressoides*. Cool, Power, and Zavarin (1991) recently showed that this species is invariant for monoterpenes and appears to have been bottlenecked. A low level of genetic variation has also been demonstrated in *Lagarostrobos franklinii* (huon pine), an endemic Tasmanian rainforest conifer (Shapcott, 1991). Alternatively, in some taxa, a low level of speciation might reflect interplay between population density and plant breeding system (Arroyo et al., 1993) and some of the more general causes of species numbers at different latitudes offered by Stevens (1989). Riveros (1991) has recently shown that many rainforest trees are dioecious or genetically self-incompatible (M. Riveros, unpublished data). Such breeding systems can slow the rate of population differentiation, although populations may have high levels of heterozygosity (Arroyo et al., 1993). Under the fairly high population densities in temperate rainforest, local inbreeding, as may occur in tropical forests, is probably uncommon, leading to a tendency for clinal variation along the major environmental gradients. Such clinal variation is evident in several species of *Nothofagus* (Donoso 1979, 1987; Ramírez, 1987), a genus in which three tested species are now known to be genetically self-incompatible (M. Riveros, unpublished data). This may explain why even *Nothofagus*, a genus that experienced large latitudinal shifts in the Pleistocene (Heusser, 1983), has not undergone much local speciation in the rainforest habitat. Hypotheses for low speciation rates in the southern South America temperate rainforest, clearly, can only be resolved with detailed genetic analyses. Detection of levels of genetic variation and minimum viable population size are not only paramount to understanding these problems, but for understanding how the rainforest will respond in relation to the enormous pressures presently inflicted on it.

Here, we have undertaken the first attempt to determine the entire vascular flora of south South American cool temperate rainforest. Without doubt, there will be additions to the rainforest flora as more detailed floristic work comes to hand. We have not referred to the extraordinarily abundant lichen flora (e.g., Galloway, 1992) and mosses of these forests, because of our lack of experience in these groups and their many systematic difficulties. What is becoming clear, however, is that the functional actors in North American and South American rainforests are turning out to belong to different taxonomic groups. Alaback (this volume) draws attention to the lack of vascular epiphytes in the north, suggesting that this niche is only filled by mosses and lichens. We have suggested here that saplings of shade-tolerant angiosperms in the undergrowth in South America and *Chusquea* will often replace the angiosperm shrubs of North American forest. There are, of course, many other functional differences related to tendencies for biotic pollination (Riveros, 1991; Riveros, Humaña, & Lanfranco, 1992) and biotic dispersal (Armesto & Rozzi, 1989) in southern South America, which have not been referred to here.

Use and conservation of any cool temperate rainforest clearly requires an intimate knowledge of the basic biology, ecological dynamics, and population genetics of the component species in a framework of climate change. Research in the future must go beyond a basic descriptive level, yet not fall into a spiral of broad generalities. Topics need to be chosen with great care in order to maximize the practical and theoretical advances that comparative studies will bring.

Acknowledgments. The original data in this paper were assembled thanks to MacArthur Foundation Grant No. 90-9929, BSP/WWF Grant No. 7506, and FONDECYT Grant No. 92-1135. We thank Donald Bran for supplying his unpublished phytosociological data for Argentina.

Appendix 8.1. Genera and Number of Species per Genus (in Parenthesis) of Vascular Plant Species Occurring in the Cool Temperate Rainforest of Southern South America

Pteridophyta

ADIANTACEAE: Adiantum (2); Pteris (1)
ASPLENIACEAE: Asplenium (3)
BLECHNACEAE: Blechnum (10)
DENNSTAEDTIACEAE: Histiopteris (1); Hypolepis (1)
DRYOPTERIDACEAE: Cystopteris (1); Elaphoglossum (2); Megalastrum (1); Polystichum (5); Rumohra (1)
EQUISETACEAE: Equisetum (1)
GLEICHENIACEAE: Gleichenia (4)
HYMENOPHYLLACEAE: Hymenoglossum (1); Hymenophyllum (19); Serpyllopsis (1); Trichomanes (1)
LOPHOSORIACEAE: Lophosoria (1)
LYCOPODIACEAE: Lycopodium (4)
POLYPODIACEAE: Grammitis (3); Pleopeltis (1); Polypodium (1)
SCHIZAEACEAE: Schizaea (1)

Gymnospermae

ARAUCARIACEAE: Araucaria (1)
CUPRESSACEAE: Austrocedrus (1); Fitzroya (1); Pilgerodendron (1)
PODOCARPACEAE: Podocarpus (2); Prumnopitys (1); Saxegothaea (1)

Angiospermae: Dicotyledoneae

AEXTOXICACEAE: Aextoxicon (1)
ANACARDIACEAE: Schinus (1)
APIACEAE: Hydrocotyle (4); Osmorhiza (2), Schizeilema (1)
APOCYNACEAE: Elytropus (1)
ARALIACEAE: Pseudopanax (2)
ASCLEPIDACEAE: Cynanchum (3)
ASTERACEAE: Acrisione (1); Adenocaulon (1); Aster (1); Baccharis (9); Chiliotrichum (2); Conyza (1); Dasyphyllum (1); Gamochaeta (2); Haplopappus (1); Hieracium (1); Hypochaeris (2); Iocenes (1); Lagenifera (3); Leptinella (1); Leucheria (5); Macrachaenium (1); Mutisia (2); Perezia (7); Senecio (12)
BERBERIDACEAE: Berberis (11)
BIGNONIACEAE: Campsidium (1); Eccremocarpus (1)
BUDDLEJACEAE: Buddleja (1)
CAMPANULACEAE: Lobelia (2)
CARYOPHYLLACEAE: Stellaria (1)
CELESTRACEAE: Maytenus (4)
CORIARIACEAE: Coriaria (1)
CORNACEAE: Griselinia (4)
CRUCIFERAE: Cardamine (1)
CUNONIACEAE: Caldcluvia (1); Weinmannia (1)
DESFONTAINIACEAE: Desfontainia (1)
DONATIACEAE: Donatia (1)
DROSERACEAE: Drosera (1)
ELAEOCARPACEAE: Aristotelia (1); Crinodendron (1)
EMPETRACEAE: Empetrum (1)
EPACRIDACEAE: Lebetanthus (1)

Appendix 8.1. *Continued*

Angiospermae: Dicotyledoneae

EREMOLEPIDACEAE: Antidaphne (1); Lepidoceras (1)
ERICACEAE: Gaultheria (8)
EUCRYPHIACEAE: Eucryphia (1)
EUPHORBIACEAE: Dysopsis (1)
FAGACEAE: Nothofagus (7)
FLACOURTIACEAE: Azara (3)
GERANIACEAE: Geranium (1)
GESNERIACEAE: Asteranthera (1); Mitraria (1); Sarmienta (1)
GOODENIACEAE: Selliera (1)
GUNNERACEAE: Gunnera (3)
HYDRANGEACEAE: Hydrangea (1)
LARDIZABALACEAE: Boquila (1); Lardizabala (1)
LAURACEAE: Persea (1)
LENTIBULARIACEAE: Pinguicula (2)
LOASACEAE: Loasa (3)
LORANTHACEAE: Desmaria (1); Notanthera (1); Tristerix (1)
MALVACEAE: Corynabutilon (1)
MISODENDRACEAE: Misodendrum (8)
MONIMIACEAE: Laurelia (1); Laureliopsis (1)
MYRTACEAE: Amomyrtus (2); Blepharocalyx (1); Luma (2); Myrceugenia (5); Myrteola (1); Tepualia (1); Ugni (2)
ONAGRACEAE: Fuchsia (1)
OXALIDACEAE: Oxalis (2)
PAPILIONACEAE: Adesmia (3); Lathyrus (1); Sophora (1); Vicia (1)
PHYTOLACCACEAE: Ercilla (2); Phytolacca (1)
PIPERACEAE: Peperomia (2)
PLANTAGINACEAE: Plantago (2)
POLYGONACEAE: Muehlenbeckia (1)
PRIMULACEAE: Lysimachia (1)
PROTEACEAE: Embothrium (1); Gevuina (1); Lomatia (3)
RANUNCULACEAE: Anemone (2); Caltha (1); Ranunculus (5)
RHAMNACEAE: Discaria (1); Rhamnus (1)
ROSACEAE: Acaena (5); Fragaria (1); Geum (2); Rubus (2)
RUBIACEAE; Galium (3); Hedyotis (1); Leptostigma (1); Nertera (1); Relbunium (1)
SANTALACEAE: Myoschilos (1): Quinchamalium (1)
SAXIFRAGACEAE: Chrysosplenium (1); Escallonia (6); Francoa (1); Ribes (4); Tribeles (1), Valdivia (1)
SCROPHULARIACEAE: Calceolaria (3); Gratiola (1), Hebe (2), Jovellana (1); Mimulus (1); Ourisia (3)
SOLANACEAE: Latua (1); Solanum (2)
THYMELAEACEAE: Ovidia (2)
TROPAEOLACEAE: Tropaeolum (1)
URTICACEAE: Pilea (2); Urtica (2)
VALERIANACEAE: Valeriana (4)
VERBENACEAE: Rhaphithamnus (1)
VIOLACEAE: Viola (9)
VITACEAE: Cissus (1)
WINTERACEAE: Drimys (2)

Appendix 8.1. *Continued*

Angiospermae: Monocotyledoneae

AMARYLLIDACEAE: Alstroemeria (1)
BROMELIACEAE: Fascicularia (1); Greigia (2)
CENTROLEPIDACEAE: Gaimardia (1)
CORSIACEAE: Arachnitis (2)
CYPERACEAE: Carex (13); Cyperus (2), Eleocharis (1); Oreobolus (1); Schoenus (2); Scirpus (1); Uncinia (8).
DIOSCORIACEAE: Dioscorea (4)
GRAMINEAE: Agrostis (2); Anthoxanthum (3); Bromus (3) Chusquea (7); Deschampsia (1); Festuca (5); Melica (1); Nasella (1); Poa (2); Relchela (1); Rhytidosperma (1); Stipa (1); Trisetum (1)
IRIDACEAE: Libertia (1), Sisyrinchium (2); Tapeinia (1)
JUNCACEAE: Juncus (4); Luzula (1); Marsippospermum (2); Rostkovia (1)
JUNCAGINACEAE: Tetroncium (1); Triglochin (1)
LILIACEAE: Astelia (1)
ORCHIDACEAE: Chloraea (1); Codonorchis (1); Gavilea (4)
PHILESIACEAE: Lapageria (1); Luzuriaga (3); Philesia (1)

References

Alaback, P.B. 1991. Comparison of temperate rainforests of the Americas. *Rev Chil Hist Nat*, 64, 399–412.

Alberdi, M., Romero, M., Rios, D., and Wenzel, H. 1985. Altitudinal gradients of frost resistence in *Nothofagus* communities of southern Chile. *Oecol Plantar*, 6, 21–30.

Armesto, J.J., and Figueroa, J. 1987. Stand structure in the temperate rain forests of Chiloé. *J Biogeog*, 14, 367–376.

Armesto, J.J., and Rozzi, R. 1989. Seed dispersal syndromes in the rainforest of Chiloé: evidence for the importance of biotic dispersal in a temperate rainforest. *J Biogeog*, 16, 219–226.

Armesto, J.J., Mitchell, J.D., and Villagrán, C. 1985. A comparison of spatial patterns of trees in some tropical and temperate forests. *Biotropica*, 18, 1–11.

Arroyo, M.T.K., Armesto, J., Squeo, F., and Villagrán, C. 1993. Global change: the flora and vegetation of Chile. In H. Mooney, E. Fuentes, and B. Kronberg (eds.), *Earth System Responses to Global Change: Contrasts between North and South Americas* (pp. 239–264). New York: Academic Press.

Arroyo, Mary T., Kalin, Cavieres L., Marticorena C., and Muñoz M. 1995. Convergence in mediterranean floras of central Chile and California: insights from comparative biogeography. In M.T. Kalin Arroyo, M. Fox, and Paul Zedler (eds.), *Ecology and Biogeography of Mediterranean Ecosystems in Chile, California, and Australia*. (pp. 43–88). New York: Springer-Verlag, 1995.

Arroyo, M.T.K., Raven, P.H., and Sarukhan, J. 1992. Biodiversity. In J.C.I. Dooge, G.T. Goodman, J.W.M. la Rivière, Marton-Lefèvre, T. O'Riordan, and F. Praderie (eds.), *An Agenda of Science for Environment and Development into the 21st Century* (pp. 205–219). Cambridge, England: Cambridge University Press.

Arroyo, M.T.K., Squeo, F., Armesto, J., and Villagrán, C. 1988. Effects of aridity on plant diversity in the northern Chilean Andes: Results of a natural experiment. *Ann Miss Bot Gar*, 75, 55–78.

Arroyo, S.C. 1975. *Lebetanthus*, género austroamericano de Epacridaceae y sus diferencias con el género *Prionotes* dè Tasmania. *Darwiniana*, 19, 312–330.
Ashworth, A.C., Markgraf, V., and Villagrán, C. 1991. Late Quaternary climatic history of the Chilean channels based on fossil pollen and beetle analyses, with an analysis of the modern vegetation and pollen rain. *J Quat Sci*, 6, 279–291.
Axelrod, D.I. 1992. What is an equable climate? *Paleogeog, Palaeoclim, Palaeoecol*, 91, 1–12.
Axelrod, D.I., Arroyo, M.T.K., and Raven, P.H. 1991. Historical development of temperate vegetation in the Americas. *Rev Chil Hist Nat*, 64, 413–446.
Bailey, H.P. 1964. Toward a unified concept of the temperate climate. *Geog Rev*, 54, 516–545.
Berry, P.E. 1989. A systematic revision of *Fuchsia* sect. *Quelusia* (Onagraceae). *Ann Miss Bot Gar*, 76, 532–584.
Brubaker, L.B. 1991. Climate change and the origin of old growth Douglas-Fir forests in the Puget Sound lowland. In *Department of Geography Publication Series, Occasional Paper No. 15* (pp. 5–18). Ontario, Canada: University of Waterloo.
Cabrera, A. 1959. Revisión del género *Dasyphyllum* (Compositae). *Rev Mus Plata*, 9, 21–100.
Cool, L.G., Power, A.B., and Zavarin, E. 1991. Variability of foliage terpenes of *Fitzroya cupressoides*. *Biochem Sys Ecol*, 19, 421–432.
Cowling, R.M., Holmes, P.M., and Rebelo, A.G. 1992. Plant diversity and endemism, In R. Cowling (ed.), *The Ecology of Fynbos: Nutrients, Fire and Diversity* (pp. 62–112). Cape Town: Oxford University Press.
Dalziel, I.W. 1992. Antarctica; a tale of two supercontinents? *Ann Rev Ea Plan Sci*, 20, 501–526.
Dawson, J.W. 1972. Pacific capsular Myrtaceae 8: *Tepualia*. *Blumea*, 22, 335–337.
Diamond, J.M. 1988. Factors controlling species diversity: overview and synthesis. *Ann Miss Bot Gar*, 75, 117–129.
Dickinson, K.J.M, Mark, A.F., and Dawkins, B. 1993. Ecology of lianoid/epiphytic communities in coastal podocarp rain forest, Hasst Ecological District, New Zealand. In R.S. Hill (ed.), *Southern Temperate Ecosystems: Origin and Diversification* (p. 35). Hobart, Tasmania.
Dimitri, M.J. 1972. *La Región de los Bosques Andino-Patagónicos: Sinópsis General*. Buenos Aires: Colección Cientifica del INTA.
Dimitri, M.J., Correa, H. 1966–1967. La flora andinopatagónica. Estudio fitosociológico de una comunidad edáfica entre Puerto Blest y Laguna Frias, del Parque Nacional Nahuelhuapi. *An Par Nac*, 11, 5–39.
Dollenz, O. 1980. Estudios fitosociológicos en el archipiélago del Cabo de Hornos. I: Relevamientos en caleta Lientur, isla Wollaston y sugidero Romanche, isla Bayly. *An Inst Patag*, 11, 225–238.
Dollenz, O. 1981. Estudios fitosociológicos en el archipiélago del Cabo de Hornos. II: Relevamientos en isla Hornos. *An Inst Patag*, 12, 173–182.
Dollenz, O. 1982a. Estudios fitosociológicos en el archipiélago del Cabo de Hornos. III: Relevamientos en la isla Deceit. *An Inst Patag*, 13, 145–151.
Dollenz, O. 1982b. Estudios fitosociológicos en las Reservas Forestales Alacalufes e Isla Riesco. *An Inst Patag*, 13, 162–170.
Dollenz, O. 1985–1986. Relevamientos fitosociológicos en la Peninsula Muñoz Gamero, Magallanes. *An Inst Patag*, 16, 55–62.
Donoso, C. 1979. Genecological differentiation in *Nothofagus obliqua* (Mirb.) Oerst. in Chile. *For Ecol Mgt*, 2, 53–66.
Donoso, C. 1984a. Estructura y dinámica de bosques del tipo forestal siempreverde en un sector de Chiloé insular. *Bosque*, 5, 82–104.

Donoso, C. 1984b. Hibridación natural entre *Nothofagus betuloides* Mirb. (Oerst) y *N. nitida* (Phil.) Krasser. *Med Amb*, 7, 9–16.
Donoso, C. 1987. Natural variation in *Nothofagus* species in Chile. *Bosque*, 8, 85–97.
Donoso, C., and Atienza, J. 1983. Hibridación natural entre especies de *Nothofagus* siempreverdes en Chile. *Bosque*, 5, 21–34.
Donoso, C., Escobar, B., and Urrutia, J. 1985. Estructura y estrategias regenerativas de un bosque virgen de Ulmo (*Eucryphia cordifolia* Cav.)—Tepa (*Laurelia philippiana* (Phil.) Looser) en Chiloé, Chile. *Rev Chil His Nat*, 58, 171–186.
Ehrendorfer, F., Silberbauser-Gottsberger, I., and Gottsberg, G. 1979. Variation on the population, racial, and species level in the primitive relic angiosperm genus *Drimys* (Winteraceae) in South America. *Plant Sys Evol*, 132, 53–83.
Elliott-Fisk, D.L. 1988. The Boreal Forest. In M.G. Barbour and W.D. Billings (eds.), *North American Terrestrial Vegetation* (pp. 33–62). Cambridge, England: Cambridge University Press.
Fisher, J. 1985. The structure and floristic composition of the rainforest of the Liverpool Range, New South Wales, and its relationships with other Australian rainforests. *Aust J Ecol*, 10, 315–325.
Franklin, J. 1988. Pacific Northwest Forests. In M.G. Barbour and W.D. Billings (eds.), *North American Terrestrial Vegetation* (pp. 103–130). Cambridge, England: Cambridge University Press.
Galli-Olliver, C. 1967. Pediplain in northern Chile and the Andean uplift. *Science*, 158, 653–655.
Galloway, D.J. 1992. Lichens of Laguna San Rafael, Parque Nacional 'Laguna San Rafael', southern Chile: Indicators of environmental change. *Gl Ecol Biogeog News*, 2, 37–45.
Heusser, C.J. 1983. Quaternary pollen record from Laguna de Tagua Tagua, Chile. *Science*, 219, 1429–1432.
Jarman, S.J., Brown, M.J., and Kantvilas, G. 1984. *Rainforest of Tasmania.* Tasmania: National Parks and Wildlike Service.
Jarman, S.J., Kantvilas, G., and Brown, M.J. 1991. *Floristic and Ecological Studies in Tasmanian Rainforest.* Tasmanian NRCP Report No. 3.
Kim, K., Jansen, R.K., Wallace, R.S., Michaels, H.J., and Palmer, J.D. 1992. Phylogenetic implications of *rbc*L sequence variation in the Asteraceae. *Ann Mis Bot Gar*, 79, 428–445.
Kuijt, J. 1988. Monograph of the Eremolepidaceae. *Sys Bot Monog*, 18, 1–60.
Landrum, L.R. 1981. The phylogeny and geography of *Myrceugenia* (Myrtaceae). *Brittonia*, 33, 105–129.
Landrum, L.R. 1988. Systematics of *Myrteola* (Myrtaceae). *Sys Bot*, 13, 120–132.
Lara, A. 1993. A 3620-year temperature reconstruction from *Fitzroya cupressoides* tree rings in southern South America. In R.S. Hill (ed.), *Southern Tempe rate Ecosystems: Origin and Diversification* (p. 58). Hobart, Tasmania.
Markgraf, V. 1989. Paleoclimates in Central and South America since 18,000 BP base on pollen and lake-level records. *Quat Sci Rev*, 8, 1–24.
Midgeley, J.J., and Bond, W.H. 1991. How important is biotic pollination and dispersal to the success of the angiosperms? *Philosophical Transactions of the Royal Society of London*, B333, 209–215.
Moore, D.M., Harbourne, J.B., and Williams, C.A. 1970. Chemotaxonomy, variation and geographical distribution of the Empetraceae. *Bot J Linn Soc*, 63, 227–293.
Oberdörfer E. 1960. *Pflanzensoziologische Studien in Chile.* Weinheim, Germany: Verlag con J. Cramer.

Peet, R.K. 1988. Forests of the Rocky Mountains. In M.G. Barbour and W.D. Billings (eds.), *North American Terrestrial Vegetation* (pp. 61–101). Cambridge, England: Cambridge University Press.
Pisano, E. 1977. Fitogeografiia de Fuego-Patagonia chilena. I: Comunidades vegetales entre las latitudes 52 y 56° S. *An Inst Patag*, 8, 121–250.
Pisano, E. 1980a. Distribución y características de la vegetación del archipiélago del Cabo de Hornos. *An Inst Patag*, 11, 191–224.
Pisano, E. 1980b. Catálogo de la flora vascular del archipiélago del Cabo de Hornos. *An Inst Patag*, 11, 151–189.
Pisano, E. 1989. Sectorización fitogeográfica del archipiélago Sud-Patagónico-Fuegino. III: Relaciones floristicas. *An Inst Patag*, 19, 5–16.
Ramírez, C. 1987. The genus *Nothofagus and its importance in Chile. Bosque*, 8, 71–76.
Ramírez, C., and Figueroa, H. 1985. Delimitación ecosociológica del bosque Valdiviano (Chile) mediante análisis estadísticos multivariados. *Stud Oecol*, 6, 69–82.
Ramírez, C., and Figueroa, H. 1987. Plant sociology of the Chilean *Nothofagus* in the hygromorphic zone. *Bosque*, 8, 121–126.
Ramírez, C., San Martin, C., Figueroa, H., MacDonald, R., and Ferrada, V. 1991. Estudios ecosociológicos en la vegetación de los ñadis de la Decima Región de Chile. *Agro Sur*, 19, 34–47.
Rapoport, E.H. 1975. *Areografía; Estrategías Geográficas de las Especies*. Mexico City: Fondo de Cultura, Económico.
Read, J., and Hill, R.S. 1989. The response of some Australian temperate rainforest trees to freezing temperatures and its biogeographical significance. *J Biogeog*, 16, 21–27.
Regal, P.J. 1982. Pollination by wind and animals: ecology of geographic patterns. *Ann Rev Ecol Sys*, 13, 497–524.
Riveros, M. 1991. Biologia reproductiva en especies vegetales de dos comunidades de la zona templada del sur de Chile, (40° S). Unpublished doctoral dissertation, University of Chile, Santiago.
Riveros, M., Humaña, A.M., and Lanfranco, D. 1992. Actividad de los polinizadores en el Parque Nacional Puyehue, X Región, Chile. *Med Amb*, 11(2), 5–12.
San Martin, J., Figueroa, H., and Ramirez, C. 1984. Fitosociología de los bosques de ruil (*Nothofagus alessandri* Espinosa) en Chile central. *Rev Chil His Nat*, 57, 171–200.
San Martin, J., Troncoso, A., and Ramirez, C. 1986. Fitosociología de los bosques de *Nothofagus antarctica* (Forst.) Oerst. en la cordillera costera de Cauquenes (Chile). *Bosque*, 7, 65–78.
Schmidt, H., and Lara, A. 1985. Descripción y potencialidad de los bosques nativos de Chile. *Amb Des*, 1, 91–108.
Schmithüsen, J. 1958. Die raumliche Ordnung der chilenischen vegetation. Forschungen in Chile. *Bonn Geog Abhand*, 15, 1–86.
Schmithüsen, J. 1960. Die Nadelholzer in den Waldgesellschaften der sudlichen Andean. *Vegetation*, 9, 313–327.
Schodde, R. 1983. A new genus in the Monimiaceae alliance (Laureles). *Parodiana*, 2, 198–299.
Shapcott, A. 1991. *Studies in the Population Biology and Genetic Variation of Huon Pine* (*Lagarostrobus franklinii*). Tasmanian National Rainforest Conservation Report 4.
Stevens, G.C. 1989. The latitudînal gradïent in geographical range: How so many species coexist in the tropics. *Am Nat*, 133, 240–256.

Stevens, G.C. 1992. The elevational gradient in altitudinal range: An extension of Rapoport's latitudinal rule to altitude. *Am Nat*, 140, 893–911.
Troncoso, A., and San Martin, J. 1988. Ampliación de área para diversas especies de plantas vasculares en la Cordillera de la Costa de la Región de Maule. *Bol Mus Hist Nat*, 41, 45–56.
Van Balgooy, M.M. 1984. *Plant Pacific Areas*, Vol. 4. Leiden, The Netherland Rijkherbarium.
Veblen, T.T. 1982. Regeneration patterns in *Araucaria araucana* forests in Chile. *J Biogeog*, 9, 11–28.
Veblen, T.T. 1985. Forest development in tree-fall gaps in the temperate rainforests of Chile. *Nati Geog Res*, Spring, 162–183.
Veblen, T.T., and Ashton, D.H. 1978. Catastrophic influences on the vegetation of the Valdivian Andes, Chile. *Vegetatio*, 36, 149–167.
Veblen, T.T., Ashton, D.H., Schlegel, F.M., and Veblen, A.T. 1978. Tree influence on the understory in a mixed evergreen-deciduous forest in Antillanca, Osorno, Chile. *Bosque*, 2, 88–104.
Veblen, T.T., Delmastro, R.J., and Schlatter, J.E. 1976. The conservation of *Fitzroya cupressoides* and its environment in southern Chile. *Environ Conserv*, 3, 291–301.
Veblen, T.T., Donoso, C., Schlegel, F.M., and Escobar, B. 1981. Forest dynamics in south-central Chile. *J Biogeog*, 8, 211–247.
Veblen, T.T., Schlegel, F.M., and Oltremari, J.V. 1983. Temperate broad-leaved evergreen forests of South America. In J.D. Ovington (ed.), *Ecosystems of the World* (pp. 5–31). Amsterdam: Elsevier.
Villagrán, C., Soto, C., and Serey, I. 1974. Estudio preliminar de la vegetación boscosa del Parque Nacional Vicente Pérez Rosales. *An Mus Hist Nat Valp*, 77, 125–154.
Villagrán, C. 1988. Expansion of Magellanic moorland during the late Pleistocene: Palynological evidence from northern Isla de Chiloé, Chile. *Quat Res*, 30, 304–314.
Villagrán, C. 1991. Historia de los bosques templados del sur de Chile durante el Tardiglacial y Postglacial. *Rev Chil Hist Nat*, 64, 447–460.
Waitt, R.B., and Thorson, R.M. 1983. The cordillera ice sheet in Washington, Idaho, and Montana. In S.C. Porter (ed.), *Late Quaternary Environments of the United States. The Late Pleistocene* (pp. 53–70). Minneapolis: University of Minnesota Press.
Williams, P.H., Humphries, C.J., and Vane-Wright, R.I. 1991. Measuring biodiversity: Taxonomic relatedness for conservation priorities. *Aust Sys Bot*, 4, 665–679.
Weigand, J., Mitchell, A., and Morgan, D. 1992. *Coastal Temperate Rainforest: Ecological Characteristics, Status and Distribution Worldwide*. Portland AOR: Ecotrust-Conservation International Occasional Paper Series No. 1.

9. A Comparative Review of Forest Dynamics and Disturbance in the Temperate Rainforests of North and South America

Thomas T. Veblen and Paul B. Alaback

In this chapter, we review and compare North and South American temperate rainforests in terms of patterns of forest dynamics and natural disturbance, stand responses to disturbances, human modifications of disturbance regimes, and the effects of recent climate variation on forest dynamics. Contemporary ecologists view vegetation change as the outcome of interacting populations constrained by fluctuating environmental conditions (Glenn-Lewin, Peet, & Veblen, 1992; Pickett, Collins, & Armesto, 1987). Disturbance accounts for much of the environmental heterogeneity in time and space. Substantial research is now devoted to characterizing, qualitatively as well as quantitatively, the spatial and temporal characteristics of disturbances, or "disturbance regimes," that are relevant to a particular population or community (Pickett & White, 1985).

Temperate rainforests are distinguished from other temperate forest types by the prevalence of evergreen trees in climates characterized by cool summer temperatures, relatively mild winters, and abundant moisture throughout the year (Alaback, 1991; Brown & Read, this volume; Veblen, Schlegel, & Oltremari, 1983). The climates of temperate rainforests are generally characterized by annual precipitation greater than 1400 mm, of which 10% or more occurs during the summer months, and by a mean temperature of less than 16°C during the warmest month (Alaback, 1991). The temperate rainforests of North and South America may be divided into three subtypes on both continents: subpolar, perhumid, and seasonal.

Although the much greater size of the North American land mass at temperate latitudes results in greater continentality at the same latitude and a more poleward location of rainforests in North America, this tripartite subdivision shows the broad similarity of climatic and vegetation gradients along the west coasts of the temperate latitudes of both continents.

Subpolar rainforests occur along the coasts at the highest latitudes. They are characterized by the most homogeneous annual distribution of precipitation and the coolest summers (Almeyda & Saez, 1958; Farr & Hard, 1987; Far & Harris, 1979). These rainforests have the wettest summers (>20% annual precipitation during summer), the coolest temperatures (<12°C mean for the warmest month), the greatest snowfall, and the least stature and diversity of the rainforests. Subpolar forests also occupy the smallest fraction of the landscape, which is also characterized by large areas of bog, tundra, and ice. In South America, subpolar forests (known as Magellanic rainforests) include the coastal forests extending from approximately 48° southward to Tierra del Fuego, as shown in Figure 9.1. In North America, subpolar forests (also known as subboreal rainforests) occur west and north of Yakutat Bay, mostly in the Prince William Sound region, 58° N–61° N, as shown in Figure 9.2.

Perhumid rainforests are the modal, or more typical, form of temperate rainforest and are also the most common type globally. Summer precipitation is generally within the range of 10% to 20% of annual rainfall, and the mean temperatures of the warmest month are usually 12°C–16°C. In North America, perhumid rainforests occur primarily in Southeast Alaska, from Yakutat Bay south to the northern tip of Vancouver Island (north of the Brooks Peninsula). In South America, this includes the area from approximately the Taitao Peninsula (ca. 47°30′ S) north to the Chilean province of Valdivia (c. 40° S).

The seasonal rainforest is a transitional type between rainforest and mesophytic temperate forest. Annual rainfall exceeds 1400 mm, but summer temperatures are often much higher (especially in the more continental climate of North America, where mean July temperatures may exceed 20°C) and summer droughts are common. Fire is often an important aspect of the ecology of these regions (Agee, 1990, 1991; Heusser, Rabassa, Brandani, & Stuckenrath, 1988; Veblen, 1985b). Species richness is greatest in seasonal rainforests, since they include a rich representation of rainforest species as well as species from adjacent biogeographic regions (e.g., the warm temperate forest zone (deciduous in South America, evergreen in North America), the mesophytic forest zone, as well as sclerophyll species from the region of Mediterranean-type climate in central Chile and northern California; Arroyo et al., this volume; Alaback, this volume). Seasonal rainforests are well represented in North America, including the coastal forests of Oregon, Washington,

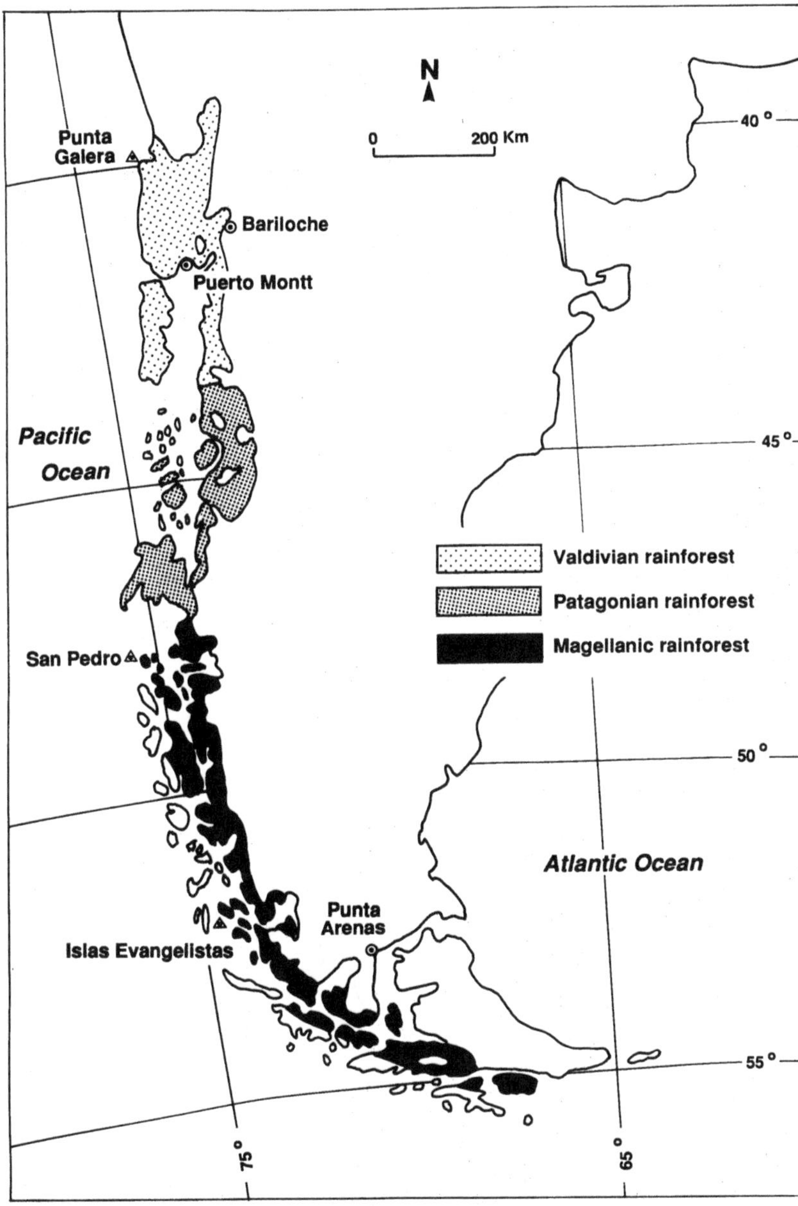

Figure 9.1. Map of the rainforests of southwestern South America. For comparison with the rainforests of northwestern North America the following climatically defined regions can be recognized: subpolar rainforests, 47°30′ to 55° S; perhumid rainforests, 40 to 47°30′ S; and seasonal rainforests, 37°45′ to 40° S. The boundary between the two latter rainforest types varies by more than two degrees of latitude according to position relative to orographic precipitation and rain shadows.

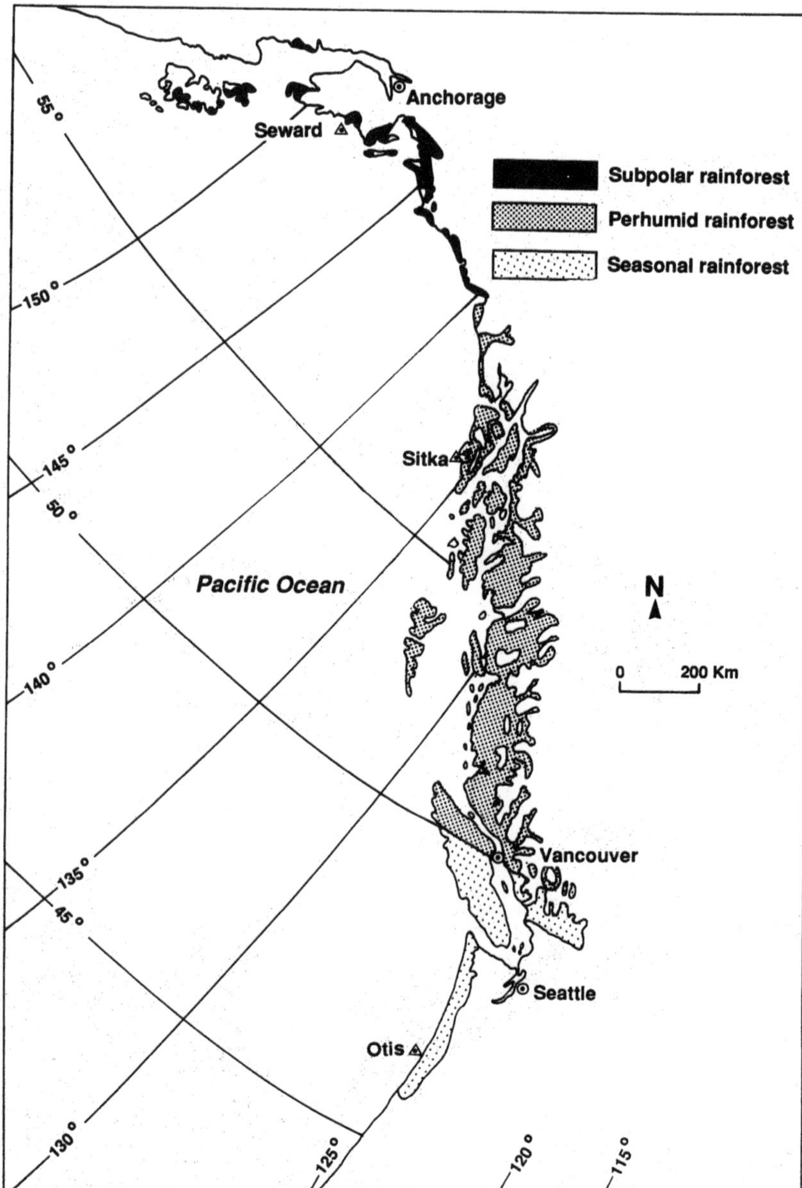

Figure 9.2. Map of the rainforests of northwestern North America.

and southern British Columbia, north to approximately the Brooks Peninsula on Vancouver Island (44° N to 50° N). In South America, this region is relatively small and includes the northern part of the Valdivian rainforest region (ca. 37°45′ S to 40° S).

The Physical Environment and Broad-Scale Vegetation Patterns

South America

The temperate rainforests of South America are associated with the west-coast marine climate of mid- to high-latitude South America (see Lawford, this volume). The west coast of South America south of approximately 37° S is characterized by mild temperature ranges and high annual precipitation, as shown in Figure 9.3. To the north is the Mediterranean-type climate, with hot and dry summers. Mediterranean-type precipitation seasonality is evident in the west-coast maritime climate as far south as approximately 42° S, where the vegetation is temperate rainforest (refer to Figure 9.3). The coastal mountains intercept the mid-latitude westerlies and produce a minor rainshadow effect on the central depression of south-central Chile, where at approximately 40° S, mean annual precipitation is typically less than 2000 mm (Almeyda & Saez, 1958). In the Andes at approximately 40° S, mean annual precipitation ranges from 3000 mm to more than 5000 mm on the windward slopes of mountains. Eastward of the Andes, the rainshadow effect becomes extremely pronounced, so that precipitation declines from approximately 3000 mm to less than 800 mm over a west-to-east distance of only 50 km. Thus, the rainforests occupy only a narrow belt on the eastern side of the Andes in southern Argentina and occur more extensively in southern Chile. South

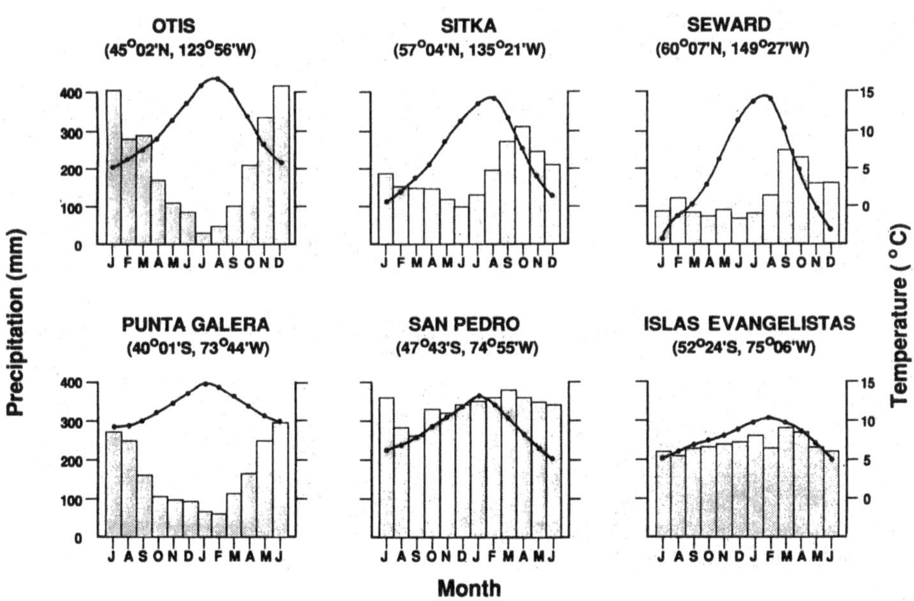

Figure 9.3. Mean monthly temperature and precipitation for selected coastal sites along the coasts of North and South America.

of approximately 42° S the seasonal distribution of precipitation is relatively uniform, with stormy conditions prevailing year-round. Along this north-to-south gradient of decreased precipitation seasonality, there also is a trend toward lower mean annual temperatures. In the far south, at approximately 52° S–55° S, cool, windy conditions prevail for most of the year. Soils are often derived from volcanic ash in the northern part of the perhumid and seasonal rainforest zones, but locally they may be developed from a broad variety of parent materials ranging from metamorphic rocks to glaciofluvial deposits. Farther south (approximately 42° S), glaciofluvial deposits predominate as soil parent materials.

The coarse-scale vegetation patterns of southern Chile and Argentina mirror the north-to-south and west-to-east climatic gradients (see Figure 9.1). Central Chile is characterized by evergreen sclerophyll shrublands and low forests. Bordering this zone on the south is the district of warm temperate deciduous *Nothofagus* forest. Toward the south, as conditions become more humid, temperate rainforests prevail south of approximately 37°45′ S.

The temperate rainforests of South America are a mixture of evergreen conifer and broadleaf tree species. Among the conifers, two Cupressids (*Fitzroya cupressoides* and *Pilgerodendron uviferum*) and three Podocarps (*Podocarpus nubigena*, *Podocarpus saligna*, and *Saxegothaea conspicua*) are important species in the rainforests, as noted in Table 9.1. Among the broadleaf elements, *Nothofagus* species are the most important trees in the rainforests. Although the three evergreen *Nothofagus* species are most typical of rainforest (occurring primarily in the subpolar and perhumid types), deciduous *Nothofagus* species also occur in areas with mean annual precipitation well over the 1400 mm threshold. The deciduous *N. obliqua* and *N. alpina* occur in the northern seasonal rainforests characterized by relatively dry summers. The deciduous *N. antarctica* occurs in poorly drained areas throughout the rainforest district. The deciduous *N. pumilio* occurs at high elevation, typically forming upper timberline on both the western and eastern sides of the Andes within the rainforest district. This belt of deciduous *N. pumilio* includes sites with mean annual precipitation >4000 mm (Veblen, Ashton, Schlegel, & Veblen, 1977); the same species as well as *N. antarctica* also dominate dry forests farther east of the Andes, where mean annual precipitation is less than 1000 mm. The occurrence of deciduous trees within the temperate rainforest district at high elevation and the occurrence of these same species outside the rainforest region at low elevations in a cold, dry climate is a pattern best developed in South America, but it does occur to a limited extent for one species in North America. *Larix lyallii* is the only deciduous tree species that can be a canopy dominant in mature forests within the seasonal rainforest zone in North America. It occurs at high elevations (>1500 m) within the rainforest zone in southwest British Columbia and adjacent Washington state and in moist subalpine sites in

Table 9.1. Common Tree Species of the Temperate Rainforests of South America. The Letters G, R, and T Refer to Most Common Mode of Regeneration (Fine-Scale Gap, Stand Replacement, or Shade-Tolerant, Respectively) of Each Species. However, Modes of Regeneration for the Same Species May Vary According to Associated Tree and Understory Species.

	Reproductive Strategy	Valdivian Rainforest	Patagonian Rainforest	Magellanic Rainforest
*Nothofagus alpina**	R	×		
*Nothofagus obliqua**	R	×		
Laurelia sempervirens	T	×		
Amomyrtus meli	G	×		
Crinodendron hookerianum	G	×		
Eucryphia cordifolia	R	×		
Persea lingue	T	×		
Aextoxicon punctatum	T	×		
Gevuina avellana	G	×		
Fitzroya cupressoides#	R	×		
Luma apiculata	T	×		
Podocarpus saligna	G	×		
Nothofagus dombeyi	R	×	×	
Nothofagus nitida	R	×	×	
Laureliopsis philippiana	T	×	×	
Amomyrtus luma	T	×	×	
Pseudopanax laetevirens	G	×	×	
Chusquea bamboos	G	×	×	
Caldcluvia paniculata	G	×	×	
Saxegothaea conspicua#	G	×	×	
Weinmannia trichosperma	R	×	×	
Podocarpus nubigena#	T	×	×	
Pilgerodendron uviferum#	R	×	×	×
Nothofagus betuloides	R, G	×	×	×
*Nothofagus antarctica**	R	×	×	×
*Nothofagus pumllio**	R, G	×	×	×
Drimys winteri	R, G	×	×	×

* Deciduous broadleaf species.
# Evergreen conifers.
All others are evergreen broadleaf trees.

the Rocky Mountains. The closely related *L. laricina* occurs in peatlands and bogs at low elevations in the boreal zone (but only in areas <1000 m) (Arno & Habeck, 1972).

The gradual decline in species richness from north to south is the basis for recognizing three major subdivisions within the rainforest district. The Valdivian rainforest occurs from about 37°45′ S to 43°20′ S, the north Patagonian rainforest from approximately 43°20′ S to 47°30′ S, and the Magellanic rainforest south of approximately 47°30′ S (Veblen et al., 1983). Although dominated by evergreen broadleaf trees and evergreen conifers, a few dominant deciduous tree species also occur in the Valdivian rainforest district (refer to Table 9.1). Bamboos (*Chusquea* spp.) typically dominate the rainforest understories in the north but are absent south of approximately 48° S. All three rainforest types are characterized by heavy loads of epiphytes; bromeliads and hymenophyllaceous ferns are common epiphytes in Valdivian and north Patagonian rainforests, whereas to the south, the epiphytes are primarily bryophytes. Lianas, both wiry ones and robust ones with stem diameters >15 cm, are typical of Valdivian and north Patagonian rainforests, which is an important contrast to North American rainforests. Following the strong west-to-east precipitation gradient from the windward side of the Andes to the rainshadow in their lee is the gradient from temperate rainforests, through mesic forests, xeric woodlands, to the Patagonian steppe of bunchgrasses and shrubs.

North America

The gradient from 44° N to 62° N along the Pacific slope of North America extends from a cool, wet climate with summer drought to perpetually wet rainforests and tundra, analogous to the one described for the equivalent latitudinal range in South America. As contrasted with South America, the gradient in North America is more gradual, and all the principal rainforest types occur approximately 7° to 10° farther north (see Figure 9.2). This latitudinal shift is due to the decrease in land area with latitude in temperate South America versus the poleward increase in land area in North America. Due to the greater continentality of the North American climate, summer temperatures are substantially greater in North America at the same latitudes. The rainforest zone is also somewhat broader in South America, owing to the greater zonal flow in the broad belt of westerly circulation over the largely landless South Pacific Ocean, although isolated pockets of marine-influenced climate and rainforest do occur as far east as the Caribou Mountains in British Columbia and west of the Bitterroot Divide in Idaho in North America (Lorain, 1988). Winters are more severe in North America, with freezing temperatures recorded at least annually throughout the whole region.

Subpolar forests are dominated by *Tsuga mertensiana*, *T. heterophylla*, and *Picea sitchensis* (Cooper, 1942; Borchers, Wattenbarger, & Ament, 1989). These forests have strong compositional affinities to the subalpine subzone of the perhumid rainforest region. Terrains in this zone have been ice-free for no more than 4000 years and tend to be much less compositionally stable than in South America. Proportionately less subpolar rainforest occurs in North America than in South America, perhaps as a result of less persistent winds and more localized ocean influence in North America. North American perhumid forests include *Picea sitchensis*–*Tsuga heterophylla* forests on well-drained sites and some mixture of *Chamaecyparis nootkatensis*, *Thuja plicata*, and *Pinus contorta* on poorer sites (Alaback & Juday, 1989; Martin, 1989). In seasonal rainforests, *Larix lyallii*, *Pseudotsuga menziesii*, *Abies amabilis*, several species of *Acer*, and *Alnus* occur in addition to all the species of the perhumid forest region (Henderson, Peter, Lester, & Shaw, 1989; Klinka, Pojar, & Meidinger, 1991; McKee, LaRoi, & Franklin, 1982; see Table 9.2). Riparian forests are much more compositionally unique in North America

Table 9.2. Common Tree Species of the Temperate Rainforests of North America. The Letters G, R, and T Refer to Most Common Mode of Regeneration (Fine-Scale Gap, Stand Replacement, or Shade-Tolerant, Respectively) of Each Species. However, Modes of Regeneration for the Same Species May Vary According to Associated Tree and Understory Species.

	Reproductive Strategy	Seasonal Rainforest	Perhumid Rainforest	Subpolar Rainforest
Abies grandis	T	×		
*Acer glabrum**	G	×		
*Acer circinatum**	G	×		
*Acer macrophyllum**	G	×		
*Alnus rubra**	R	×		
Pseudotsuga menziesii	R	×		
Larix lyallii#	G	×		
Abies amabalis	T	×	×	
Taxus brevifolia	T	×	×	
Chamaecyparis nootkatensis	G	×	×	
Thuja plicata	T	×	×	
Picea sitchensis	R, G	×	×	×
Pinus contorta	G	×	×	×
*Populus trichocarpa**	R	×	×	×
Tsuga heterophylla	T	×	×	×
Tsuga mertensiana	G		×	×

* Deciduous broadleaved tress. All others are conifers, and are usually evergreen.
# Deciduous conifer.

than in South America and are usually dominated by *Picea*, *Alnus*, or *Pseudotsuga* (Alaback & Sidle, 1986; Gagnon & Bradfield, 1986; McKee et al., 1982).

The Coast Range delimits the rainforest region from mesophytic forest throughout the whole region, although it is relatively low in elevation and with a less dramatic gradient west to east in the Oregon Coast Range and the Willapa Hills in southwestern Washington than farther north. In contrast to the central depression of Chile, where only slight depression in precipitation occurs and rainforest largely persists, interior valleys east of the Coast Range in North America have as little as 300 mm to 1000 mm of precipitation and are dominated by fire-dependent grassland, semiarid, or mesophytic forest types. Isolated fragments of seasonal rainforest occur inland at higher elevations on the western slopes of the Cascades and in the Caribou mountains of central British Columbia and in the Bitterroots of northern Idaho, although these areas tend to have more seasonal precipitation and greater fire frequency than their coastal counterparts (Habeck, 1988; Lorain, 1988; Spies & Franklin, 1988). West of Prince William Sound, the forest becomes first a pure Sitka spruce type, then gives way to wet coastal tundra. Heavy winds and cool growing-season temperatures delimit this western extension of the rainforest zone (Farr & Hard, 1987).

A broadly analogous pattern of increasing summer precipitation and total precipitation and declining temperatures with latitude occurs along the coast of North America. Precipitation has greater seasonality at higher latitudes than in South America, with peak rainfall typically occurring in the fall. An equally significant and more consistent gradient of temperature characterizes this north–south transect, as shown in Figure 9.3. This gradient in temperature encompasses a wider range than in South America and is closely related to forest yield and productivity (Farr & Harris, 1979). Seasonal variation of temperature is greatest in the north, whereas diurnal variation is greatest in the south. Fog plays a key role in depressing spring and summer temperatures, particularly in the seasonal rainforest zone. Along the Oregon coast, the moderating influence of the ocean is most constricted, often to within only 1 km of the coast, and is most pervasive in southeast Alaska, where moderated temperatures may extend 150 km inland.

Patterns of Stand Development

Patterns of stand development in North and South American rainforests are the consequence of both coarse-scale disturbances producing whole-stand replacement patterns and small treefalls resulting in fine-scale gap dynamics. In landscapes dominated by a high rate of coarse-scale disturbance, patterns of stand development are well described by Oliver's (1981) general model of whole-stand replacement, consisting of the following four stages: stand initiation, stem exclusion, understory reinitiation, and old growth.

In South American temperate rainforests, coarse-scale disturbances are particularly important, and stand development patterns conform with Oliver's (1981) model of whole-stand replacement. Following coarse-scale disturbances such as landslides, fire, glacial activity, and volcanic ash deposition, shade-intolerant tree species such as *Nothofagus* species, *Weinmannia trichosperma*, and/or *Eucryphia cordifolia* form even-aged cohorts. Over extensive areas, relatively old (>300 years) forests are dominated by these same shade-intolerant species, but young or small trees of these shade-intolerant species are absent or scarce in these old stands. In contrast, shade-tolerant tree species (such as *Laurelia sempervirens*, *Laureliopsis philippiana*, *Persea lingue*, *Aextoxicon punctatum*, and numerous myrtaceous trees) are typically abundant and occur as all-aged populations in the old stands (Armesto & Figueroa, 1987; Donoso, Escobar, & Urrutia, 1985; Donoso, Grez, Escobar, & Real 1984; Veblen, 1985b; Veblen & Ashton, 1978; Veblen, Donoso, Schlegel, & Escobar, 1981). In the absence of disturbance, there is a gradual successional trend toward dominance by these shade-tolerant species. In landscapes with a high frequency of coarse-scale disturbances, this successional trend, however, is not completed, and the shade-intolerant trees (especially *Nothofagus*) remain as the dominants of the oldest stands. In other landscapes with lower rates of coarse-scale disturbance, such as the coastal nonvolcanic mountains of south-central Chile at approximately 41° S and the island of Chiloé where mass movements are less important, extensive areas are dominated by the shade-tolerant tree species and forest dynamics follow a pattern of fine-scale treefall gap dynamics (Armesto & Fuentes, 1988; Veblen et al., 1981).

In South American temperate rainforests, the pattern of stand development varies according to the type of stand-devastating disturbance. Disturbances that remove nearly all the preexisting vegetation and expose bare mineral soil (such as mass movements, floods, glaciers, and volcanism) result in stand initiation and stem exclusion stages consisting solely of even-aged cohorts of shade-intolerant tree species (typically *Nothofagus* species). Shade-tolerant tree species and tall understory *Chusquea* bamboos typically do not establish in abundance until the understory reinitiation stage. As the initial cohort of shade-intolerant species become senescent, die, and create gaps, these gaps are preempted by the advance regeneration of the shade-tolerant tree species and the proliferation of bamboos. If the frequency and scale of disturbance are sufficiently high, the shade-intolerant colonists do not die before the next stand-devastating disturbance occurs, so a true old-growth stage (or steady-state stand) is not attained. Where the stand-initiating disturbance does not remove the understory (e.g., a blowdown), the response is dominated by the accelerated growth of plants that survive the disturbance rather than by new establishment. In this case, it is mostly small individuals of shade-tolerant trees and the *Chusquea* bamboos that increase their growth rates. Shade-intolerant species such as *N. dombeyi* may also regenerate in small numbers following this kind of canopy disturbance that does not entirely remove the understory. In blowdowns of mixed-species rainforest, scattered *N. dombeyi* may establish on nurse logs and survive to maturity in low numbers in competition with the bamboos and the released advance regeneration of shade-tolerant species such as *Laureliopsis philippiana*.

In the old-growth forests of more stable habitats (e.g., the coastal mountains of Chiloé and south-central Chile), forest dynamics are controlled mainly by the occurrence of small treefall gaps (Armesto, Figueroa, 1987; Armesto & Fuentes, 1988; Veblen et al., 1981). The most shade-tolerant tree species are able to regenerate beneath a continuous canopy cover, but their growth rates are enhanced by the creation of small canopy gaps (Armesto & Fuentes, 1988; Veblen, 1985b). Somewhat less shade-tolerant species, both subcanopy and main canopy in potential stature, are totally dependent on treefall gaps for successful regeneration (Armesto & Fuentes, 1988). Again, preemption of canopy gaps by rapidly growing *Chusquea* bamboos may have a major inhibitory influence on tree regeneration, even of the most shade-tolerant tree species. Interference from bamboos sometimes results in a slow rate of gap filling. Tree species with some mode of vegetative reproduction (e.g., root suckering, basal sprouting) typically have an advantage in filling gaps (Veblen et al., 1981).

In North American temperate rainforests, broadly analogous patterns of stand development occur. An interesting contrast is the tendency for primary succession to include a sequence of dominance by different life forms (from bryophytes to forbs, shrubs, and then mixtures of all these life forms plus trees) following disturbance (Cooper, 1937). In contrast, at most South American sites, dominance by tree life forms seems to develop earlier, perhaps because of the prevalence of more nutrient-rich volcanic soils or milder winter temperatures. In the perhumid zone of North America, the pace and direction of development have been shown to be highly dependent on soil fertility and drainage: Rapid stand initiation and a long stage of stem exclusion occur on moderately productive sites, whereas on nutrient-poor sites, stand initiation is slower and the stem exclusion phase is shorter or nonexistent (Demeo, Martin, & West, 1992). On nutrient-rich, well-drained sites in the seasonal rainforest zone, by contrast, the stem exclusion stage is shorter than on moderately productive sites, because of the rapid development of upper-canopy layers and early self-thinning (Harcombe, 1986). In South America, although site factors such as drainage also play a major role in controlling the pace and direction of stand development, research on the role of soil fertility is scant.

In North America, coarse- and fine-scale disturbances are both important in controlling forest dynamics, especially in steep terrain, along the mainland or exposed outer coast of the perhumid zone and throughout the seasonal rainforest zone (Edmonds, Thomas, & Maybury, 1993; Ott, 1994; Spies & Franklin, 1988). Fire typically is the dominant catastrophic agent in seasonal rainforests, whereas wind, flooding, avalanches, and landslides are the key agents in perhumid rainforests. In many cases, successional trajectories are very long, with directional change continuing to occur 4 to 5 centuries following disturbance (Alaback, this volume; Spies, 1991).

Riparian habitats constitute a special case, due to the high frequency of relatively large flood events and mudslides (debris torrents). The most common scenario in large coastal streams appears to be where a large mudslide or debris torrent creates a lense of material that forms a new terrace on which *Alnus rubra* and *Picea sitchensis* establish over a 200- to 800-year period. After the initial 100 to 200 years of stand development, gap processes become important in creating fine-scale structure. The riparian stands along large streams are more likely to be characterized by whole-stand replacement, due to the higher frequency of catastrophic floods. Less catastrophic disturbance (e.g., blowdown) often results in local dominance by *Rubus* species or *Alnus* species, except on raised organic surfaces such as old windthrow mounds or logs (Beaudry, 1989; Clement, 1985; Schrader, 1991). In contrast, the riparian stands along smaller streams tend toward continual recruitment, although some of these smaller systems may be more susceptible to landslides and debris torrents.

In North America, the species that respond to different disturbances vary significantly with elevation and topographic position. For example, at sea level, *Tsuga heterophylla* is an agressive competitor and excludes all other tree species on productive sites (Alaback, 1982; Stewart, 1986). At higher elevations or on poorer sites, the forest is more open, and trees such as *Picea sitchensis*, in the perhumid zone, or *Abies* species, in the seasonal zone, *Pinus contorta*, *Thuja plicata*, and *Chamaecyparis nootkatensis* can compete more aggressively. Near timberline, *Picea sitchensis* can regenerate in single-tree treefall gaps, because of the openness of the canopy. *Tsuga mertensiana*, which is less shade tolerant than *T. heterophylla*, may form nearly pure stands maintained by gap processes at high elevations where stands are very open and have a persistent snowpack (Demeo et al., 1993; Lertzman, 1992; Martin, 1989; Matson & Boone, 1984). Analogously, in South America the regeneration of *Nothofagus dombeyi* is almost entirely dependent on coarse-scale disturbances in the more species-rich rainforests but regenerates following fine-scale treefalls in species-poor mesophytic forests (Veblen, 1989).

In North America on upland sites there is no equivalent understory dominant to the *Chusquea* bamboos of South America. However, in riparian sites and in burned or logged sites in the seasonal rainforest region *Rubus spectabilis* can play a role similar to that played by *Chusquea* in that establishment of *Picea*, *Pseudotsuga* and *Tsuga* is retarded for 50–100 yr. or more (Barber, 1976; Beaudry, 1989; Ruth, 1968; Schrader, 1991). *Alnus rubra* can also retard successional development, especially after severe soil disturbance in both the perhumid and seasonal rainforest zones (Binkley, 1984; Newton, el Hassan, & ZavitKovski, 1968). These aggressive deciduous species generally do not have the degree of persistence in mature forest as does *Chusquea* in South America perhaps because of the more open multi-layered canopies that typify Valdivian rainforests.

Natural Disturbances

Disturbances of Tectonic Origin

Located at a boundary between two major crustal plates, the southern Andean region is an area of great volcanic and tectonic activity and, therefore, of geologically related ecological disturbances. For example, on the wet western side of the Andes, in the Valdivian rainforest district, strong earthquakes such as the 1960 Valdivian earthquake measuring 8.5 on the Richter scale, periodically, in a single event, trigger thousands of mass movements in the Andes. In addition on mass movements triggered by earthquakes, many other mass movements are associated with prolonged and torrential rainfall in this region of glacially oversteepened slopes. Bare sites created by mass movements are favorable to the establishment of shade-intolerant species, such as *Nothofagus* spp., *Weinmannia trichosperma*, and *Fitzroya cupressoides* that otherwise tend to be replaced by the numerous shade-tolerant tree species of the Chilean rainforests (Lara, 1991; Veblen & Ashton, 1978; Veblen et al., 1981). On the eastern side of the Andes in Argentina, earthquakes periodically kill entire stands due to the intense shaking of unconsolidated debris fans and even on stable substrates they cause increased rates of treefalls (Veblen, Kitzberger, & Lora, 1992a).

Historical data as well as dendroecological evidence indicate that earthquakes of a magnitude similar to that of 1960 have frequently affected tree establishment and tree growth in the rainforest district (Veblen et al., 1992a). In relatively species-rich Valdivian and Patagonian rainforests, disturbance by earthquake-caused landslides can prevent the successional replacement of shade-intolerant spp. (e.g., *Nothofagus* spp., *Weinmannia trichosperma*, and *Fitzroya cupressoides*) by more shade-tolerant species. In the species-poor Magellanic rainforest and the pure evergreen *Nothofagus* forests of northern Patagonia, past earthquakes have had a significant effect on tree mortality and growth patterns, but in the absence of the shade-tolerant tree species that are common in the Chilean rainforests, coarse-scale disturbances are not *required* for the continued dominance of stands by *Nothofagus*. Throughout the coastal areas of southern Chile and in the Andes, subsidence associated with earthquakes has caused extensive inundation and tree mortality in coastal forests (Musters, 1871; Rothkugel, 1916). Inundated coastal forests occur throughout the southern Chilean Archipelago from just south of Chiloé to Tierra del Fuego (Goodall, 1979).

In North America, earthquake-caused landslides and mudslides are of less regional importance to rainforests than they are in South America. Successional patterns on landslide or mudflow surfaces are similar to those that occur on glacial deposits (see the next section), with *Picea* and *Alnus* dominating in the perhumid and subpolar zones and *Pseudotsuga* or *Abies* dominating in the seasonal zone. Once the surfaces are stabilized, succession is more rapid and more commonly includes trees at the early stages than in postglacial successions (Frenzen, Krasney, & Rigney, 1988; Halpern and Harmon, 1983). Colonization of colluvial fans at the foot of landslides is rapid, whereas succession is slow or radically altered on steep escarpments near the tops of landslides (Miles & Swanson, 1986; Smith & Commandeur, 1986). In the perhumid zone, landslides are more frequent and widespread on the mainland, but are generally localized events on islands. Only 2 to 3% of the land area is affected by landslides, but many of these impacts are of regional ecological significance because they may destroy critical salmonid habitat (Swanston & Marion, 1991). Notable exceptions occur in the Queen Charlotte Islands in British Columbia, where rugged terrain and high intensity storms lead to widespread slope failures (Schwab, 1983). Expansible clay soils derived from volcanic ash in the Cascades, loose volcanic ash, and interbedded mudstone and siltstones are also prone to frequent slope failures following intense storms in the seasonal rainforest zone (Swanston & Swanson, 1976). These slope failures often dominate the terrain in the southern Cascades. Logging and associated road construction dramatically increase probability of slope failure (Schwab, 1983; Swanston & Swanson, 1976). Slope failures are particularly frequent when slopes exceed 70% in high rainfall zones. In general, slope failures in logged and roaded regions are smaller, occur at lower elevations, develop on gentler gradients, and tend to travel shorter distances than those in natural undisturbed landscapes (Swanston & Marion, 1991).

There is some evidence that earthquakes, although infrequent, may still have important implications for stand dynamics in the perhumid and subpolar rainforest zones in North America in as much as they may trigger massive slope failures. In the Queen Charlotte Islands, for example, surveys of landslides suggest that a large portion of the landslides that are now visible on the terrain occurred at about the same time, approximately 85 years ago, and were most likely triggered by earthquakes (Steve Chatwin, B.C. Ministry of Forests Research Branch, Victoria, B.C. personal communication). There are several large landslides in northern Southeast Alaska that appear to be of the same age, and may also have been the result of an earthquake in the area (Douglas Swanston, U.S.D.A., Forest Service, Forestry Sciences Laboratory, Juneau, AK, personal communication).

Volcanism

In South America, volcanic ash deposition is important in creating conditions favorable to the establishment of shade-intolerant tree species (especially *Nothofagus* spp.) that otherwise tend to be successionally replaced. *Nothofagus* spp. are frequently observed growing on recent volcanic ash and lava deposits, but specific studies of *Nothofagus* stand development patterns on recent volcanic deposits are lacking. At relatively high elevations in areas of high rainfall, *Fitzroya cupressoides* often is the first tree species to colonize volcanic ash or lava, or may cocolonize with *Nothofagus betuloides* (Lara, 1991; Veblen, Delmastro, & Schlatter, 1976). At mid elevations in competition with Shade-tolerant tree species, the regeneration of *F. cupressoides* may be dependent on volcanism and/or landslides (Lara, 1991).

Volcanic ash deposition is not common in North American perhumid rainforests, but is common in the southern seasonal rainforest zone and the western subpolar rainforest zone. Only at Kruzof Island near Sitka and Blue Lake in northern Misty Fjords National Monument are there effects of recent (post-Pleistocene) volcanic eruptions within the perhumid zone in Alaska (Brew, 1988). At sites on volcanic deposits, forest productivity appears to be low and the rate of stand development appears slow as compared to adjacent, older terrains. Volcanic terrains also have a strong tendency towards bog or fen formation (Klinger, 1988).

The most intensive work on post volcanic succession was done following the 1980 eruption of Mount St. Helens in the southern end of the seasonal rainforest zone (del Moral & Bliss, 1993). Despite the severity of the eruption, preexisting vegetation played a significant role in the regeneration process (Franklin, Modmation, Swanson, & Sedill, 1985). In high elevation locations snow diminished the effects of ash on plants (Antos & Zobel, 1985). In other cases, rhizomatous plants were able to grow up through the ash. Several species were able to germinate and establish on surfaces with 10 cm of ash or less. On sites with heavier ash deposition, water stress, low nutrient status, and poor substrate stability limited regeneration success. Early colonizing species mostly consisted of wind dispersed small seeded species, such as *Anaphalis* and *Epilobium*. *Lupinus* was the principal nitrogen-fixing species, which was an early colonist of ash surfaces. In contrast to glacial succession, bryophytes and cyanobacteria played a very limited role and were not notable components of the succession until 6 to 10 years after the eruption. Most of the old-growth forests and many plant communities on Mount St. Helens originated after the last major eruption (Kruckeberg, 1987). In contrast to South America it does not appear that the eruption will result in a major change in tree species composition. The same species which would dominate after stand replacing wildfire (*Pseudotsuga*, *Abies*) appear to dominate after ash deposition.

As with glacial succession, it appears that the early stages of postvolcanic succession are slow and sensitive to stochastic factors. During dry summers, for example, colonization is limited and mortality may also include established plants from previous years. Succession is slowest on isolated sites and most rapid where intact mature communities are located nearby. Limited summer rainfall combined with unstable surfaces has led to limited plant colonization on Mount St. Helens (generally less than 10% plant cover 11 years after the eruption). Alpine areas were particularly slow to recover because of poor dispersal of taxa there, but also because of heavy unstable ash deposits. Lahars and mudflows appear to revegetate in part more quickly than the upland areas because of the reduced impacts of ash, but often are extremely heterogeneous due to textural differences in exposed surfaces and differences in water movement and substrate stability. In many cases early pioneer species are not facilitators directly but appear to limit colonization of climax species (del Moral & Bliss, 1993). *Lupinus* appeared to be a major nitrogen-fixing pioneer species, but often inhibited colonization by other species. It was a facilitator mostly after the clones died. Unfortunately few studies have been done of postvolcanic succession in a perhumid or subpolar rainforest climatic zone in either hemisphere, and no studies have been made of the early stages of postvolcanic succession in South America. It is unclear how well these observations on postvolcanic succession would apply to a wet summer climate or other types of volcanoes in the seasonal rainforest zone.

Disturbances Associated with Glacial Processes

In the South American rainforest district, glacial activity is an important agent of disturbance creating apportunities for establishment of shade-intolerant tree species. For example, in northern Patagonia on glacier-covered Mt. Tronador (41°10′ S), *Nothofagus dombeyi* establishes in abundance on glacial deposits at low elevations in moist habitats shortly after the deposits are exposed (Lawrence & Lawrence 1959; Veblen, Ashton, Rubulis, Lorenz & Cortes, 1989a). Even-aged *N. dombeyi* stands often develop on moraines that have become stabilized. *Nothofagus dombeyi, Weinmannia trichosperma*, and *Eucryphia cordifolia* are even capable of establishing and forming dense stands on in-transit moraines where mass movements have deposited debris on top of the glacial ice (Veblen et al., 1989a). At higher elevations and on drier substrates, *N. antarctica* and *N. pumilio* may dominate the colonization of moraines and outwash deposits.

Further south, at San Rafael Glacier (46°40′ S), Heusser (1964) and Lawrence and Lawrence (1959) described the establishment of *Nothofagus nitida*, *N antarctica*, and *N. betuloides* on moraines and outwash. In southern Patagonia, establishment of *N. betuloides*, *N. pumilio* and *N. antarctica* on moraines and outwash has been reported for numerous glacial valleys, (Armesto, Casassa & Dollenz, 1992; Pisano, 1978; Pisano & Dimitri, 1973). Nitrogen-fixing herbs and shrubs, such as *Gunnera chilensis*, *Coriaria ruscifolia*, and *Discaria* spp., are common on freshly exposed glacial deposits as well as landslide and riparian sites. Although they undoubtedly play an important role in modifying nitrogen availability at these sites, this role has not yet been investigated.

Extensive flooding is another disturbance phenomenon associated with glacial processes in the rainforest district. In southern Patagonia, for example, rapid glacial melting and partial melting of ice blocks in lakes periodically flood large areas of *Nothofagus* forest (Peña & Escobar, 1983). In comparison with drier climates, postglacial succession is generally more rapid in rainforest climates due to mild summer temperatures and lack of drought. In North America. *Alnus* and *Picea* establish quickly following glacial retreat in the perhumid and supolar zones. *Abies lasiocarpa* is a common colonist in drier parts of the perhumid zones or at high elevations in the seasonal zone. Dense even-aged *Picea* or *Abies* forest develops within 200 years of deglacation. These forests are distinguished from forests growing on older terrains by less dense canopies and the presence of deciduous hardwoods (*Populus trichocarpa*, *Alnus*, or *Betula*); they are also characterized by Rosaceous shrubs and semisaprophytic herbs, such as *Pyrola*, and the absence of Ericaceous shrubs and *Tsuga*. Ericaceous understory shrubs are the slowest to colonize postglacial substrates (after at least 300 years). *Tsuga* colonizes pure *Picea* forests approximately 300 years or more following deglaciation. There are many possible paths of succession following deglaciation, however, few of these alternative pathways have been studied in detail (Fastie, 1990; Reiners, Worley & Lawrence, 1971)

The first work on postglacial succession in the temperate rainforest zone focused on the eastern arm of Glacier Bay, Alaska (e.g., Cooper, 1937; Crocker & Major, 1955, Reiners et al., 1971). A simple chronosequence was used to characterize this succession from bryophytes (e.g., *Rhacomitrium* spp.), through a shrub stage (*Salix* spp.), to a *Picea sitchensis* forest, and finally to a steady-state *Tsuga heterophylla* forest. *Alnus sinuata* is widely recognized for playing an important role succession following deglaciation by fixing atmospheric nitrogen (Crocker & Major, 1955). Following establishment of *Alnus* forests, *Picea* seedlings grow slowly in the understory, eventually overtopping the short-lived *Alnus*. The nitrogen-rich soils then allow for rapid growth of *Picea* and the establishment of a pure *Picea* forest within two centuries of deglaciation (Cooper, 1937). Each of the principal stages of this succession is associated with significant changes in soil chemistry and structure (Crocker & Major, 1955).

Recent work on postglacial succession in North America has emphasized plant life-history strategies and competitive relationships to examine whether succession truly follows the facilitation model of Connell and Slatyer (1977). Preliminary results of some of this work suggest that postglacial succession is a far more complex and multifaceted process than is usually presented. For example in Glacier Bay, the original work was done only on the southern and eastern arm of the Bay where glacial retreat occurred rapidly (Cooper, 1937). Farther up the eastern arm and in the western arm, where a wide variety of parent materials predominate, a wide range of plant successional trajectories develop following deglaciation (Fastie, 1990). The speed of deglaciation and physical processes associated with it such as scour, fill, flooding, sediment texture, and other edaphic changes lead to divergent successional pathways. For example *Pinus contorta* plays a key role in early succession on sandy sites or dry sites. *Sheperdia* is also characteristic of these sites, especially on limestone. On others *Picea* comes in first with *Salix* and *Populus*. On coarse textured soils there is only limited colonization by *Alnus* resulting in lower concentrations of nitrogen and a slower successional pattern. Late seral species can also occur at the earliest stages of succession, imposing some constraints on the facilitation model and on the generality of Cooper's descriptions of the succession.

In Washington, revegetation following glaciation follows a different pattern (Oliver, Adams & Zasoski, 1985). As in southern Chile establishment of trees and shrubs proceeds rapidly following deglaciation. Distinct stages as described by Cooper (1937) do not generally occur. *Alnus sinuata* is characteristic of the early stages of succession but *Populus* is absent. *Vaccinium* commonly colonizes in the early stages of the succession in Washington, but is only part of the succession in Glacier Bay 300 years or more following deglaciation. Primary forest in Washington is much less compositionally distinctive, with *Tsuga* being a part of the succession from the earliest stages. But in both regions old growth forest generally develops within 400 to 600 years.

Disturbance by Snow Avalanches

In the southern Andes in subalpine forests of *N. pumilio* and/or *N. betuloides*, infrequent but stand-devastating snow avalanches typically result in the development of even-aged stands (Veblen et al., 1981). Where avalanches occur with greater frequency, however, they may prevent *N. pumilio* from successionally replacing *N. antarctica*. Although *N. pumilio* is longer lived and appears to be a better competitor, *N. antarctica's* typically more prostrate, shrubby form and effective vegetative reproduction (through adventitious roots on damaged branches) give it an advantage at sites frequently disturbed by avalanches (Veblen et al., 1977).

Because of the cooler winters in North America, on the mainland and northern part of the perhumid and subpolar rainforest regions, avalanches are a widespread and more frequent disturbance agent than in analogous parts of the Andes. Temperatures are often near freezing during winter in the perhumid and subpolar rainforest regions which allows for accumulation of wet heavy snow on steep slopes. Forests on avalanche slopes are usually distinguished from nearby old-growth forest by the presence of pure even-aged early successional species such as *Picea sitchensis* or *Pseudotsuga* with *Alnus* and *Rubus spectabilis* mixtures (Martin, 1989). These are open forests with a dense shrub layer usually composed of *Alnus*, *Rubus*, *Ribes*, and *Oplopanex*. In many cases the species composition of these forests is very similar to that of a riparian floodplain forest. These forests will succeed to *Tsuga* or *Abies* in approximately 100 to 200 years if not subject to additional avalanches. Understory plant communities vary with the frequency of avalanches and can be used to predict frequency of avalanches (Cushman, 1976). In Washington succession following avalanches often is rapid and frequently includes the resprouting or release of *Vaccinium* spp. and *Abies amabilis* from the original forest

(Oliver et al., 1985). In this case forest development is much less compositionally distinctive than it is in coastal Alaska.

Disturbance by Wind

The mid- to high-latitude zone of western South America is well known for its strong winds, and blowdowns are a common disturbance throughout the temperate rainforest zone (Agostini, 1941; Eskuche, 1973). Windstorms on Tierra del Fuego, causing the windthrow of individual trees as well as the blowdown of entire stands, are particularly important in controlling the structure of *Nothofagus* forests (Puigdefábregas, del Barrio, Gallart et al., in press; Rebertus & Veblen 1993a, 1993b). On Tierra del Fuego, most old-growth forests of *Nothofagus pumilio* and *N. betuloides* have >10% gap area (surface area beneath treefall-caused openings) and complete turnover rates through gap processes (i.e., the time required for treefalls to disturb an entire stand) range from 300 to 500 years (Rebertus & Veblen, 1993b), which is similar to more species-rich Valdivian rainforests in Chile (Veblen, 1985b). Uprooting and stem breakage are the predominant modes of treefall. Most gaps involve multiple treefalls and typically exhibit two or more events of gap creation and subsequent expansion (Rebertus & Veblen, 1993b).

On Tierra del Fuego, under certain circumstances of wind exposure, topographic position, and substrate, gradual gap enlargement results in a wave-like pattern similar to well-documented fir waves in the northeastern United States (Puigdefábregas et al., in press). Well-developed mortality waves in *N. betuloides* forest occur at right angles to the coast along the Beagle Channel and Isla de los Estados, where strong coastal winds and possibly salt spray may be the driving forces. These mortality waves are conspicuous because of the bleached white dead-standing trees and logs pointing in a leeward direction. Further inland, waves do occur, but they are less conspicuous and more irregular in shape and movement (Rebertus & Veblen, 1993a). Large-scale blowdowns, sometimes involving hundreds of hectares, are a conspicuous feature in the Fuegian landscape. Indeed, some areas, where soils are thin, are a complete mosaic of blowdowns of different ages.

Farther north in the Valdivian rainforest region, fine-scale treefalls have been studied on the Island of Chiloé (Armesto & Figueroa, 1987; Armesto & Fuentes, 1988; Veblen, 1985b) and in the Andes in the Province of Valdivia (Veblen, 1985b). These studies have shown that the shade-intolerant species such, as *Nothofagus dombeyi* and *Weinmannia trichosperma*, typically do not regenerate in gaps created by single or small groups of trees. The exception is where a large nurse log allows a *N. dombeyi* seedling to establish above the bamboo-dominated understory. If the seedling's roots develop a secure and sustaining anchorage in the soil before the nurse log decays, this regeneration strategy may be successful (Veblen, 1985b). In general, seedling establishment on logs and stumps is important throughout the South American rainforest region. The most shade-tolerant tree species, such as *Laureliopsis philippiana*, are able to regenerate beneath a continuous canopy cover but their growth rates are enhanced by the creation of small canopy gaps (Armesto & Fuentes, 1988; Veblen, 1985b). Somewhat less shade-tolerant species, both subcanopy and main canopy in potential stature, are totally dependent on treefall gaps for successful regeneration (Armesto & Fuentes, 1988). Tree species with some mode of vegetative reproduction (e.g., root suckering, basal sprouting) typically have an advantage in filling gaps (Veblen et al., 1981).

In North America, windstorms are frequent and cause extensive impact throughout the rainforest region (Harris, 1989; Ruth & Harris, 1979). Windthrow severity limits stand development on parts of the outer Pacific coast, but in most regions stem breakage (a combination of fungal attack and wind) or fungal mortality is of greater frequency than uprooting (Alaback, unpublished data; Lertzman & Krebs, 1991; Ott, 1994; Taylor, 1990). Maintenance dynamics are common throughout the region, but stand-replacement level windthrow occurs periodically as well, especially along exposed coastlines. Snow breakage is also a key disturbance agent. It can be particularly important in young even-aged stands because of the susceptibility of tall thin stems to wet heavy snow in early spring (Alaback & Tappeiner, 1991).

Preliminary studies of treefall gaps in North and South American temperate rainforests indicate broadly similar patterns of gap sizes, turnover times, and percentages of the forest area in the state of a gap (Armesto & Fuentes, 1988; Lertzman, 1992; Ott, 1994; Rebertus & Veblen, 1983b; Spies, Franklin, & Klopsch, 1990; Taylor, 1990; Veblen, 1985b). The major difference in treefall gap parameters, is the larger percentage of uprooted trees in South American forests (Figure 9.4). Possibly this reflects a greater resistance of the angiosperm trees (the wood of which generally have greater specific densities) to bole breakage, higher intensity and more frequent storms in the Andes, greater incidence of root decay fungi in North America, and/or deeper more paludified soils in North America.

Disturbance by Fire

In South America, the abundance of fire-adapted sprouting species in the drier deciduous *Nothofagus* forests (e.g. *N. antarctica*, *N. obliqua*, *N. alpina*, *Lomatia hirsuta*, *Schinus patagonicus*) suggests that fire has been an important component of these ecosystems for many thousands of years. However, in the wettest temperate rainforests, prior to human occupation, natural fires ignited by lightning or volcanism were rare events. Within the rainforest zone, natural fires appear to occur most

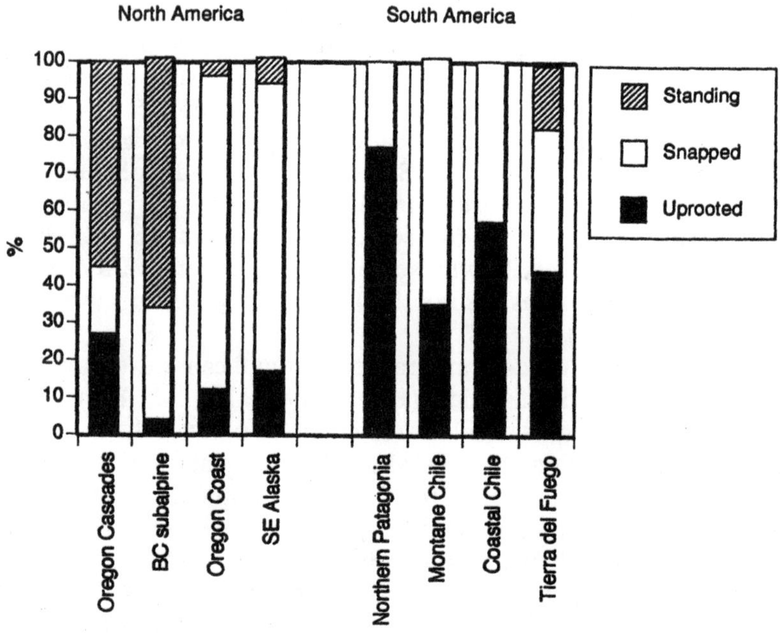

Figure 9.4. Comparison of causes of mortality of gap makers in old-growth forests in North and South America. Data are based on surveys of recently formed canopy gaps. Snapped = tree bole broken (from fungal decay and wind); Windthrow = tree blown over and rootthrow mound created. Data derived from Lertzman & Krebs, 1991, Ott, 1994; Spies et al., 1990; Taylor, 1990; Rebertus & Veblen, 1993b; Veblen 1985a, 1989.

frequently near the northern and eastern limits of these forests (Heusser et al., 1988; Veblen et al., 1992a).

In the rainforest zone where *Chusquea* bamboos occur, they resprout vigorously following fire and may impede tree regeneration until they flower synchronously and die. If the preburn forest had a dense canopy with only sparse bamboo in the understory, then there is relatively little interference from bamboo, and even-aged stands of the shade-intolerant tree species (especially *Nothofagus* spp.) will develop (Veblen, 1985a). In drier forests, where the bamboos are scarce or absent, as long as seed sources are available, all *Nothofagus* spp. typically regenerate in abundance and develop into even-aged stands following stand-devastating fire (Pisano & Dimitri, 1973; Roig, Dollenz, Faggi & Méndes, 1985; Veblen & Lorenz, 1987). Except for *N. antarctica*, *N. obliqua*, and *N. alpina*, vegetative reproduction following fire is rare for rainforest trees.

In North America, natural fire is also rare in perhumid and subpolar rainforest regions except in severe rain shadow locales such as Lynn

Canal (in the Skagway and Haines region north of Juneau; Noste, 1969). *Pinus contorta* subs. *contorta* and *Betula papyrifera* occur where fires are frequent (and nowhere else within the rainforest zone). In the perhumid zone, continuously wet and thick organic horizons prevent effective heat penetration from forest fires into the soil layer. As a result, fires often do not lead to dramatic changes in plant species composition. *Thuja* and *Picea* appears to regenerate better and outcompete *Tsuga* more effectively following fire. Much of the effect appears to be more one of eliminating advance regeneration of *Tsuga* than creating more favorable growth conditions for *Thuja* or *Picea*.

In the seasonal rainforest zone in North America fire is infrequent (100 to 300 years), but often catastrophic in scale. The widespread occurrence of *Pseudotsuga menziesii*, in the southern portion of this region, for example, is usually attributable to catastrophic fire (Agee, 1991; Harcombe, 1986; Juday, 1976). Formerly extensive even-aged old-growth forests of *Pseudotsuga* covered large portions of the Oregon Coast Range and are attributable to just a few catastrophic fires. In the Olympic Mountains of Washington, fire is a critical factor on the eastern slope of the mountains (east of the Bailey Range) where *Pseudotsuga* forest may also dominate (Fonda & Bliss, 1969). Likewise fire frequencies in the Cascades are generally much higher than along the coast both because of increased lightning frequencies and lower fuel moisture. From the western Olympics northwards to the northern limit of the seasonal rainforest type, stands originated by fire become progressively more localized or infrequent, and are usually associated with excessively drained, rocky sites, exposed southern slopes, or other anomalously warm or dry microsites (Gagnon & Bradfield, 1987).

Disturbance by Native Herbivores

Although South American temperate rainforests are generally depauperate of large mammalian herbivores, in pre-European times native guanacos (*Lama guanicoe*) were common along the ecotone between *Nothofagus* forests and the Patagonian steppe. Guanacos were the principal game animal of Patagonian aboriginals and were described by nineteenth century explorers as occurring in immense herds (Musters, 1871; Prichard, 1902). Today, in some *N. pumilio* stands on Tierra del Fuego, guanacos are abundant and seriously impede *Nothofagus* regeneration (A. Rebertus, personal communication). Other large herbivores that may have affected forest composition include huemul deer (*Hippocamelus bisulcus*) and small pudu deer (*Pudu pudu*), both hunted by Indian populations (Cox, 1863; Machoni, 1732). Pudu deer occur principally in the rainforests of southern Chile and adjacent areas of Argentina, while huemul deer favor ecotonal areas near alpine timberline or the Patagonia steppe (Povilitis, 1978).

Another native herbivore affecting some *Nothofagus* forests is the defoliating caterpillar, *Ormiscodes cinnamomea* (Gentili & Gentili, 1988) which periodically defoliates areas of many square kilometers in northern Patagonia. Although outbreaks result in relatively little tree mortality, but incomplete recovery has led to an increases in trees with partial "crown dieback." Thus, periodic insect defoliation followed by incomplete recovery may contribute to the abundance throughout Patagonia of *Nothfagus* with partially dead crowns, which formerly was attributed to recent climate change (Auer, 1939, 1958; Kalela, 1941). In the Punta Arenas region wood borers (*Chilecomadia* sp.) attack *Nothofagus pumilio*, but the extent of the damage has not been investigated. In northern Patagonia we have also observed that *N. pumilio* with partial crown dieback which has been attacked by wood-boring insects.

In North America, in the outer coastal environment, where mild winters prevail high native deer (*Odocoileus hemonius* var, *sitkensis* and var. *columbiana*) populations persist and result in reduced shrub and tree seedling populations. In mainland and more protected island regions within the perhumid zone, deer seldom reach sufficient population levels to significantly alter successional pathways (Klein, 1985; Lewis, 1992). In the seasonal rainforest region logging and fire have created abundant deer habitat. Parks and preserves, such as the Olympic rainforest, develop a park-like structure from the heavy browsing of deer and elk (*Cervus elaphus*). *Thuja* and *Tsuga* generally decline in seedling recruitment compared to *Picea* under heavy browsing pressure, but reductions in elk populations do not necessarily correspond to increases in *Tsuga* seedling recruitment (Harmon & Franklin, 1983). Highly palatable shrubs such as *Vaccinium parvifolium* and *Ribes* spp. may also decline in heavily browsed sites. Planted tree seedlings generally have to be protected from deer until they reach above their browsing height. Porcupines (*Erethixon dorsatum*) can also significantly affect forest structure by killing dominant trees especially in young even-aged forests thereby creating small canopy gaps. Mountain beaver (*Alpodontia rufa*) may also play an important role in limiting tree regeneration in the seasonal rainforest zone.

Insect outbreaks can also have important localized effects in temperate rainforest especially during successive warm or dry years in which over-winter survival is unusually high. In the perhumid zone black-headed budworm (*Acleris gloverana*) and the hemlock sawfly (*Neodiprion tsugae*) are considered the most serious defoliating insects. Outbreaks are infrequent, and tend to be of greatest consequence in even-aged or managed forests (Ruth & Harris, 1979). In the seasonal zone Sitka spruce weevil (*Pissodes sitchensis*) is one of the key factors limiting the growth of *Picea sitchensis* inland from the coastal fog belt (Ruth & Harris, 1979). The spruce budworm (*Choristoneura* spp.) can also cause significant defoliation during warm or dry years or after a large accumulation of organic debris from fire or windthrow. In comparison with interior forests, however, insect outbreaks appear to be of lesser ecological significance to coastal forests.

Human Modifications of Disturbance Regimes

Impacts of Introduced Animals

Although European settlement prior to the mid-nineteenth century was rare in the rainforest districts of Chile and Argentina, domestic livestock arrived at least two centuries earlier as escapees from coastal settlements (Fonck, 1900; Martinic, 1985). Throughout southern Chile and Argentina, forests have been heavily used to graze livestock for at least 60 years and in many cases for over a century (Cunill, 1974; Goodall, 1979; Roig et al., 1985). Although there are some inaccessible areas where livestock are not found, these account for only a small percentage of rainforest area.

Exotic deer (mostly the red deer, *Cervus elaphus*, but also fallow deer, *Dama dama*, and axis deer, *Axis axis*), upon which a lucrative sport hunting industry depends, have attained high population densities and have had a deleterious impact on native vegetation in southern Chile and particularly in Argentina (Ramilo, 1985; Ramirez, Godoy, Eldridge, & Pacheco, 1985; Veblen, Mermoz, Martin, & Ramilo, 1989b). Deer and livestock have an inhibitory effect on the regeneration of common arboreal species (Martín, Mermoz, & Gallopín, 1985; Ramirez et al., 1981; Veblen et al., 1989b; Veblen, Mermoz, Martín, & Kitzberger, 1992b).

European hares (*Lepus capensis*) and wild boars (*Sus scrofa*) have colonized nearly all suitable habitats of mainland Patagonia (Daciuk, 1978). Because of their rooting activities, wild boars are responsible for localized alteration of grasslands and forest openings, especially in the Andean parts of the Argentine provinces of Rio Negro and Chubut (Daciuk, 1978). European hares occur throughout continental Patagonia primarily in grasslands where they often attain high population densities and have had a significant impact on the vegetation. Although hares and boars do not occur on Tierra del Fuego, introduced rabbits (*Oryctolagus cuniculus*) and beavers (*Castor canadensis*) are severe pests there (Goodall, 1979; Lizarralde, Escobar, & Bianciotto, 1990). Their damage includes both cutting of trees and flooding of large areas of Nothofagus forest.

In North America, there are few instances of exotic introductions of herbivores that have been so successful as to significantly affect native vegetation. Many huntable species have been introduced to regions that are slightly out of their natural range. Impacts of species introductions are of much less importance in North America than in South America. For example mountain goats were introduced into the Olympic Mountains where they have had localized impacts on alpine vegetation. Pine marten and various squirrel species were introduced to islands that did not have them for historical biogeographical reasons. Squirrels may be affecting availability of conifer seed or altering competitive relationships with birds.

One of the most significant examples of animal introductions affecting forest processes in the perhumid rainforest region is the introduction of Sitka black-tailed deer to the Queen Charlotte Islands in the early 1990s. The Queen Charlotte Islands are an isolated group of islands 100 km off of the coast of central British Columbia, well known as a partial glacial refugium (Calder & Taylor, 1968). No natural predators of deer occurred on the Queen Charlotte Islands. Deer are not native to the islands although an isolated population of caribou occurred there historically. The forest vegetation has been radically altered by the introduced deer (Pojar & Banner, 1984). The affected forests retain only a small remnant of the woody shrub stratum. Only bryophytes, ferns and unpalatable herbs persist in the most heavily affected areas. Shrubs develop into normal-sized plants only when rooted above the level at which deer can browse, or when the shrub branches are above browsing height. *Chamaecyparis* and *Thuja* regeneration is severely constrained by deer because of its high nutritive value and palatability (Hennon, 1992), and even *Picea* can be browsed. Because of the paucity of undergrowth (both tree and shrub) these forests are park-like.

Elk were introduced about 10 years ago in southeast Alaska in one small region (southern Etolin Island). The population is small and stable, and local game managers do not believe there is a danger of a major impact. If the population grows, however there will be major concern because of the affinity of elk for riparian habitats, which are fragile habitat for many native plant and animal species and are the most diverse and productive habitats in most watersheds. Effects of elk on forage and habitat selection for deer may be significant as well. Deer have also been introduced north of their natural range in Prince William Sound and Kodiak Island. Populations remain small and unstable in the Prince William Sound area, but on Kodiak Island populations have expanded steadily since their introduction in 1924 and are continuing to move westward (Smith, 1979). The Kodiak Island populations differ from the other Alaska populatons in their ability to tolerate winters in nonforested habitats (natural shrublands), beyond the western range of *Picea* and all other conifers.

Anthropogenic Fire

Charcoal particles in sedimentary samples from northern Patagonia (c. 39° S) to Tierra del Fuego, document the occurrence of fire for at least several thousand years (Heusser, 1987; Heusser et al., 1988). Although more charcoal is found in association with pollen assemblages indicating xeric open vegetation, the joint occurrence of charcoal with forest taxa imply some incidence of prehistoric fire in forests. Despite the uncertainty of the relative importance of natural ignitions versus human-set fires in prehistoric times, there is no doubt that with the arrival of aboriginal humans the frequency of forest burning increased. In northern Patagonia, early explorers described Indians using fires to hunt rheas and guanacos in the steppe adjacent to forests (Cox, 1863; Fonck, 1900; Musters, 1871). They also described vast areas of burnt rainforest in the eighteenth and nineteenth centuries that they attributed to Indian-set fires (Cox, 1863; Fonck, 1900).

During the nineteenth century and early twentieth century period of white settlement of southern Chile and Argentina, vast areas of rainforest were deliberately burned to increase pasture and to permit agriculture (Cunill, 1974; Willis, 1914). In many areas this period of burning is reflected by the abundance today of extensive 70 to 100 year-old even-aged *Nothofagus* forests (Veblen & Lorenz, 1987). In the Chonos and Guaitecas Archipelagoes and on the Taitao Peninsula, extensive areas of forest were intentionally burned to facilitate access for extraction of the fire-resistant timber of *Pilgerodendron uviferum* (Grosse, 1979).

In the perhumid region of North America, fire was used by native populations whose settlements mostly were in rainshadow climates. Fire was used to fell large trees and to clear burial grounds, but the effects were localized. In recent decades, some fires ignited by chainsaws and related activities have resulted in small fires, especially in clearcut areas (Noste, 1969). Fire is used extensively in the seasonal rainforest region in British Columbia and in the Pacific Northwest to facilitate planting of nursery-grown tree seedings on commercial forestland and to decrease shrub competition with trees (Ruth & Harris, 1979). Indigenous peoples used fire extensively in the seasonal zone to maintain more open habitats for game species, particularly on dry south-facing slopes and valley bottoms. Human-caused fires may cause tremendous damage in the seasonal zone because of the large fuel accumulations which occasionally reach low moisture status during summer months.

Logging

Although the aboriginal population had a significant impact on the rainforests of South America through their use of fire, timber exploitation was insignificant until European colonization. In the Valdivian rainforest district, by the seventeenth century, Spanish shipyards were using large quantities of *Fitzroya cupressoides*. Massive clearing and burning of the forests of the Central Depression of south-central Chile began in the mid-nineteenth century with the arrival of German colonists (Cunill, 1974). Commercial logging became significant after completion of the railroad linked south-central Chile to central Chile in 1912. Until the advent of chipping in the 1980s, logging in the Andean forests of the Valdivian rainforest region involved mainly selective logging of the best-formed individuals of the most valuable species (especially *Nothofagus* spp. and *Fitzroya cupressoides*). The high-grading of these forests has resulted in relatively little regeneration of these shade-intolerant species, so that the timber quality of large areas of native forest in the Valdivian rainforest region has steadily declined over the past half century. Large areas of Andean rainforests were logged and burned, and in some cases converted to cattle pasture. Where regeneration has been protected from livestock, however, large areas of second-growth *Nothofagus*-dominated stands have developed. The patterns of stand development of these *Nothofagus*-dominated second-growth stands are similar to those of stands which originate following natural fire, landslides, or floods (Veblen, 1985a). Much second-growth native forest of substantial timber potential has been converted to plantations of exotic conifers (especially *Pinus radiata* because of the incentives of reforestation legislation (Lara & Veblen, 1992). Today, production of wood chips throughout the Valdivian and Patagonian rainforest regions is decimating vast areas of native forest.

Large-scale clearcutting comparable in scale to that practiced in the Pacific Northwest has occurred in some Chilean *Fitzroya cupressoides* forests. Following clearcutting in the Andes, the regeneration of *F. cupressoides* has been practically nil and sites exploited in the mid-1960s are dominated by a mixture of shrubs and economically unimportant tree species (Lara, 1991; Veblen et al., 1976). Regeneration on logged terrains in the Coastal Range of Chile, however, is more successful, but a general decline in *Fitzroya* is still common (Donoso, Sandoval, Grez, & Rodriguez, 1993; Parker & Donoso, 1993; Veblen & Ashton, 1982). The greater regeneration success in the Coastal Range following logging may reflect the presence of some seed trees in contrast to the large clearcuts in the Andes.

In North America, the deforestation and settlement of the west coast was one of the most rapid in history, and has contributed to increasing atmospheric carbon (Harmon, Ferrell, & Franklin, 1990). In the mid-latitudes of the Pacific Northwest and northern California most of the forests in valley bottoms and the most productive sites were cleared between the mid-1800s and the 1940s. The mid-elevation sites were mostly cleared during the economic expansion of the 1950's through the late 1960's (Sedell & Duval, 1985). Much of the logging from the 1980s through the present has been on high elevation sites and less productive or more remote sites. In British Columbia and Alaska, logging of sea level sites began on an industrial scale in the 1950s and 1960s, and peaked in the early 1990s (Sedell & Duval, 1985; U.S.D.A., Forest Service, unpublished data). Logging harvest volume is expected to decline over the next four to five decades throughout the Pacific Northwest and north coast due to overharvesting and a resulting imbalance in forest plantation age structure. The greatest extent of old-growth forest now occurs in the perhumid zone in British Columbia and Alaska and in relatively small fragments in southern British Columbia, Washington, and Oregon.

Commercial forestry plantations were established from the 1930s through the 1960s. Problems with poor seed selection and poor regeneration techniques plagued early attempts at forest plantations. Genetic improvements, better planting, and nursery techniques improved results from the 1970s to the present. Intensive forestry, including thinning, fertilization, and pest control, have only been implemented in the past two decades, and still exist primarily on private industrial land in Oregon and Washington. Throughout the seasonal rainforest zone *Pseudotsuga* has been planted even along the coast, leading to a less species rich forest than would occur after natural disturbance. Natural forest regeneration is relied upon rather than planted nursery stock for the great majority of lands within the perhumid rainforest region.

The mild wet climate in the rainforest zone allows for a rapid establishment of forest stands following logging or fire. *Tsuga* regeneration is usually rapid with significant survival from the previous stand (Alaback & Tappeiner, 1991; Harris, 1974). Herbs and shrubs resprout and dominate sites 5 and 15 years after logging, respectively. By stand age 20 to 30 years most understory plants and all tree seedlings are shaded out. In the perhumid zone on productive sites this depauperate stem exclusion stage (*sensu* Oliver, 1981) may persist for 100 years or more (Alaback, 1982). By stand age 150 to 250 years, the understory reinitiation stage develops, first with shade-tolerant shrubs like *Vaccinium*, then eventually with evergreen herbs or short shrubs such as *Cornus canadensis*, *Rubus pedatus* and *Tiarella trifoliata*. In two to four centuries, mature forests develop with a rich diversity of understory species and overstory canopy structure. The conversion of old-growth to young second-growth vegetation thus results in a significant decline in diversity and habitat value for deer and other wildlife species (Alaback, 1982; Hanley, Robbins, & Spalinger, 1989; Wallmo & Schoen, 1980). In the Pacific Northwest this pattern is less dramatic because of the less dense canopies of *Pseudotsuga* and greater solar radiation intensities. Understory plants colonize more rapidly during the aggradation stages of forest development and ruderal species play a much larger role in the initial stages of the succession than in the perhumid zone (Bailey & Poulton, 1968; Schoonmaker & McKee 1988; Spies & Franklin 1989).

The Effects of Recent Climate Variation on Disturbance Regimes and Forest Dynamics

The effects of recent (i.e., <1000 years B.P.) climate variations in southern Chile and Argentina are intimately interrelated with the effects of anthropogenic disturbances such as fire and livestock (Veblen & Lorenz, 1988; Veblen & Markgraf, 1988; Veblen et al., 1992a). In South America, *Nothofagus*-dominated forests generally occur at sites that are suboptimal for tree growth due to climatic or edaphic factors. Thus, along a west-to-east trans-Andean gradient at c. 40° S from moist to a dry climate, the Valdivian rainforest of south-central Chile is replaced by monotypic stands of *Nothofagus*, which in turn give way to xeric woodlands and Patagonian steppe. Similarly, along north-to-south as well as along elevational gradients from warm to cooler temperatures, relatively species-rich temperate rainforests are replaced by *Nothofagus* forests, then by alpine or subantarctic tundra. The occurrence of *Nothofagus* forests in these ecotones between forest and nonforest vegetation makes them highly susceptible to the effects of climate fluctuation (Markgraf, 1983, 1987).

In general for northern Patagonia, both tree-ring records and glacier variations indicate intervals of below average temperatures for 1300 to 1380, 1500 to 1700, and around 1840 A.D. In contrast, above average temperatures occurred in 1100 to 1300, 1850 to 1890, and 1905 to 1920 A.D. (Mercer, 1965, 1970; Rothlisberger, 1986; Villalba, 1990; Villalba, Boninsegna, & Gobos, 1989). These climate variations are likely to have had an influence on the occurrence of fire and on forest dynamics generally. For example, preliminary data indicate there is a marked concentration of *Austrocedrus chilensis* mortality from c. 1911 to 1919 which was a period of unusually low summer precipitation and high mean annual temperature (Villalba & Veblen, unpublished). Similarly, the peak of anthropogenic forest burning in northern Patagonia coincides with drier periods in the late nineteenth and early twentieth centuries (Veblen et al., 1992a).

The pollen record from the Pacific coast of North America, especially along the northern coast and interior, indicates a significant warming trend since the last major ice advance (e.g., Brubaker, 1988; Heusser, 1960). Many tree species in Alaska are physiologically adapted to grow much further north or west than the limit of their present day distribution (e.g., *Abies amabllis*, *A. lasiocarpa*, *Picea glauca*, *P. sitchensis*). *Picea sitchensis* continues to expand its range westward into the wet coastal tundra areas and shrublands at the western extreme of its distribution in Alaska (Griggs, 1934).

Subtle changes in climate in the southern part of the rainforest district correspond with dramatic changes in northern areas including glacial surges or rapid recessions and rapid timberline movement. For example, during the Little Ice Age approximately 200 years ago all of Glacier Bay was filled with ice. Presently ice only remains in the head of the Bay, more than 95 km from the mouth of the Bay. Partly because of these massive fluctuations in glaciers and climate, species in the northern part of the rainforest district tend to be habitat generalists and are less likely to be uniquely adapted to current climatic conditions (Kellison & Weir, 1987).

Two centuries ago or more in the seasonal rainforest zone some fire-adapted tree species such as *Pseudotsuga menziesii* colonized coastal and high-elevation sites which now are dominated by cool moist-adapted species such as *Tsuga* or *Abies* (Jan Henderson, U.S.D.A., Forest Service, Olympic National Forest, personal communication). Natural regeneration following logging at mid-elevations in the Oregon Cascades may also result in major shifts in dominant species, perhaps also suggesting a somewhat different climate two to five centuries ago when dominant trees from the original forest established. Similarly, in southern Oregon reforestation following logging on southwest-facing slopes is often slow or unsuccessful unless manipulation of competing vegetation, shade cards and other intensive reforestation techniques are applied, suggesting a possible climatic shift from when these forests established two to three centuries ago (Minore, 1978; Perry, Amaranthus, Borchers et al., 1989).

Since the late 1970s a 0.1 degree per decade warming has been documented in interior Alaska, providing a possible model of the effects of global warming on northern landscapes (Trenberth, 1990). Frequency of severe snow accumulation has declined dramatically in the perhumid rainforest zone, resulting in dramatic increases in large mammal populations which were formerly constrained by extreme winter conditions. This has resulted in reduced tree regeneration by the most palatable species, such as *Chamaecyparis nootkatensis* and *Thuja plicata* (Hennon, 1992). Frequency of snow avalanches has also dramatically declined in coastal regions. This has resulted in a rapid colonization of alpine tundra by *Tsuga mertensiana*, or of *Cassiope*- or *Carex*-dominated meadows by *Rubus spectabilis*.

Conclusions

The disturbance regimes of the temperate rainforests of North and South America show some important similarities related to their locations in areas of often rugged topography and great tectonic activity, and in cool moist climates. In comparison with other types of forest ecosystems, fire is a relatively infrequent natural disturbance in the subpolar and perhumid rainforests. In the seasonal rainforests, however, fire often is a major factor in their dynamics. In both North and South American temperate rainforests, windstorms and landslides are important agents of disturbance. Towards higher latitudes and altitudes, disturbance by glacial activity is also important.

In general, many characteristics of the temperate rainforests of the west coasts of the Americas reflect the interplay between fine-scale and coarse-scale disturbances. Both regions have long-lived forest dominants, a rich epiphyte layer, and complex canopy structure reflecting both large-scale and small-scale disturbances. The relative importance of disturbances of different scales varies significantly, and to a large degree predictably, from one habitat type to another. Despite the taxonomic differences between the temperate rainforests of North and South America, the general patterns of stand development following disturbance are broadly similar. These patterns are well described by general models of whole-stand replacement (e.g., Oliver, 1981) and fine-scale treefall gap dynamics (e.g., Watt, 1947). Many areas have a complex history that includes disturbances at many spatial scales often occurring on the same terrain but at different points in time. The structure, diversity, and functioning of these forests appears to be closely tied to the character of these disturbances and their spatial and temporal patterns.

Although the extensive data required to compare the extent and frequency of most disturbances in North versus South American temperate rainforests are not available, the incomplete information on natural disturbance regimes in the two areas suggests some significant differences. Although both areas occur in regions of great tectonic activity, the South American temperate rainforests appear to be more extensively affected by geologic disturbances such as volcanism and earthquake-triggered mass movements. Disturbance by volcanism is less significant in North American temperate rainforests because volcanoes mostly occur in the southeastern part of the seasonal rainforest zone or west of the subpolar region (Aleutians, Alaska Range, Katmai Range). In contrast, disturbance by snow avalanches and wet snowfalls appears greater in North America. This reflects the cooler winters and greater snow accumulation associated with the more continental North American climate.

A major difference exists between North and South American temperate rainforests in terms of the degree of human modification mediated through introductions of large herbivores. Introduced mammals have had only a minor impact in the north but have become a dominant influence in large parts of South American temperate rainforests. Introduced livestock, deer, and beavers have and continue to play a major role in modifying forest composition and structure in South America. The greater impact of large introduced herbivores in South American temperate rainforests mainly reflects the much greater densities of the animals. However, in South American forests the dominant plant species appear to be less resistant to browsing which may reflect the paucity of large native herbivores in these forests.

Both North and South American temperate rainforests may be particularly susceptible to the effects of climate variation. Relative to other forest formations, their distributions are relatively restricted and at their poleward limits they occur in habitats where temperature plays a particularly important role in determining tree survival. Thus, global warming is likely to have a significant ecological impact at ecotones which are mainly controlled by low temperatures. Similarly, ecotones mainly controlled by moisture availability (i.e., at their eastern and equatorward limits) are likely to be greatly affected by changes in fire regimes which will inevitably accompany global warming. Given the relatively minor role of natural fires over most of the distribution of temperate rainforests, climatically induced increases in fire occurrence are likely to have major influences on the structure, composition, and functioning of these ecosystems.

Acknowledgments. Support for the research on which this review is based was provided by the Council on Research and Creative Work of the University of Colorado, the National Geographic Society, the USDA Forest Service Pacific Northwest Research Station, the Ecological Processes Program, and the National Science Foundation. Some of the material on South America was drawn from collaborative work in progress with Thomas Kitzberger, Alan Rebertus, and Ricardo Villalba, whose contributions are gratefully acknowledged. The manuscript was considerably improved by comments of the reviewers K.P. Lertzman and T.A. Spies.

References

Agee, J.K. 1990. The historical role of fire in Pacific Northwest forests. In J.D. Walstad, S. Radosevich, and D. Sandberg (eds.), *Natural and Prescribed Fire in Pacific Northwest Forests* (pp. 25–38). Corvallis, OR: Oregon State University Press.

Agee, J.K. 1991. Fire history of Douglas-fir forests in the Pacific Northwest. In L.F. Ruggiero, K.B. Aubry, A.B. Carey, and M.H. Huff (eds.), *Wildife and Vegetation of unmanaged Douglas-fir forests* (pp. 25–33). Portland. OR: U.S.D.A., Forest Service, Pacific Northwest Research Station, General Technical Report PNW-GTR-285.

Agostini, A.M. 1941. *Andes Patagónicos: Viajes de Exploración a la Cordillera Patagónica Austral*. Buenos Aires: Imprenta de Luis L. Gotelli.

Alaback, P.B. 1982. Dynamics of understory biomass in Sitka spruce-western hemlock forests of southeast Alaska. *Ecology*, 63, 1932–1948.

Alaback, P.B. 1991. Comparative ecology of temperate rainforests of the Americas along analogous climatic gradients. *Rev Chil His Nat*, 64, 399–412.

Alaback, P.B., and Juday, G.P. 1989. Structure and composition of old-growth forests in the Research Natural Areas of Southeast Alaska. *J Nat Areas Assoc*, 9, 27–39.

Alaback, P.B., and Sidle, R.C. 1986. Biomass, structure and nutrients of riparian vegetation on a small watershed on Chichagof Island, southeast Alaska. In D.L. Correll (ed.), *Watershed Research Perspectives. Proceedings of the Symposium, June 10–12, 1985* (pp. 135–165). Edgewater, MD: Smithsonian Environmental Research Center.

Alaback, P.B., and Tappeiner II, J.C. 1991. Response of western hemlock (*Tsuga heterophllya*) and early huckleberry (*Vaccinium ovalifolium*) seedlings following forest windthrow. *Can J For Res*, 21, 534–539.

Almeyda, A.E., and Saez F.S. 1958. *Recopilación de Datos Climáticos de Chile y Mapas Sinópticos Respectivos*. Santiago: Ministerio de Agricultura.

Antos, A.J., and Zobel, D.B. 1985. Recovery of forest understories buried by tephra from Mount St. Helens. *Vegetatio*, 64, 105–114.
Armesto, J.J., and Figueroa, J. 1987. Stand structure and dynamics in the temperate rain forests of Chiloé Archipelago, Chile. *J Biogeogr*, 14, 267–376.
Armesto, J.J., and Fuentes, E.R. 1988. Tree species regeneration in a mid-elevation, temperate rain forest in Isla de Chiloé, Chile. *Vegetatio*, 74, 151–159.
Armesto, J.J., Casassa, I., and Dollenz, O. 1992, Age structure and dynamics of Patagonian beech forests in Torres del Paine National park, Chile. *Vegetatio*, 98, 13–22.
Arno, S.F., and Habeck, J.R. 1972. Ecology of alpine larch (*Larix lyallii* Parl.) in the Pacific Northwest. *Ecol Monog*, 42, 417–450.
Auer, V. 1939. Der Kampf zwischen Wald und Steppe auf Feuerland. *Petermanns Mitteilungen*, 6, 193–197.
Auer, V. 1958. The pleistocene of Fuego-Patagonia. Part II. The history of the flora and vegetation. *Ann Acad Scient Fenni*, Series A, 50, 1–239.
Bailey, A.W., and Poulton, C.T. 1968. Plant communities and environmental interrelationships in a portion of the Tillamook burn, northwest Oregon. *Ecology*, 49, 1–13.
Barber, H.W.J. 1976. *An Autecological Study of Salmonberry* (Rubus spectabilis *Pursh*) in Western Washington. Unpublished doctoral dissertation, University of Washington, Seattle, WA.
Beaudry, P.G. 1989. Hydrology of the Skeena River floodplains. II: Flood hazard classification for silviculture. In E.B. Alexander (ed.), *Watershed '89: A Conference on the Stewardship of Soil, Air, and Water Resources* (pp. 173–178). Juneau, AK: U.S.D.A., Forest Service.
Binkley, D. 1984. Importance of size density relationships in mixed stands of Douglas-fir and red alder. *For Ecol Mgt*, 9, 81–88.
Borchers, S.L., Wattenbarger, J., and Ament, R. 1989. Forest plant associations of Montague Island, Chugach National Forest: Preliminary results. In E.B. Alexander (ed.), *Watershed 89: A Conference on the Stewardship of Soil, Air, and Water Resources* (pp. 29–46). Juneau, AK: U.S.D.A., Forest Service.
Brew, D.A. 1988. Latest Mesozoic and Cenozoic igneous rocks of Southeast Alaska—A synopsis. Menlo Park, CA: U.S.D.A., U.S. Geological Survey, Open File Report 88-405.
Brubaker, L.B. 1988. Vegetation history and anticipating future vegetation change. In J.K. Agee and D.R. Johnson (eds.), *Ecosystem Management for Parks and Wilderness* (pp. 41–61). Seattle: University of Washington, Press.
Calder, J.A., and Taylor, R.L. 1968. *Flora of the Queen Charlotte Islands. Part I: Systematics of the Vascular Plants*. Ottawa: Canada Department of Agricultural Research Branch Monogr 4(1), Queens Printer.
Clement, J.E. 1985. Floodplain succession on the west coast of Vancouver Island. *Can. Field Nat.*, 99, 34–39.
Connell, J.H., and Slatyer, R.O. 1977. Mechanisms of succession in natural communities and their role in community stability and organization. *Am Nat*, 111, 1119–1144.
Cooper, W.S. 1937. The recent ecological history of Glacier Bay: An initial report upon a long term study. *Ecology*, 4, 355–365.
Cooper, W.S. 1942. Vegetation of the Prince William Sound Region, Alaska; with a brief excursion into post-Pleistocene climatic history. *Ecol Monog*, 12, 1–22.
Cox, G. 1863. Viajes a las regiones septentrionales de Patagonia 1862–1863. *An Univ Chile*, 23, 3–239, 437–509.
Crocker, R.L., and Major, J. 1955. Soil development in relation to vegetation and surface age at Glacier Bay, Alaska. *J Ecol*, 43, 427–448.

Cunill, P. 1974. Geografia del deterioro del paisaje chileno. In L. Capuro (ed.), *Ecología: Ciencia Básica para la Supervivencia del Hombre* (pp. 125–140). Valparaíso, Chile: Editorial Universitarios de Velparaíso.
Cushman, M. 1976. *Vegetation Composition as a Predictor of Major Avalanche Cycles, North Cascades, Washington*. Unpublished masters thesis, University of Washington, Seattle, WA.
Daciuk, J. 1978. Notas faunísticas y bioecológicas de Península Valdés y Patagonia. IV: Estado actual de las especies de mamíferos introducidos en la subregión araucana (Rep. Argentina) y del grado de coacción ejercido en el ecosistema. *An Parq Nac*, 14, 105–130.
del Moral, R., and Bliss, L.C. 1993. Mechanisms of primary succession: Insights resulting from the eruption of Mt. St. Helens. In M. Begon and A.H. Fitter (eds.), *Advances in Ecological Research* (pp. 1–66). New York: Academic Press.
Demeo, T., Martin, J., and West, R. 1993, *Forest Plant Association Guide: Ketchikan Area, Tongass National Forest*. Juneau, AK: U.S.D.A., Forest Service, Alaska Region, No. R10-MB-210.
Donoso, G., Grez, R., Escobar, B., and Real, P. 1984. Estructura y dinámica de bosques del tipo forestal siempreverde en un sector de Chiloé insular. *Bosque*, 5, 82–104.
Donoso, C., Escobar, B., and Urrutia, J. 1985. Estructura y estrategias regenerativas de un bosque virgen de Ulmo (*Eucryphia cordifolia* Dav.) y Tepa (Laurelia philippiana Phil.) Looser en Chiloé, Chile. *Rev Chil Hist Nat*, 58, 171–186.
Donoso, C., Sandoval, V., Grez R., and Rodriguez, J. 1993. Dynamics of *Fitzroya cupressoides* forests in southern Chile. *J. Veg. Sci.*, 4, 303–312.
Edmonds, R.L., Thomas, T.B., and Maybury, K.P. 1993. Tree population dynamics, growth, and mortality in old-growth forests in the western Olympic Mountains, Washington. *Can J For Res* 23, 512–519.
Eskuche, U. 1973. Estudios fitosociológicos en el norte de Patagonia. I. Investigación de algunos factores de ambiente en comunidades de bosque y de chaparral. *Phytocoenología*, 1, 164–213.
Farr, W.A., and Hard, J.S. 1987. Multivariate Analysis of Climate along the Southern Coast of Alaska—Some Forestry Implications. Portland, OR: U.S.D.A., Forest Service, Pacific Northwest Research Station, Research Paper PNW-RP-372.
Farr, W.A., and Harris, A.S. 1979. Site index of Sitka spruce along the Pacific coast related to latitude and temperatures. *For Sci*, 25, 145–153.
Fastie, C. 1990. Inference and verification in chronosequence studies at Glacier Bay. In A.M. Milner and J.D. Wood, Jr. (eds.), *Proceeding of the Second Glacier Bay Science Symposium September, 19–22, Glacier Bay Lodge, Alaska* (pp. 147–149). Anchorage, AK: U.S.D.I., Park Service.
Fonck, F. 1900. *Viajes de Fray Francisco Menéndez a Nahuelhuapi*. Valparaiso, Chile: C.F. Niemeyer.
Fonda, R.W., and Bliss, L.C. 1969. Forest vegetation of the montane and subalpine zones, Olympic Mountains, Washington. *Ecolog Monog*, 39, 271–301.
Franklin, J.F., MacMahon, J.A., Swanson, F.J., and Sedell, J.R. 1985. Ecocsystem responses to the eruption of Mt. St. Helens. *Nat Geog Soc Res*, 1, 198–216.
Frenzen, P.M., Krasney, M.E., and Rigney, L.P. 1988. 33 years of plant succession on Kautz Creek mudflow, Mount Rainier National Park, Washington. *Can J Bot* 66, 130–137.
Gagnon, D., and Bradfield, G.E. 1986. Relationships among forest strata and environment in southern coastal British Columbia. *Can J For Res*, 16, 1264–1271.

Gentili, M., and Gentili, P. 1988. Lista comentada de los insectos asociados a las especies sudamericanas del género *Nothofagus*. Monografías de la. Academia Nacional de Ciencias Exactas, *Físicas y Naturales*, 4, 85–106.

Glenn-Lewin, D.C., Peet, R.K., and Veblen, T.T. (eds.). 1992. *Plant Succession: Theory and Prediction*, London: Chapman and Hall.

Goodall, R.N.C. 1979. *Tierra del Fuego*. Buenos Aires: Ediciones Shanamaiim.

Griggs, R.F. 1934. The edge of the forest in Alaska and the reason for its position. *Ecology*, 15, 80–96.

Grosse, J.A. 1979. Colonización en Aisén y Chiloé Continental. *Trapananda*, 3, 3–5.

Habeck, J.R. 1988. Old-growth forests in the northern Rocky Mountains. *Nat. Areas J*, 8, 202–211.

Halpern, C.B., and Harmon, M.E. 1983. Early plant succession on the Muddy River mudflow, Mount St. Helens. *Am Midl Nat* 110, 97–106.

Hanley, T.A., Robbins, C.T., and Spalinger, D.E. 1989. *Forest Habitats and the Nutritional Ecology of Sitka Black-Tailed Deer: A research Synthesis with Implications for Forest Management*. Portland, OR: U.S.D.A., Forest Service, Pacific Northwest Research Station, General Technical Report PNW-230.

Harcombe, P.A. 1986. Stand development in a 130-year old spruce-hemlock forest based on age structure and 50 years of mortality data. *For Ecol Mgt*, 14, 41–58.

Harmon, M.E., and Franklin, J.F. 1983. Age distribution of western hemlock and its relation to Roosevelt elk populations in the South Fork Hoh River valley, Washington. *NW Sci* 57, 249–255.

Harmon, M.E., Ferrell, W.K., and Franklin, J.F. 1990. Effects on carbon storage of conversion of old-growth forests to young forests. *Science*, 247, 699–702.

Harris, A.S. 1974. Clearcutting, reforestation and stand development on Alaska's Tongass National Forest. *J For*, 72, 330–337.

Harris, A.S. 1989. *Wind in the Forests of Southeast Alaska and Guides for Reducing Damage*. Portland, OR: U.S.D.A., Forest Service, Pacific Northwest Research Station, General Technical Report PNW-GTR-244.

Henderson, J.A., Peter, D.H., Lesher, R.D., and Shaw, D.C. 1989. *Plant Associations of the Olympic National Forest*. Portland, OR: U.S.D.A., Forest Service, Pacific Northwest Region, R6 Ecol. Technical Paper, 001-88.

Hennon, P.E. 1992. *Ecology and Silviculture of Yellow-Cedar in Southeast Alaska: November 1992 Information Exchange at Sitka, Alaska*. Juneau AK: U.S.D.A., Forest Service, Forest Health Management Report, General Technical Report R10 TP-24.

Heusser, C.J. 1960. *Late-Pleistocene Environments of North Pacific North America*. Washington, DC: American Geographical Society.

Heusser, C.J. 1964. Some pollen profiles from the Laguna de San Rafael Area, Chile. In *Ancient Pacific Floras: 10th Pacific Science Congress Series* (pp. 95–115). Honolulu: University of Hawaii Press.

Heusser, C.J. 1987. Fire history of Fuego-Patagonia. *Quat South Am Ant Penin*, 5, 93–109.

Huesser, C.J., Rabassa, J., Brandani, A., and Stuckenrath, R. 1988. Late-Holocene vegetation of the Andean *Araucaria* region, Province of Neuquén, Argentina. *Mountain Res Devel*, 8, 53–63.

Juday, G.P. 1976. *The Location Composition and Structure of Old-Growth Forests of the Oregon Coast Range*. Unpublished doctoral dissertation, Oregon State University, Corvallis, OR.

Kalela, E.K. 1941, Über die Holzarten und die durch die klimatischen Verhältnisse verursachten Holzartenwechsel in den Wäldern Ostpatagoniens. *Ann Acad Scient Fenn*, Seres A, 2, 5–151.

Kellison, R.C., and Weir, R.J. 1987. Breeding strategies in forest tree populations to buffer against elevated atmospheric carbon dioxide levels.In W.E. Shands and J.S. Hoffman (eds.), *The Greenhouse Effect, Climate Change and U.S. Forests* (pp. 285–294) Washington, DC: The Conservation Foundation.

Klein, D.R. 1965. Ecology of deer range in Alaska. *Ecolog Monog*, 35, 259–284.

Klinger, L.F. 1988. *Successional Change in Vegetation and Soils of Southeast Alaska*. Unpublished Doctoral dissertation, University of Colorado, Boulder, CO.

Klinka, K., Pojar, J., and Meidinger, D.V. 1991. Revision of biogeoclimatic units of coastal British Columbia. *Northwest Sci*, 65, 32–47.

Kruckeberg, A.R, 1987. Plant life on Mount St. Helens before the 1980 eruption. In D.E. Bilderback, (ed.), *Mount St. Helens 1980: Botanical Consequences of the Explosive Eruptions*, (pp. 3–21). Berkeley: University of California Press.

Lara, A. 1991. *The Dynamics and Disturbance Regimes of* Fitzroya cupressoides *Forests in the South-Central Andes of Chile*. Unpublished doctoral dissertation, Boulder: University of Colorado.

Lara, A., and Veblen, T.T. 1992. Forest plantation in Chile: A Successful model? In A.S. Mather (ed.), *Afforestation: Policy, Planning, and Progress*. London: Pinter.

Lawrence, D.B., and Lawrence, E.G. 1959. *Glacier Variations in Southern South America*. Technical Report, Office of Naval Research Contract 641-(04) to the American Geographical Society.

Lertzman, K.P. 1992. Patterns of gap-phase replacement in a subalpine, old-growth forest. *Ecology*, 73, 657–669.

Lertzman, K.P., and Krebs, C.S. 1991. Gap-phase structure of a subalpine old-growth forest. *Can J For Res*, 21, 1730–1741.

Lewis, S.W. 1992. *Relationships between Deer and Vegetation on Coronation Island, Southeastern Alaska*. Unpublished masters thesis, University of Alaska, Fairbanks.

Lizarralde, M., Escobar, J., and Biancciotto, O. 1990. El castor (*C. canadensis*) en Tierra del Fuego: Effectos de la alteración ambiental y su aprovechamiento como recurso natural. *Libro del PEOAF*, Museo del Fin del Mundo, Ushuaia, Argentina.

Lorain, C.C. 1988. *Floristic History and Distribution of Coastal Disjunct Plants of the Northern Rocky Mountains*. Unpublished masters thesis, University of ldaho, Moscow.

Machoni, A. 1732. *Las Siete Estrellas de la Mano de Jesús, Tratado Histórico*. Colegio de Asumpción, Córdoba, Spain.

Markgraf, V. 1983. Late and postglacial vegetational and paleoclimatic changes in subantarctic, temperate, and arid environments in Argentina. *Palynology*, 7, 43–70.

Markgraf, V. 1987. Paleoclimates of the southern Argentine Andes. *Curr Res Pleistocene*, 4, 150–157.

Martin, J.R. 1989. *Vegetation and Environment in Old Growth Forests of Northern Southeast, Alaska: A Plant Association Classification*. Unpublished Masters thesis, Arizona State University, Tempe.

Martín, C.D., Mermoz, M., and Gallopín, G. 1985. *Impacto de la Ganadería en la Cuenca del Rio Manso Superior*. Report, Administración de Parques Nacionales, Buenos Aires, Argentina.

Martinic, B.M. 1985. La Ocupación y el impacto del hombre sobre el territorio. In O. Boelcke, D.M. Moore, and F.A. Roig (eds.), *Transecta Botánica de la Patagonia Austral* (pp. 81–94). Buenos Aires: Consejo Nacional de Investigaciones Científicas y Técnicas.

Matson, P.A., and Boone, R.D. 1984. Natural disturbance and nitrogen mineralization: wave-form dieback of mountain hemlock in the Oregon Cascades. *Ecology*, 65, 1511–1516.

McKee, A.W., LaRoi, G., and Franklin, J.F. 1982. Structure composition and reproductive behavior of terrace forests south fork Hoh River, Olympic National Park. In E.E. Starkey, J.F. Franklin, and J.W. Matthews (eds.), *Ecological Research in National Parks of the Pacific Northwest* (pp. 22–29). Corvallis, OR: U.S.D.I., National Park Service, Coop Park Studies Unit.

Mercer, J.H. 1965. Glacier variations in southern Patagonia. *Geog Rev*, 55, 390–413.

Mercer, J.H. 1970. Variations of some Patagonian glaciers since the Late-Glacial: II. *Am J Sci*, 269, 1–25.

Miles, D.W.R., and Swanson, F.J. 1986. Vegetation composition on recent landslides in the Cascade Mountains of western Oregon. *Can J For Res*, 16, 739–744.

Miller, A. 1976. The climate of Chile. In W. Schwerdtfeger (ed.), *World Survey of Climatology. Vol. 12: Climates of Central and South America* (pp. 113–131). Amsterdam: Elsevier.

Minore, D. 1978. *The Dead Indian Plateau: A Historical Summary of Forestry Observations and Research in a Severe Southwestern Oregon Environment.* Portland, OR: U.S.D.A., Forest Service, Genenal Technical Report PNW-72.

Musters, G.C. 1871. *At Home with Patagonians: A Year's Wandering over Untrodden Ground from the Straits of Magellan to the Rio Nergo*. London: John Murray.

Newton, M., el Hassan, B.A., and Zavitkovski, J. 1968. Role of red alder in western Oregon forest succession. In: J.M. Trappe, J.F. Franklin, R.F. Tarrant, and G.M. Hansen (eds.), *Biology of Alder. Proc. Symp. 40th Ann. Meeting Northwest Sci. Assn., April 14–15. 1967, Pullman, WA* (pp. 73–84). Portland, OR: U.S.D.A., Forest Service.

Noste, N.V. 1969. *Analysis and Summary of Forest Fires in Coastal Alaska.* U.S.D.A., Forest Service, Inst North For.

Oliver, C.D. 1981. Forest development in North America following major disturbances. *For Ecol Mgt*, 3, 153–168.

Oliver, C.D., Adams A.B., and Zasoski, R.J. 1985. Disturbance patterns and forest development in a recently deglaciated valley in the northwestern Cascade Range of Washington, U.S.A. *Can J For Res*, 15, 211–232.

Ott, R.A. 1994. *Small-Scale Natural Disturbance and Its Influence on Vegetation Structure and Diversity in Upland Western Hemlock-Sitka* Spruce Forests in Southeast Alaska. Unpublished doctoral dissertation, University of Alaska Fairbanks.

Parker, T., and Donoso, C. 1993. Natural regeneration of *Fitzroya cupressoides* in Chile and Argentina. *For Ecol Mgt*, 59, 63–85.

Peña, H., and Escobar, F. 1983. *Análisis de las Crecidas del Rio Paine, XII Región*. Buenos Aires: Publicación Interna NO. 83/7, Ministerio de Obras Públicas, Dirección de Aguas.

Perry, D.A., Amaranthus, M.P., Borchers, J.C., Borchers, S.L., Brainerd, R.E. 1989. Bootstrapping in ecosystems. *BioScience*, 39, 230–237.

Pickett, S.T.A., and White, P.S. 1985. *The Ecology of Natural Disturbance and Patch Dynamics*. New York: Academic Press.

Pickett, S.T.A., Collins, S.L., and Armesto, J.J. 1987. Models, mechanisms and pathways of succession. *Bot Rev*, 53, 335–371.

Pisano, V.E. 1978. Establecimientos de *Nothofagus betuloides* (Mirb.) Blume (Coigue de Magallanes) en un valle en proceso de deglaciación. *An Inst Patag*, 9, 107–128.

Pisano, E., and Dimitri, M.J. 1983. Estudio ecológico de la región continental sur del area andino-patagónica. *An Inst Patag*, 4, 207–271.
Pojar, J., and Banner, A. 1984. Old-growth forest and introduced black-tailed deer on the Queen Charlotte Islands, British Columbia. In W.R. Meehan, I.R. Merrell, Jr., and T.A. Hanley (eds.), *Fish and wildlife Relationships in Old-Growth Forests. Proc. Symp., 12–15 April 1982, Juneau, Alaska* (pp. 247–257). Morehead City, NC: Am Institut of Fishery Research and Biology.
Povilitis, A. 1978. *The Chilean Huemul Project (1975–76): Huemul Ecology and Conservation*. Unpublished doctoral dissertation, dissertation, Colorado State University, Fort Collins, CO.
Prichard, H. 1902. Field-notes upon some of the larger mammals of Patagonia, made between September 1900 and June 1901. *Proc Zool Soc (London)*, 1, 272–277.
Puigdefábregas, J.O., Bianciotto, del Barrio G., Gallart, F., Clotet, N., Alogia, M., and Ferres, L. In press. Respuesta del bosque subantártico frente a perturbaciones naturales. In J. Puigdefábregas, J. Frangi, F.G. Novo, and A. Vila (eds.), *The Natural Systems of Tierra del Fuego and their Occupation by Man*. Madrid: Consejo Superior de Investigación Y Ciencia.
Ramilo, E. 1985. *Situación y Manejo del Ciervo Colorado en los Parques Nacionales Lanín y Nahuel Huapi*. Report, Administración de Parques Nacionales, Buenos Aires.
Ramirez, C., Godoy, R., Eldridge, W., and Pacheco, N. 1981. Impacto ecológico del ciervo rojo sobre el bosque de olivillo en Osorno, Chile. *An Mu Hist Nat Valparaiso*, 14, 195–210.
Rebertus, A.J., and Veblen, T.T. 1993a. Partial wave formation in old-growth Nothofagus forests on Tierra del Fuego, Argentina. *Bull Torrey Bot Club*, 120, 461–470.
Rebertus, A.J., and Veblen, T.T. 1993b. Structure and treefall gap dynamics of old-growth Nothofagus forests on Tierra del Fuego, Argentina. *J Veg Sci*, 4, 641–653.
Reiners, W.A., Worley, I.A., and Lawrence, D.B. 1971. Plant diversity in a chronosequence at Glacier Bay, Alaska. *Ecology*, 52, 55–59.
Roig, F.A.J. Anchorena, Dollenz, O., Faggi, A.M., and Méndez, E. 1985. Las Comunidades Vegetales de la Transecta Botánica de la Patagonia Austral. In O. Boelcke, D.M. Moore, and F.A. Roig (eds.), *Transecta Botánica de la, Patagonia Austral* (pp. 350–519). Buenos Aires: Consejo Nacional de Investigaciones Científicas y Técnicas.
Rothkugel, M. 1916. *Los Bosques Patagónicos*. Buenos Aires: Ministerio de Agricultura.
Rothlisberger, F. 1986. *10000 Jahre Glestschergeschichte der Erde*. Salzburg, Austria: Verlag Sauerlander.
Ruth, R.H. 1968. *Differential Effect of Solar Radiation on Seedling Establishment under a Forest Stand*. Unpublished doctoral dissertation, Oregon State University, Corvallis.
Ruth, R.H., and Harris, A.S. 1979. *Management of Western Hemlock-Sitka Spruce Forest for Timber Production*. Portland, OR: U.S.D.A., Forest Service, General Technical Repout PNW-88.
Schoonmaker, T.D., and McKee, A. 1988. Species composition and diversity during secondary succession of coniferous forests in the western Cascade mountains of Oregon. *For Sci*, 34, 960–979.
Schrader, B.A. 1991. *Vegetation and Conifer Seedling response to Clearcutting of Alluvial Spruce Sites in Southeast Alaska*. Unpublished masters thesis, University of Washington, Seattle.

Schwab, J.W. 1983. *Mass Wasting: October-November 1978 Storm, Rennell Sound, Queen Charlotte Islands, British Columbia*. Victoria: British Columbia Ministry of Forests, Research Note 91.

Sedell, J.R., and Duval, W.S. 1985. *Water Transportation and Storage of Logs. Influence of Forest and Rangeleand Management on Anadromous Fish Habitat in Western North America 5*. Portland, OR: U.S.D.A., Forest Service, Pacific Northwest Experiment Station, General Technical Report PNW-186.

Smith, R.B. 1979. History and current status of Sitka black-tailed deer in the Kodiak Archipelago. In O.C. Wallmo and J.W. Schoen (eds.), *Sitka black tailed deer, Proceedings of a Conference, February 22–24, 1978, Juneau, Alaska* (pp. 184–195). Juneau, AK: U.S.D.A., Forest Service, Pacific Northwest Research Station.

Smith, R.B., and Commandeur, P.R. 1986. *Soils, Vegetation and Forest Growth on Landslides and Surrounding Logged and Old-Growth Areas on the Queen Charlotte Islands*. Victoria: British Columbia Ministry of Forests, Land Management Report No. 41.

Spies, T.A. 1991. Plant species diversity and occurrence in young, mature, and old-growth Douglas-fir stands in Western Oregon and Washington. In L.F. Ruggiero, K.B. Aubry, A.B. Carey, and M.H. Huff (eds.), *Wildlife and Vegetation of Unmanaged Douglas-fir forests*. Portland, OR: U.S.D.A., Pacific Northwest Research Station, General Technical Report PNW-GTR-285.

Spies, T.A., and Franklin, J.F. 1988. Old growth and forest dynamics in the Douglas-fir region of western Oregon and Washington. *Nat Areas J*, 8, 190–201.

Spies, T.A., and Franklin, J.F. 1989. Gap characteristics and vegetation response in coniferous forests of the Pacific Northwest. *Ecology*, 70(3), 543–545.

Spies, T.A., Franklin, J.F., and Klopsch, M. 1990. Gap characteristics and vegetation response in coniferous forests of the Pacific Northwest. *Can J For Res*, 20, 649–658.

Stewart. G.H. 1986. Population dynamics of a montane conifer forest, western Cascade Range, Oregon. *Ecology*, 67, 534–544.

Swanston, D.N., and Marion, D.A. 1991. Landslide response to timber harvest in Southeast Alaska. In S.S. Fan and U.H. Kuo (eds.), *Proceedings of the Fifth Federal interagency sedimentation conference, March 18–21, 1991, Las Vegas, Nevada* (pp. 49–56). Washington, DC: Federal Energy Regulatory Commission.

Swanston, D.N., and Swanson, F.J. 1976. Timber harvesting mass erosion, and steepland forest geomorphology in the Pacific Northwest. In Dr. Coates (ed.), *Geomorphology and Engineering* (pp. 199–221). Stroudsburg, PA: Hutchison and Ross.

Taylor, A.H. 1990. Disturbance and persistence of Sitka spruce (*Picea sitchensis* (Bong) Carr.) in coastal forests of the Pacific Northwest, North America. *J Biogeog*, 17, 47–58.

Trenberth, K.E. 1990. Recent observed interdecadal climate changes in the northern hemisphere. *Bull Am Meteor Soc*, 71, 988–993.

Veblen, T.T. 1985a. Forest development in tree-fall gaps in the temperate rain forests of Chile. *Nat Geog Res*, 1, 161–184.

Veblen, T.T. 1985b. Stand dynamics in Chilean Nothofagus forests. In S.T.A. Pickett and P.S. White (eds.), *The Ecology of Natural Disturbance and Patch Dynamics* (pp. 35–51). New York: Academic Press.

Veblen, T.T. 1989. Tree regeneration in response to gaps along a transandean gradient. *Ecology*, 70, 543–545.

Veblen, T.T., and Ashton, D.H. 1978. Catastrophic influences on the vegetation of the Valdivian Andes. *Vegetatio*, 36, 149–167.

Veblen, T.T., and Ashton, D.H. 1982. The regeneration status of *Fitzroya cupressoides* in the Cordillera Pelada, Chile. *Biol Conserv*, 23, 141–161.
Veblen, T.T., and Lorenz, D.C. 1987. Post-fire stand development of *Austrocedrus-Nothofagus* forests in Patagonia. *Vegetatio*, 73, 113–126.
Veblen, T.T., and Lorenz, D.C. 1988. Recent vegetation changes along the forest/steppe ecotone in northern Patagonia. *Ann Assoc Am Geog*, 78, 93–111.
Veblen, T.T., and Markgraf, 1988. Steppe expansion in Patagonia? *Quat Res*, 30, 331–38.
Veblen, T.T., Ashton, D.H., Rubulis, S., Lorenz, D.C., and Cortes, M. 1989a. Nothofagus stand development on in-transit moraines, Casa Pangue Glacier, Chile. *Arctic Alpine Res*, 21, 144–155.
Veblen, T.T., Ashton, D.H., Schlegel, F.M., and Veblen, A.T. 1977. Plant succession in a timberline depressed by volcanism in south-central Chile. *J Biogeog*, 4, 275–294.
Veblen, T.T., Delmastro, R.J., and Schlatter, J.E. 1976. The conservation of *Fitzroya cupressoides* and its environment in southern Chile. *Environ Conser*, 3, 291–301.
Veblen, T.T., Donoso, C. Schlegel, Z.F.M., and Escobar, B. 1981. Forest dynamics in south-central Chile. *J Biogeog*, 8, 211–247.
Veblen, T.T., Kitzberger, T., Lara, A. 1992a. Disturbance and forest dynamics along a transect from Andean rain forest to Patagonian shrublands. *J Veg, Sci*, 3, 507–520.
Veblen, T.T., Mermoz, M., Martín, C., and Kitzberger, T. 1992b. Ecological impacts of introduced animals in Nahuel Huapi National Park, Argentina. *Conser Biol*, 6, 71–83.
Veblen, T.T., Mermoz, M., Martín, C., and Ramilo, E. 1989b. Effects of exotic deer on forest regeneration and composition in northern Patagonia. *J Appl Ecol*, 26, 711–724.
Veblen, T.T., Schlegl. F.M., and Oltremari, J.V. 1983. Temperate broad-leaved evergreen forests of South America. In J.D. Ovington (ed.), *Temperate Broad-Leaved Forests* (pp. 5–31). Amsterdam: Elsevier.
Villalba, R. 1990. Climatic fluctuations in northern Patagonia during the last 1000 years as inferred from tree ring records. *Quat Res*, 34, 346–360.
Villalba, R., Boninsegna, J.A., and Cobos, D.R. 1989. A tree-ring reconstruction of summer temperature between A.D. 1500 and 1974 in western Argentina. *Third International Conference of the Southern Hemisphere Meteorology and Oceanography* (pp. 196–197). Buenos Aires.
Wallmo, O.C., and Schoen, J.W. 1980. Response of deer to secondary succession in southeast Alaska. *For Sci*, 26, 448–462.
Watt, A.S. 1947. Pattern and process in the plant community. *J Ecol*, 35, 1–22.
Willis, B. 1914. El *Norte de la Patagonia*. Buenos Aires: Dirección de Parques Nacionales.

10. Patterns of Terrestrial Vertebrate Diversity in New World Temperate Rainforests

Peter L. Meserve

Temperate rainforests occur on the western margins of North and South America between approximately 40°–60° N–S latitude. These occur primarily in the Pacific Northwest region of the United States and Canada and in southern Chile, with extensions into neighboring precordilleran Argentina. The climate of these regions is characterized by high, predominantly winter precipitation as rain and/or snow, averaging 1500 mm to 3000 mm or more annually, and mild maritime conditions. Depending on whether the forests are perhumid or seasonal, summers may also be wet, or relatively dry, but not hot. In both hemispheres, volcanism, orogenic uplifts, and periodic Pleistocene glaciations have played a major role in shaping the modern landscape. The southern temperate forest region has also been affected by marine transgressions, particularly in the early and mid-Cenozoic. Today, the temperate forests of both regions are essentially isolated from coterminous regions by high mountain ranges (the Cascades and Rocky Mountains, for North America, and the Andes, for South America) that produce significant rainshadows to the east and, at the same time, ameliorate conditions to the west. As recently as the Pliocene, the southern temperate forests were much more extensive than at present, extending east through much of what is now semiarid Patagonian steppes. Forest vegetation may have also extended northward during glacial maxima, occupying the area now dominated by Mediterranean scrub or matorral (Vuilleumier, 1985; but see Markgraf, 1991, and Markgraf et al., 1992, for a different view). In southern Chile, the pattern of forest advance and retreat paralleling glaciation and deglaciation episodes during the Pleistocene did not result in extensive speciation among indigenous animal species, probably because large areas of the mainland were uncovered by ice, and the forest remained relatively continuous although displaced latitudinally (Markgraf, 1991; Markgraf et al., 1992; Vuilleumeir, 1985). On the other hand, as southern forest areas were isolated from other forests to the north, their faunas became increasingly characterized by depauperization and endemism. Today, these forests are effectively separated by at least 1100 km of arid high-altitude puna and Patagonian steppes from the nearest forested areas to the north and northeast.

Methods

The terrestrial vertebrate faunas of the southern temperte forests were reanalyzed by Meserve and Jaksic (1991) in regional comparisons with available data for Pacific Northwest forests from earlier workers (e.g., Brown, 1985; Cei, 1979; Duellman, 1979a, 1979b; Formas, 1979; Glanz, 1982; Harris, 1984; Hershkovitz, 1972; Lacher & Mares, 1986; Pearson & Pearson, 1982; Ruggiero, Aubry, Carey, & Huff, 1991; Vuilleumier, 1985; and references therein). The results of these analyses are summarized here. In addition, we include data for rainforests in coastal British Columbia and southeastern Alaska, using range maps and species lists in regional guides (i.e., Banfield, 1974; Cook, 1984; Hall, 1981; MacDonald & Cook, 1993; Nussbaum, Brodie, & Storm, 1983; Stebbins, 1985). Only total numbers of species are considered, supplemented by local estimates or composite inventories at specific sites over several years. Except for the data in Ruggiero et al. (1991), which specifically concern Pacific Northwest Douglas fir forests, the lists are limited in their ability to resolve questions regarding differences in point or between-habitat diversity; but as summary data, they are valuable for overall comparisons of southern and northern temperate forests. Information is also available on mammalian habitat use and diet from the same reanalyses. In this chapter, we ignore avian species, because such data are presented in several recent papers (i.e., Armesto, Smith-Ramírez, & Sabag, this volume; Jaksic & Feinsinger, 1991; Ralph, 1985; Ralph, Paton & Taylor, 1991; Vuilleumier, 1985; Willson, De Santo, Sabag, & Armesto, this volume).

After describing some salient features emerging from these data, we discuss three specific aspects with respect to small-mammal species density, distribution, and demography in the southern temperate forest region. These include: 1) some general characteristics of small-mammal assemblages in primary-growth and secondary-growth rainforests and remnant forest islands; 2) the progressive decline in small-mammal species number at specific sites proceeding southward from the Valdivian rainforest into the north Patagonian forests; and 3) the dynamics of the small-mammal assemblages at the southern temperate forest/Patagonian steppe interface, and the northern margin of temperate evergreen forests in Chile. The relevance of these data to the current status of temperate rainforests and their biota in Chile will also be considered.

Results

Table 10.1 presents the results of tabulations of numbers of amphibians and reptiles for selected collecting localities in southern Chile and the Pacific Northwest. Sites in the Pacific Northwest support more herptiles overall than southern Chile (19–32 species versus 13–18 species, respectively) for all but the most polarward sites. The total number of species present in the regions is 45 and 38 species, for Pacific Northwest and Chilean temperate rainforests respectively. Whereas salamanders make up a majority of Pacific Northwest amphibians (18/26 species, or 69.2%),

Table 10.1. Numbers of Herptiles (Amphibians and Reptiles) in Temperate Evergreen Forests of Chile and the Pacific Northwest

	Amphibians	Reptiles
Chile		
Valdivian rainforest		
Angol; 37°45′ S	8	10
N Valdivia; 39°48′ S	11	7
La Picada; 41°02′ S	10	5
S Lago Llanquihue; 41°36′ S	10	3
Isla Chiloé; 42°–43° S	8	5
Northern Patagonian rainforest		
Puerto Aisén; 45°45′	5	1
Pacific Northwest		
Northwestern California; 41° N	15 (12–14)	16 (8–12)
Oregon Cascades; 43°–45° N	14 (9–13)	18
Oregon Coast Range; 43°–45° N	11 (10–11)	12
Southern Washington Cascades; 46°–47° N	13	6
Coastal British Columbia; 49°–50° N	14	6
Southeastern Alaska; 55°–57° N	6	1

[1]Data for California, Oregon, and Washington from Aubry and Hall (1991), Bury et al. (1991), Corn and Bury (1991a), Gilbert and Allwine (1991a), and Welsh and Lind (1991); that for British Columbia and southeastern Alaska from Nussbaum et al. (1983), Cook (1984), and Stebbins (1985). Data for South America from Meserve and Jaksic (1991). Numbers in parentheses indicate point estimates for specific sites sampled over 1 to 2 years.

all Chilean amphibians are anurans ($N = 28$ species), and 82% of these are in one family, the Leptodactylidae. For South America as a whole, leptodactylids make up only 38% of the amphibians (Duellman, 1979a). Similar to salamanders, leptodactylids have exploited the moist terrestrial environments in the southern temperate forests. Some 14/28 (50%) species of amphibians have highly localized distributions within the southern temperate forest region (defined as occuring in only one latitudinal region, as identified in Table 10.1) compared to 9/26 (35%) in the Pacific Northwest. These species may be considered to be essentially endemic. Lizards make up a majority of southern temperate forest reptiles (8/10 species, or 80%) and all but one are in the genus *Liolaemus*, whereas snakes comprise a majority of Pacific Northwest reptiles (12/19 species, or 63%) regionally. Finally, 15/38 (40%) reptile species have highly localized distributions within the southern temperate forest region versus 9/45 (20%) species in the Pacific Northwest. Also of interest is the general decline in numbers of herptiles proceeding toward the pole in both regions (see Table 10.1). Bury, Corn, & Aubry (1991) note the decline in numbers of reptiles northward to the southern Washington Cascades; this pattern continues into British Columbia and southeastern Alaska. Whereas amphibians also decline in number, a greater proportion of this decline is due to the salamanders.

Clearly, continental drift and isolation have contributed to these intercontinental differences. For example, leptodactylids are considered to be an ancient Gondwanaland element (Cracraft, 1974), and available evidence suggests that they dispersed northward from South America into Central America via island hopping (Duellman, 1979a). Cei (1979) has suggested an austral origin for primitive members of this family, and pockets of diversity in the Argentine Patagonian steppes may represent remnants from periods of forest expansion eastward. Similarly, different species complexes of the lizard genus *Liolaemus* may have originated from northern and southern Argentina (Cei, 1979).

The impoverishment of the southern temperate forest mammal fauna is evident in Table 10.2. The Chilean (and Argentine) forests include a maximum of 29 native species in contrast to 69 in the Pacific Northwest. Most of these differences are due to the lack of insectivores in South America (a total of 11 species), a poorer chiropteran fauna (6 versus 13 species), and the absence of sciurid and microtine rodents (a total of 17 species). Subtracting this difference of 34 species from the 69 species in the Pacific Northwest yields 35 species, more similar to the total for the southern temperate forests. Some 7/29 (24%) species are endemic to the southern temperate forests (versus 10/69 species, or 14%, for the Pacific Northwest). Specific forests and sites in the Pacific Northwest also have more small-mammal and chiropteran species than those in southern Chile. There appears to be a less dramatic decrease in numbers of mammal species polarward as compared to herptiles.

Table 10.2. Numbers of Native Mammal Species in Temperate Evergreen Forests in Chile and the Pacific Northwest

	Small Mammals	Bats	Other Species
Chile			
Valdivian rainforest			
San Martín; 39°38′ S	9	6	11
Antillanca; 40°47′ S	9	6	12
La Picada; 41°02′ S	9	6	12
Pacific Northwest			
Northwestern California; 41° N	21 (10–14)	—	14
Oregon Cascades; 43°–45° N	19 (12–16)	—	—
Oregon Coast Range; 43°–45° N	19 (14–18)	8	—
Southern Washington Cascades; 46°–47° N	22	12	17
Coastal British Columbia; 49°–50° N	27	12	18
Southeastern Alaska; 55°–57° N	17	5	11

[1] Data for California, Oregon and Washington from Aubry et al. (1991), Corn and Bury (1991b), Gilbert and Allwine (1991b), Ralph et al. (1991), Thomas and West (1991), and West (1991). Data for British Columbia and southeastern Alaska from Banfield (1974), Hall (1981), and MacDonald and Cook (1993). Data for South America from Meserve and Jaksic (1991), Meserve et al. (1991b), and Martínez and Rau (unpublished data). Numbers in parentheses indicate point estimates for specific sites trapped over 1 to 2 years. Dashes (—) indicate data not available.

The origins of the southern forest mammal fauna are complex. Many larger mammals, particularly the carnivores, are widespread species or closely related to forms that invaded South America at the time of the establishment of the Pliocene land bridge. Others, such as the sigmodontine rodents, which invaded considerably earlier (Hershkovitz, 1972), are either closely related to Andean forms or to those from the Patagonian steppes. Pearson (1982) suggested that a wave of species that radiated primarily in the southern Bolivian altiplano utilized the Andes as a corridor to invade more southerly areas of South America.

In habitat use, mammals show a higher proportion of fossorial/semifossorial species and a lower proportion of arboreal/climbing species than in the Pacific Northwest, as shown in Table 10.3. The lack of arboreal forms was noted by Glanz (1982). The majority of arboreal/climbing species in the Pacific Northwest is made up almost entirely of sciurids, which are lacking in the southern temperate forests. Southern temperate forests possess more carnivores and frugivores and less insectivores, proportionally, than do forests of the Pacific Northwest. The lesser importance of insectivores in southern forests is due to a lower number of bats as well the absence of shrews and moles (Insectivora), which alone constitute

Table 10.3. Ecological Characteristics of Native Mammals in Temperate Evergreen Forests in Chile and the Pacific Northwest

Functional Habit	Fossorial/ Semifossorial	Aquatic	Climbing/ Arboreal	Terrestrial/ Scansorial	Volant
Chile	14%	7%	7%	55%	17%
Pacific Northwest	6%	7%	12%	57%	18%

Diets	Carniv.	Insect.	Frug.	Gran.	Omniv.	Herb.	Mycoph.
Chile	28%	28%	10%	3%	10%	21%	0% (10%)*
Pacific Northwest	16%	35%	0%	4%	16%	23%	6% (12%)*

[1] Data from Meserve and Jaksic (1991).
* = partially mycophagous.

16% (11/69) of the mammal species in Pacific Northwest forests. Although there are no principally mycophagous species in southern forests, unlike the Pacific Northwest, 10% of the mammals are at least partially mycophagous, similar to the percentage for the Pacific Northwest.

Conclusions

This overview of patterns in species densities and ecological characteristics of herptile and mammalian faunas of the southern temperate forests identifies several general patterns. First, for all but the amphibians, faunas are conspicuously depauperate in southern temperate forests. Even for amphibians, one order (Anura) make up all, and one family (Leptodactylidae) virtually all (82%) living representatives. A second salient feature is that endemism is much higher among terrestrial vertebrates in southern forests—between 24% and 50%, depending on taxa. This suggests a major role of isolation and past historical events on the composition of the indigenous faunas, as noted previously.

Further insight into the processes that have affected the southern faunas may be gained from recent studies on small-mammal assemblages in southern Chile. Total small-mammal species numbers are similar in secondary-growth coastal rainforests (i.e., San Martín) and primary-growth Valdivian rainforests in Chile (see Table 10.2) as well as temperate rainforests in neighboring Argentina (Meserve et al., 1991b). However, small-mammal faunas of Chilean forests tend to be dominated by widespread sigmodontine rodent species unlike Argentine forests; these include *Akodon olivaceus* and *Oligoryzomys longicaudatus*, which occur in the northern Chilean semiarid region as well as in southern Patagonia. Among Chilean forests, there also appear to be differences in the demography of resident species. For example, two undisturbed primary-growth Valdivian forests located approximately 45 km apart in the precordilleran Andes

(Antillanca and La Picada, Table 10.2) had very similar trends in small-mammal numbers/ha over 2 years (1983–1984; see Figure 10.1); fluctuations were characterized by low-amplitude changes in monthly densities (coefficient of variation = 0.3–0.5). In contrast, small-mammal numbers tend to vary one to two orders of magnitude or more by season in secondary-growth forests such as San Martín, with forest populations dropping to virtually zero in dry summer months (Murúa & González, 1986; Murúa, González, & Meserve, 1986; unpublished data; *C.V.* = 1.2 for 1980–1982). In addition, whereas 90% or more of the captures are of the above-noted species in San Martín, those in Antillanca and La Picada are more evenly distributed among up to five species, including, in addition, *Abrothrix longipilis*, *Ab. sanborni*, and *Geoxus valdivianus*. Other species at least occasionally represented in live traps include *Irenomys tarsalis*, *Dromiciops gliroides*, and *Rhyncholestes raphanurus*. All but the last species are known from snap-trapping efforts in San Martín (Meserve & Kelt, unpublished), but, with the exception of *Ab. longipilis*, rarely occur in live-trap censuses in the forest.

Similarly, Kelt (1989) trapped six forest habitat islands ranging from 0.25 ha to 20 ha in the central valley west of La Picada and Antillanca during 1985–1986; this area was largely cleared by the mid-1800s (Hartwig, 1991). A mean of four-small mammal species/island were encountered, but 96.2% of the 397 captures were of three species (*A. olivaceus*, *Ab.*

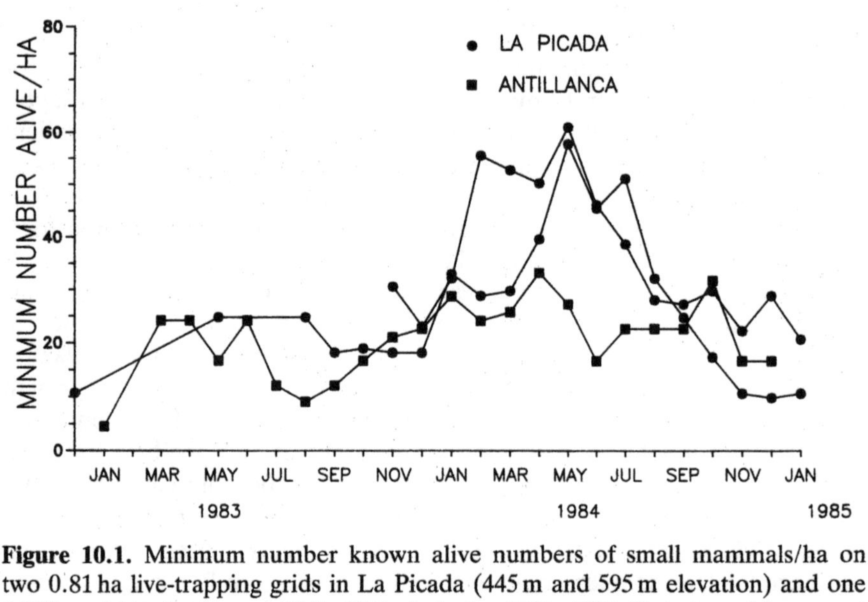

Figure 10.1. Minimum number known alive numbers of small mammals/ha on two 0.81 ha live-trapping grids in La Picada (445 m and 595 m elevation) and one 0.36 ha grid in Antillanca, Chile (820 m) during 1983–1984. Data from Meserve et al. (1991b), and D.R. Martínez and J.R. Rau (Unpublished data). Latitudes given in Table 10.2.

longipilis, and *O. longicaudatus*). Captures of the remaining four species constituted only 3.8%, as compared to about 39% for five remaining species in a similar sampling effort conducted in La Picada in 1983 (Patterson, Meserve, & Lang, 1989). Therefore, small forest islands appear to support a more depauperate small-mammal fauna than contiguous, intact forests.

Meserve, Kelt, & Martínez (1991a) sampled 15 primary-growth temperate rainforest sites between approximately 41° S and 45°45′ S latitude, using a standarized trapping protocol during 1985–1987. The transect followed the Carretera Austral (Southern Highway) as it was being constructed through formerly undisturbed Valdivian and north Patagonian rainforest. Small-mammal species number declined progressively, from a maximum of seven species in La Picada to a minimum of three near the southern end of the transect in Chile Chico; latitude was the single most predictive variable explaining species number in stepwise regressions. Thus, point diversity appears to decline progressively to the south, and widespread species such as *A. olivaceus* and *O. longicaudatus* predominate at more southerly sites (73% of the captures for the five southern sites; Meserve et al., 1991a). While it may be tempting to conclude that this is due to a peninsular effect as rainforest becomes more narrowly restricted in the southern terminous of South America, examination of vegetation maps (e.g., Hartwig, 1991) shows that the most conspicuous narrowing of hygrophytic forests occurs near the Patagonian icefields (ca. 47° S), well to the south of the transect. Furthermore, virtually all small-mammal species occurring at the northern end of the transect have been historically documented at the southern end with exception of the marsupials (*Rhyncholestes raphanurus* and *Dromciops gliroides*), which do not occur south of 42° S on the mainland (Kelt & Martínez, 1989; Meserve et al., 1991a). Clearly, other factors must be operating here. We suggest that widespread extirpation of forests in the Puerto Aisén region, which was settled much earlier than the intervening forest sampled by the transect, has already resulted in local depauperization. Although the central valley region also has suffered massive deforestation between 37° S and 42° S, coastal forests still retain most species, perhaps because they are relatively continuous and/or secondary-growth forests such as at San Martín that have recovered sufficiently (ca. 100 yrs). Even in the latter case, however, most small mammals are comprised of two species (*A. olivaceus* and *O. longicaudatus*), unlike the precordilleran forests in Antillanca and La Picada but similar to north Patagonian forests sampled by Meserve et al. (1991a).

Finally, Johnson, Franklin, Iriarte (1990) describe an apparent "dilemma" of high small-mammal species numbers at the interface of the southern forests and steppes in Chilean Patagonia (ca. 50°38′ S). Here a total of 38 species of mammals are reported, of which 16 species are native rodents. This is in contrast to Valdivian forests, where rodents make up only 9 of a total 27 species (see Table 10.2). However, it should be noted that at least seven species recorded by Johnson et al. (1990) are typical Patagonian steppe species. Elsewhere, there is almost complete faunal turnover between temperate forest and Patagonian steppe small mammals at lower latitudes (i.e., at ca. 41° S and 46° S latitude (Kelt, 1989; Pearson & Pearson, 1982). Virtually no Patagonian steppe species penetrate the temperate forest region, and only a few widespread forest elements occur on the steppe margins. Hence, the apparent "dilemma" noted in southern Patagonia reflects a unique mixing of faunas in a region where the two vegetational communities show a high degree of intergradation. Whereas 38% of the Patagonian fauna are of Valdivian and Magellanic forest origin, 38% are primarily of Andean affinity, and 28% are from the Patagonian steppes (Johnson et al., 1990).

As discussed earlier, historical and geographic factors may explain the relatively depauperate and highly endemic character of terrestrial vertebrates in southern temperate rainforests. In addition, recent events involving the large-scale extirpation of temperate forests in Chile may also have had significant effects. Estimates of extant forest remaining in Chile vary; Gilroy (1992), Hartwig (1991), and Schmidt & Lara (1985) provide estimates of 6,250,000 ha, 6,970,000 ha, and 7,312,000 ha, respectively, of extant forest remaining in Chile. However, estimates from GIS yielded an estimate of the original area of 11,675,100 ha, with only about 25% left unlogged (Alaback in Kellogg, 1992). Brun (1975; cited in Veblen & Ashton, 1978) gave an estimate of 16% remaining in 1952. A total of 5,447,903 ha were listed by Veblen, Schlegel, & Oltremari (1983) as being in national parks, monuments, or reserves; however, not all of this is forest, much is in relatively inaccessible areas, and several large reserves have since been removed from protected status by executive decree.

At the northern end of the Chilean temperate forest region, Muñoz-Pedreros, Murúa, & González (1990) describe small mammal assemblages in exotic pine (*Pinus radiata*) plantations north of Concepción, where Mediterranean scrub and evergreen sclerophyll forests formerly dominated. Although, as expected, small-mammal assemblages include typical widespread species, such as *A. olivaceus* and *O. longicaudatus*, and Mediterranean scrub/semiarid elements, such as *Phyllotis darwini* and the marsupial *Thylamys elegans*, most notable is the importance of a large octodontid rodent, *Octodon bridgesi*, heretofore considered rare and uncommon in Chile (c.f., Redford & Eisenberg, 1992). In view of the rapid rate at which native rainforest is disappearing in southern Chile, it seems at least plausible that if pine plantations become progressively more expansive in areas formerly dominated by temperate evergreen forests, typical Mediterranean small-mammal elements will expand their southern limits accordingly.

It is perhaps fortunate that many terrestrial vertebrates discussed here have relatively broad geographic ranges within the southern temperate rainforest region, a characteristic that applies to many of Chile's vertebrates generally (Cody et al., 1977). Nevertheless, a large number of herptiles and mammals typical of southern temperate rainforests are in the "Red List of Chilean Terrestrial Vertebrates" (Glade, 1988) under "Endangered," "Vulnerable," "Rare," or "Insufficiently known" categories. Given the already depauperate nature of southern temperate vertebrate faunas, it may be expected that continuing extirpation of forests and increasing isolation of remaining stands will ultimately have irreversible effects on local and regional diversity. The apparent depauperization of small-mammal assemblages in north Patagonian forests in continental Chile Chico as well as in remnant forest islands in the central depression suggests that such effects have probably already occurred.

Acknowledgments. I thank Eduardo Fuentes and Rick Lawford for the invitation to participate in the AMIGO meetings held in Victoria, British Columbia, Canada, September 21–26, 1992. Thanks are also extended to the following people for assistance and the sharing of information used in the preparation of this paper: Joseph Cook, P. Stephen Corn, Ramón Formas, Douglas A. Kelt, David R. Martínez, Hymen Marx, Roberto Murúa, Juan Carlos Ortiz, Donald W. Thomas, and Stephen D. West. Support for various aspects of fieldwork conducted in southern Chile during 1980–1987 came from U.S. National Science Foundation Grants INT-8018631 and INT-8302724, an American Republics award from the Fulbright Foundation, a fellowship from the Organization of American States, and the Graduate School, Northern Illinois University, DeKalb, Illinois.

References

Aubry, K.B., and Hall, P.A. 1991. Terrestrial amphibian communities in the southern Washington Cascade Range. In L.F. Ruggiero, K.B. Aubry, A.B. Carey, and M.H. Huff (eds.), *Wildlife and Vegetation of Unmanaged Douglas-Fir Forests* (pp. 327–338). Portland, OR: U.S.D.A., Forest Service, Pacific Northwest Research Station, General Technical Report PNW-GTR-285.

Aubry, K.B., Crites, M.J., and West, S.D. 1991. Regional patterns of small mammal abundance and community composition in Oregon and Washington. In L.F. Ruggiero, K.B. Aubry, A.B. Carey, and M.H. Huff (eds.), *Wildlife and Vegetation of Unmanaged Douglas-Fir Forests* (pp. 285–294). Portland, OR: U.S.D.A., Forest Service, Pacific Northwest Research Station, General Technical Report PNW-GTR-285.

Banfield, A.W.F. 1974. *The Mammals of Canada*. Toronto: National Museum of Natural Sciences, National Museums of Canada, University of Toronto Press.

Brown, E.R. (ed.) 1985. *Management of Wildlife and Fish Habitats in Forests of Western Oregon and Washington. Parts 1 and 2*. Portland, OR: U.S.D.A., Forest Service, Pacific Northwest Region, General Technical Report R6-F&WL 192.

Brun, R. 1975. Estructura y potencialidad de distintos tipos de bosque nativo en el sur de Chile. *Bosque* (Valdivia), 1, 6–17.

Bury, R.B., Corn P.S., and Aubry, K.B. 1991. Regional patterns of terrestrial amphibian communities in Oregon and Washington. In L.F. Ruggiero, K.B. Aubry, A.B. Carey, and M.H. Huff (eds.), *Wildlife and Vegetation of Unmanaged Douglas-Fir Forests* (pp. 340–350). Portland, OR: U.S.D.A., Forest Service, Pacific Northwest Research Station, General Technical Report PNW-GTR-285.

Cei, J.M. 1979. The Patagonian herpetofauna. In W.E. Duellman (ed.), *The South American Herpetofauna: Its Origin, Evolution, and Dispersal* (pp. 309–339). Lawrence, KS: Museum of Natural History, University of Kansas, Monograph No. 7.

Cody, M.L., Fuentes, E.R., Glanz W., Hunt, J.H., and Moldenke, A.R. 1977. Convergent evolution in the consumer organisms of mediterranean Chile and California. In H.A. Mooney (ed.), *Convergent Evolution in Chile and California* (pp. 144–192). Stroudsburg, PA: Dowden, Hutchinson and Ross, Inc.

Cook, F.R. 1984. *Introduction to Canadian Amphibians and Reptiles*. Ottawa: National Museum of Natural Sciences, National Museums of Canada.

Corn, P.S., and Bury, R.B. 1991a. Terrestrial amphibian communities in the Oregon Coast Range. In L.F. Ruggiero, K.B. Aubry, A.B. Carey, and M.H. Huff (eds.), *Wildlife and Vegetation of Unmanaged Douglas-Fir Forests* (pp. 305–317). Portland, OR: U.S.D.A., Forest Service, Pacific Northwest Research Station, General Technical Report PNW-GTR-285.

Corn, P.S., and Bury, R.B. 1991b. Small mammal communities in the Oregon Coast Range. In L.F. Ruggiero, K.B. Aubry, A.B. Carey, and M.H. Huff (eds.), *Wildlife and Vegetation of Unmanaged Douglas-Fir Forests* (pp. 241–254). Portland, OR: U.S.D.A., Forest Service, Pacific Northwest Research Station, General Technical Report PNW-GTR-285.

Cracraft, J. 1974. Continental drift and vertebrate distribution. *Ann Rev Ecol Sys*, 5, 215–261.

Duellman, W.E. 1979a. The South American Herpetofauna: A Panoramic View. In W.E. Duellman (ed.), *The South American herpetofauna: Its Origin, Evolution, and Dispersal* (pp. 1–28). Lawrence, KS: Museum of Natural History, University of Kansas, Monograph No. 7.

Duellman, W.E. 1979b. The herpetofauna of the Andes: Patterns of distribution, origin, differentiation, and present communities. In W.E. Duellman (ed.), *The South American Herpetofauna: Its Origin, Evolution, and Dispersal* (pp. 371–459). Lawrence, KS: Nuseum of Natural History, University of Kansas, Monograph No. 7.

Formas, J.R. 1979. La herpetofauna de los bosques temperados de Sudamérica. In W.E. Duellman (ed.), *The South American Herpetofauna: Its Origin, Evolution, and Dispersal* (pp. 341–369). Lawrence, KS: Museum of Natural History, University of Kansas, Monograph No. 7.

Gilbert, F.F., and Allwine, R. 1991a. Terrestrial amphibian communities in the Oregon Cascade Range. In L.F. Ruggiero, K.B. Aubry, A.B. Carey, and M.H. Huff (eds.), *Wildlife and Vegetation of Unmanaged Douglas-Fir Forests* (pp. 319–324). Portland, OR: U.S.D.A., Forest Service, Pacific Northwest Research Station, General Technical Report PNW-GTR-285.

Gilbert, F.F., and Allwine, R. 1991b. Small mammal communities in the Oregon Cascade Range. In L.F. Ruggiero, K.B. Aubry, A.B. Carey, and M.H. Huff (eds.), *Wildlife and Vegetation of Unmanaged Douglas-Fir Forests* (pp. 257–269). Portland, OR: U.S.D.A., Forest Service, Pacific Northwest Research Station, General Technical Report PNW-GTR-285.

Gilroy S. 1992. Disturbing the ancients. *Buzzworm: Env J*, 4, 38–43.
Glade, A.A. (ed.) 1988. Red list of terrestrial vertebrates. Santiago: Corporación Nacional Forestal.
Glanz, W.E. 1982. Adaptive zones of Neotropical mammals: A comparison of some temperate and tropical patterns. In M.A. Mares and H.H. Genoways (eds.), *Mammalian Biology in South America* (pp. 95–110). Linesville, PA: Special Publication Series 6, Pymatuning Laboratory of Ecology, University of Pittsburgh.
Hall, E.R. 1981. *The Mammals of North America*. 2 Vol. New York: Wiley.
Harris, L.D. 1984. *The Fragmented Forest*. Chicago: University of Chicago Press.
Hartwig, F. 1991. *Chile, Desarrollo Forestal Sustentable*. Santiago: Editorial Los Andes.
Hershkovitz, P. 1972. The recent mammals of the Neotropical region: A zoogeographical and ecological review. In A. Keast, F.C. Erk, and B. Glass (eds.), *Evolution, Mammals, and Southern Continents* (pp. 311–431). Albany: State University of New York Press.
Jaksic, F.M., and Feinsinger, P. 1991. Bird assemblages in temperate forests of North and South America: Comparative diversity, guild structure, and resource use. *Rev Chil Hist Nat*, 64, 491–510.
Johnson, W.E., Franklin, W.L., and Iriarte, J.A. 1990. The mammalian fauna of the northern Chilean Patagonia: A biogeographical dilemma. *Mammalia*, 34, 458–469.
Kellogg, E. (ed.) 1992. Coastal temperate rain forests. Portland, OR: Ecotrust, Occasional Paper Series No. 1.
Kelt, D.A. 1989. *Biogeography and assemblage structure of small mammals across a transition zone in southern Chile*. Unpublished master's thesis, Northern Illinois University, DeKalb, IL.
Kelt, D.A., and Martínez, D.R. 1989. Notes on the distribution and ecology of two marsupials endemic to the Valdivian forests of southern South America. *J Mammal*, 70, 220–224.
Lacher, T.E., and Mares, M.A. 1986. The structure of Neotropical mammal communities: An appraisal of current knowledge. *Rev Chil Hist Nat*, 59, 121–134.
MacDonald, S.O., and Cook, J.A. 1993. *The Small Mammals of Southeast Alaska: An Update*. Fairbanks: University of Alaska Museum.
Markgraf, V. 1991. Late Pleistocene environmental and climatic evolution in southern South America. *Bamb Geog Schr Bd*, 11, 271–281.
Markgraf, V., Dodson, J.R., Kershaw, A.P., McGlone, M.S., and Nicholls, N. 1992. Evolution of late Pleistocene and Holocene climates in the circum-South Pacific land areas. *Cli Dyn*, 6, 193–211.
Meserve, P.L., and Jaksic, F.M. 1991. Comparisons of terrestrial vertebrate assemblages in temperate rainforests of North and South America. *Rev Chil Hist Nat*, 64, 511–535.
Meserve, P.L., Kelt, D.A., and Martínez, D.R. 1991a. Geographical ecology of small mammals in continental Chile Chico, South America. *J Biogeog*, 18, 179–187.
Meserve, P.L., Lang, B.K., Murúa, R., Muñoz, A., and González, L.A. 1991b. Characteristics of a terrestrial small mammal assemblage in a temperate rainforest in Chile. *Rev Chil Hist Nat*, 64, 157–169.
Muñoz-Pedreros, A, Murúa, R., and González, L.A. 1990. Nicho ecológico de micromamíferos en un agroecosistema forestal de Chile central. *Rev Chil Hist Nat*, 63, 267–277.
Murúa, R., and González, L.A. 1986. Regulation of numbers in two rodent species. *Rev Chil Hist Nat*, 59, 193–200.

Murúa, R., González, L.A., and Meserve, P.L. 1986. Population ecology of *Oryzomys longicaudatus philippii* (Rodentia, Cricetidae) in southern Chile. *J Anim Ecol*, 55, 281–293.

Nussbaum, R.A., Brodie, Jr., E.D., and Storm, R.M. 1983. *Amphibians and Reptiles of the Pacific Northwest*. Moscow, ID: University of Idaho Press.

Patterson, B.D., Meserve, P.L., and Lang, B.K. 1989. Distribution and abundance of small mammals along an elevational transect in temperate rainforests of Chile. *J Mammal*, 70, 67–78.

Pearson, O.P. 1982. Distribucíon de pequeños mamíferos en el altiplano y los desiertos del Peru. In P. Salinas (ed.), *Zoología Neotropical* (pp. 263–284). Mérida, Venezuela: Actas del VIII Congreso Latinoamoricano de Zoología.

Pearson, O.P., and Pearson, A.K. 1982. Ecology and biogeography of the southern rainforests of Argentina. In M.A. Mares and H.H. Genoways (eds.), *Mammalian Biology in South America* (pp. 129–142). Linesville, PA: Special Publication Series 6, Pymatuning Laboratory of Ecology, University of Pittsburgh.

Ralph, C.J. 1985. Habitat association patterns of forest and steppe birds of northern Patagonia, Argentina. *Condor*, 87, 471–483.

Ralph, C.J., Paton, P.W.C., and Taylor, C.A. 1991. Habitat association patterns of breeding birds and small mammals in Douglas-fir/hardwood stands in northwestern California and southwestern Oregon. In L.F. Ruggiero, K.B. Aubry, A.B. Carey, and M.H. Huff (eds.), *Wildlife and Vegetation of Unmanaged Douglas-Fir Forests* (pp. 379–393). Portland, OR: U.S.D.A., Forest Service, Pacific Northwest Research Station, General Technical Report PNW-GTR-285.

Redford, K.H., and Eisenberg, J.F. 1992. *Mammals of the Neotropics: The Southern Cone. Vol. 2*. Chicago: University of Chicago Press.

Ruggiero, L.F., Aubry, K.B., Carey, A.B., and Huff, M.H. (eds.) 1991. *Wildlife and Vegetation of Unmanaged Douglas-Fir Forests*. Portland, OR: U.S.D.A., Forest Service, Pacific Northwest Research Station, General Technical Report PNW-GTR-285.

Schmidt, H., and Lara, A. 1985. Descripción y potencialidad de los bosques nativos de Chile. *Amb Des*, 1, 91–108.

Stebbins, R.C. 1985. *Western Reptiles and Amphibians* (2nd ed.). Boston: Houghton Mifflin.

Thomas, D.W., and West, S.D. 1991. Forest age associations of bats in the southern Washington Cascade and Oregon Coast ranges. In L.F. Ruggiero, K.B. Aubry, A.B. Carey, and M.H. Huff (eds.), *Wildlife and Vegetation of Unmanaged Douglas-Fir Forests* (pp. 295–303). Portland, OR: U.S.D.A., Forest Service, Pacific Northwest Research Station, General Technical Report PNW-GTR-285.

Veblen, T.T., and Ashton, D.H. 1978. Catastrophic influences on the vegetation of the Valdivian Andes, Chile. *Vegetatio*, 36, 149–167.

Veblen, T.T., Schlegel, F.M., and Oltremari, J.V. 1983. Temperate broad-leaved evergreen forests of South America. In J.D. Ovington (ed.), *Temperate Broad-Leaved Evergreen Forests* (pp. 5–31). Amsterdam: Elsevier Science Publishers.

Vuilleumier, F. 1985. Forest birds of Patagonia: Ecological geography, speciation, endemism, and faunal history. In P.A. Buckley, M.S. Foster, E.S. Morton, R.S. Ridgely, and F.G. Buckley (eds.), *Neotropical Ornithology* (pp. 255–304). Washington, D.C.: Ornithological Monographs No. 36, American Ornithologists' Union.

Welsh, Jr., H.H., and Lind, A.J. 1991. The structure of the herpetofaunal assemblage in the Douglas-fir/hardwood forests of northwestern California and southwestern Oregon. In J.F. Ruggiero, K.B. Aubry, A.B. Carey, and M.H.

Huff (eds.), *Wildlife and Vegetation of Unmanaged Douglas-Fir Forests* (pp. 395–413). Portland, OR: U.S.D.A., Forest Service, Pacific Northwest Research Station, General Technical Report PNW-GTR-285.

West, S.D. 1991. Small mammal communities in the southern Washington Cascade Range. In L.F. Ruggiero, K.B. Aubry, A.B. Carey, and M.H. Huff (eds.), *Wildlife and Vegetation of Unmanaged Douglas-Fir Forests* (pp. 269–283). Portland, OR: U.S.D.A., Forest Service, Pacific Northwest Research Station, General Technical Report PNW-GTR-285.

11. Avian Communities in Temperate Rainforests of North and South America

Mary F. Willson, Toni L. De Santo, Carlos Sabag, and Juan J. Armesto

The climate and topography of the temperate rainforest biomes of North and South America have many similarities (e.g., Alaback, 1991). However, the biogeographic and phylogenetic relationships of the avian inhabitants of these biomes differ greatly, as does the structure of their habitats. There is no particular reason to expect great similarities in avian community structure in the two regions, but a comparison of communities can be instructive by setting the stage for more directly functional and predictive approaches (e.g., Jaksic & Feinsinger, 1991; Meserve & Jaksic, 1991; Willson, 1991). Here we examine first the abundance, diversity, guild structure, and selected biological attributes of bird communities in coastal temperate rainforests of Alaska and Chile. We then broaden the comparison, using previously published information for avian communities of the south temperate forests of Argentina and Chile and the north temperate forests in North America, specifically those in the northwestern part of that continent and those of the midwestern United States, to try to make more general statements. Our goals are: 1) to describe differences and similarities in avian communities in these forests, focusing on diversity, abundance, and selected biological attributes; 2) to assess the similarity of the bird community of the Chilean rainforest to the Alaskan coastal rainforest and to Canadian interior forests; and 3) to begin to examine some sources of the differences in community structure.

Methods

Census information consists of point-counts near Juneau and Haines, Alaska (58° N–59° N) and on Isla Grande de Chiloé (42° S), Chile. Counting stations, located along transects in the study sites, were separated by >150 m in Alaska (Willson & Comet, unpublished manuscript) and >100 m in Chiloé (Willson, De Santo, Sabag, & Armesto, 1994). The counts did not include nocturnal birds, shorebirds, or waterbirds. The data reported here refer to regularly occurring species, meaning those that were recorded on more than one census day; each site was sampled at least four times during the breeding season. All counts were made in 1992 (Alaska: May and June; Chile: November and December), during the morning hours. Each point-count lasted 8 minutes, and most study sites had at least eight points. On average, four points were sufficient to record 85% of the regularly occurring species observed at each site in both Chile and Alaska, and 98% (range 93%–100%) of such species were obtained by six points. *Diversity of species* is defined as species richness, for our purposes in this chapter, and all abundance data are relative abundances, in terms of the number of birds/point/day. All references to "abundance" have to do with the number of birds detected; we did not examine possible regional differences in detectability. Documentation of the Chilean censuses is reported in Willson, De Santo, Sabag, and Armesto (1994), and that of the Alaska censuses is reported in Willson and Comet (unpublished manuscript).

For purposes of comparisons of diversity and abundances in Alaska and Chile for this chapter, we use only study sites that occupied at least 50 ha, and most were much larger. The floristically diverse, broadleaf Chilean forests are dominated by *Nothofagus*, *Drimys*, *Weinmannia*, and myrtaceous trees (Arroyo, Riveros, Peñaloza, Cavieres, & Faggi, this volume). *Picea sitchensis* and *Tsuga heterophylla* dominate the Alaskan coastal conifer forests, and *Populus trichocarpa* comprises the canopy of the Alaskan deciduous forests in which point-counts were done.

We also ask whether the Chilean forest bird communities are notably similar in structure to those of the Alaskan coastal forest, by comparing them to the communities in Canadian forests at similar latitudes. For this purpose, bird censuses in *Picea glauca*, *Pinus contorta*, *Populus tremuloides*, and *Salix*/*P. trichocarpa*/*Alnus* stands in British Columbia and Yukon are used.

In addition to point-counts, mist-net censuses of understory birds were conducted in both areas, both in some of the study sites used for point-counts and in other sites. Mist nets were operated in the same kinds of forests as the point-counts, with the addition of some stands of *Alnus* in Alaska. A total of 380 mist-net hours were spread over three sites in Chiloé (8 nets/site). In Alaska, about 680 mist-net hours were accumulated in two deciduous sites and 635 net-hours in a conifer site near Juneau (at least 10 nets/site); capture rates for an additional conifer site near Ketchikan were obtained from Canterbury. Most of the nets were 30 mm or 36 mm mesh size, and 12 m by 2 m in net size. Mist-net data are used here only to compare capture rates of birds in understory.

The broader comparisons, using data in the literature, included forests with diverse species composition and a variety of stand size. South temperate data were derived from Cody (1970), Ralph (1985), and Vuilleumier (1972), with a latitudinal range of about 38° S–42° S (plus one site at 30° S). Midwestern U.S. data for deciduous forest were obtained for Wisconsin (Bond, 1957; using only stands >100 ha, to ensure that some large tracts were included) and Illinois (Willson, 1974; using each stand only once, to avoid pseudoreplication), and for coniferous forest in Minnesota and Michigan (from Willson, 1986); latitudinal range was about 40° N–49° N. In a general way, the midwestern U.S. data provide a partial control for simple latitudinal effects such as gradients in species diversity, but more importantly, they are a ready source of comparative information that allowed us to attempt these broad contrasts. Data for forests of the north and west in North America came from Spindler and Kessel (1980; coniferous and deciduous), Flack (1976; deciduous), Theberge (1976; coniferous and deciduous), Kessler and Kogut (1985; coniferous), with a latitudinal range of about 49° N–63° N. These studies were used only for comparisons of total species present and of guild structure among regions and forest types. Differences in census methods among these studies mean that all comparisons must be taken as preliminary estimates.

Guild assignments for each species were based on our observations in Alaska and Chile and on published information (e.g., Harrison, 1975, 1979, for nest locations in North America; Holmes, Bonney, & Pacata, 1979, and Manuwal, 1991, for foraging of some species; Flack, 1976, for aspen forests; Johnson, 1967, for Chilean birds). Root (1967) defined a *guild* as "a group of species that exploit the same class of environmental resources in a similar way." The guilds used for our comparisons represent broad categorizations of patterns of habitat use; fine tuning of categories (e.g., distinguishing subcanopy from canopy foragers, or ground from understory foragers, or insectivores from various types of omnivores in Chile) is not yet possible. Nevertheless, our guilds capture the major distinctions that emerged from more detailed and quantitative studies elsewhere, in which vertical stratum and proximity to tree trunks provided the major distinctions among guilds (Holmes, Bonney, & Pacala, 1979; Holmes & Recher, 1986).

Although Holmes and colleagues dealt with foraging guilds, here we include nest-sites as well, given that nesting biology has been shown to have important consequences for community structure (Martin, 1988a, 1988b). The guilds include the following broad categories: 1) big-tree users, nesting or foraging on trunks and large branches; 2) users of covered nests, including those nesting in tree cavities as well as in other holes or building dome nests; 3) understory users, including those that nest or forage principally in the understory (<3 m or so); some birds were assigned to an "understory" guild on the basis of common foraging and/or nesting locations, but males of these species often sing from the canopy (e.g., *Passerina cyanea*, *Ixoreus varius*, *Turdus migratorius*, and *Catharus guttatus*, to name a few in North America, and *Turdus falklandii* in Chile); 4) aerial feeders, including species that often feed on the wing, either hawking or hover-gleaning insects; 5) canopy seed predators, including species known to extract seeds from fruits or cones for a major portion of their diet; 6) leaf-gleaners, including insectivorous species that feed while moving along or between branches. These categories are a first approximation, pending an increased information base, especially for Chile. Many of the guilds used in this chapter are based on the location of a particular resource, because of our interest in what might happen if the availability of such resources were altered (Willson, De Santo, Sabag, & Armesto, 1994). Number 2 above is a clear exception, in that it depends in part on particular types of constructed nests, rather than a resource. Nevertheless, we include it, because nesting habits can have a profound effect on patterns of reproductive success, which in turn can influence species presence and abundance.

In this preliminary assessment, a given species can belong to more than one of these guilds, and the sum of all the guilds need not be 100% of the community. The guild assignment of most midwestern species can be obtained from Willson (1974), and those for Alaska and adjacent Canada are from Flack (1976) and Willson (unpublished). For Chiloé, some guild assignments are shown in Willson, De Santo, Sabag, and Armesto (1994); in addition, the canopy seed predators are *Enicognathus leptorhynchus*, *Columba araucana*, and *Phrygilus patagonicus*; aerial feeders include *Elaenia albiceps*, *Pyrope pyrope*, *Colorhamphus parvirostris*, and *Tachycineta leucopyga* (both *Elaenia* and *Pyrope* also consume much fruit); and leaf-gleaners include *Anairetes parulus*, *Colorhamphus parvirostris*, *Eugralla paradoxa*, *Scytalopus magellanicus*, and *Sylviorthorhynchus desmursii*.

Statistical Methods

Direct comparisons of diversity, abundance, and guild structure for our 1992 data for Chile and Alaska are made with one-way parametric ANOVAs, Kruskal-Wallis nonparametric one-way ANOVAs, and Mann-Whitney U tests. In these comparisons, as in those described in the following paragraph, we analyze the data first by the proportion of regular species in each guild and then by the proportion of individuals that belong to each guild. We use the terms *speciose*, *diverse*, and *species rich* interchangeably in referring to proportionate or relative guild structure by species, and the term *abundance* to mean proportions of individuals.

For the broader comparison among regions, we divided both northern and southern forests into two types (conifer and broadleaf), because experience in Alaska has suggested important biological differences between them and because other studies (e.g., James & Rathbun, 1981) have found differences in avian community structure between the two kinds of forest. We search for patterns, using a Model I two-way parametric ANOVA (with arcsin-transformation of percents) or region (south, midwest, northwest) *X* forest type (broadleaf, conifer). This approach allows us to detect interactions between the chosen factors and the effects of each factor. In some cases, directional differences are apparent despite a significant interaction term, and these can be treated, properly, along with the results of ANOVAs that lack significant interaction terms (Sokal & Rohlf, 1981). In addition, we can make predictions, based on what is known or believed about the basic biology of each kind of forest, about the direction of differences between communities of these forests. Post-hoc tests following the ANOVA permit a ranking of all the regional forest types, which is then examined for trends. The predictive approach is stronger in some ways, because it permits the negation of some possible explanations. However, it has the obvious frailty that the prediction is only as good as the information on which it is based; negation can occur because the background logic and information—and hence the prediction—are faulty. Nevertheless, it can be a useful tool in this preliminary phase as a means of examining some possible sources of differences among the communities.

Comparison of Alaskan and Chilean Rainforests: Results

Diversity and Abundance

Site diversity (the number of regularly occurring species at each site) differed with marginal significance among the three forest types (refer to Table 11.1; one-way ANOVA, $F = 3.64$, $p = 0.065$). However, point diversity (the average number of regularly occurring species per point) differed significantly among forest types: point diversity was highest in Alaskan deciduous forest and lowest in Alaskan conifer stands; Chiloé was intermediate and not significantly different from either (see Table 11.1; one-way ANOVA, $F = 4.0$, $p = 0.05$; Tukey post-hoc test). On average, point diversity in Chiloé accounted for a similar proportion of site diversity (68%) as in the other sites (74%, 80%), probably indicating similar spatial heterogeneity in avian distributions within sites in all three kinds of forest.

The greatest average number of birds per census point per day was found in Chilean forests, and the lowest, in Alaskan coniferous forests; Alaskan deciduous forests were intermediate (Table 11.1; one-way

Table 11.1. Avian Species Diversity and Abundance in Temperate Rainforests of Southeast Alaska and Chile, Using Regularly Occurring Species (Defined as Those Recorded on at Least 2 Census Days at a Site). Some species are migratory in the interior but not on the coast of Alaska (e.g., *Troglodytes troglodytes*, *Regulus satrapa*), and are here counted as nonmigratory.

Location, forest type	Site Diversity (Spp./site)	Point Diversity (Mean No. spp./point)	Abundance (Mean No. birds/point/day)	Percentage Long-distance Migrants
Southeast Alaska				
Deciduous	18, 18, 18	15.9, 13.3, 14.1	13.0, 11.7, 9.3	83, 83, 72
Average	18.0	14.4	11.3	79
Conifer	14, 16, 15, 20, 12, 18	7.9, 11.6, 10.8, 13.[illegible] 10.4, 13.0	5.9, 7.0, 11.9, 8.7, 7.5, 7.3	38, 50, 67, 53, 55, 41
Average	15.8	11.2	8.1	50.1
Chiloé, Chile				
Broadleaf, Mature	19, 20, 16	14.8, 13.2, 11.7	15.8, 11.8, 16.4	16 (21),[1] 20 (25),[1] 17 (17)
Broadleaf, Secondary	22	12.0	14.3	14 (18)
Average	19.3	12.9	14.6	16.8 (20.3)

[1] Including *Columba araucana*.

ANOVA, $F = 12.44$, $p = 0.002$; Tukey post-hoc test showed that the significant difference was found between Chiloé and Alaskan coniferous forest). Mist-net samples from the understory, however, suggest either that the denser populations of Chiloé occur chiefly in the upper strata of the forest, or that the understory birds are more difficult to capture in mist nets. The catch rate of birds in understory in Chiloé (0.13–0.22 birds per mist-net hour) was uniformly lower than that for Alaskan deciduous stands (0.36–1.05); the samples for Alaska coniferous stands (0.06–0.15) were extremely low (one-way anova, $F = 6.81$, $p = 0.037$).

Composition

A number of taxonomic families are shared by the avifaunas of Alaska and Chiloé forests, and both regions hold species of families and subfamilies not shared with the other region (e.g., Chiloé: Psittacidae, Furnariidae, Rhinocryptidae; Alaska: Paridae, Sittidae, Certhiidae, Bombycillidae, Vireonidae, Emberizidae—Parulinae, Muscicapidae—Sylviinae). More interesting are genera shared by these widely separated regions: *Picoides*, *Colaptes*, *Turdus*, *Zonotrichia*, *Carduelis*, *Tachycineta*, and *Troglodytes*. Chile even has one species that occurs in much of North America, although not in Alaskan coastal forests (*Troglodytes aedon*).

The most common species in the Alaskan deciduous stands were *Wilsonia pusilla*, *Vermivora celata*, and *Dendroica coronata*, ranking in the top three for abundance in the majority of sites. In Alaska coniferous stands, the most common species (top three at most sites) were *Ixoreus varius*, *Troglodytes troglodytes*, and *Dendroica townsendii*. Five of these six most common species are summer residents (and long-distance migrants) in Alaska, and four belong to a single subfamily (Emberizidae—Parulinae).

The most common species at all sites on Chiloé was the tyrannid *Elaenia albiceps*, a summer resident and long-distance migrant (Johnson, 1967; Sabag, 1993). The chucao (*Scelorchilus rubecula*; Rhinocryptidae) ranked in the top three at three sites but was probably overestimated because of its loud, ringing call. Also commonly ranked in the top three were the only lowland hummingbird in the region (*Sephanoides galeritus*) and a furnariid (*Aphrastura spinicauda*).

Notable Biological Differences

One major difference between the avifaunas of southeast Alaska and Chiloé is the high proportion of southeast Alaskan species that are long-distance migrants (see Table 11.1). Migrants comprise 38% to 67% of the regularly occurring species in Alaskan coastal conifer forest and as much as 72% to 83% of the species in deciduous forest. In contrast, long-distance migrants constitute less than 20% of the regularly occurring

Chilote forest-bird community (or <25% if *Columba araucana* is included as a long-distance migrant; we lack information on the migratory patterns of Chilote populations of *C. araucana*). Several additional species in both southeast Alaska and Chiloé make local seasonal movements.

A second major difference between North American and Chilean rainforests is found in the clutch sizes of many species of birds. Many species of breeding birds in Chiloé usually lay clutches of two to three eggs, and only a few regularly lay at least four eggs (Johnson, 1967). In contrast, many North American birds commonly lay clutches of at least four eggs; most of the data derive from studies below the 49th parallel (Ehrlich, Dobkin, & Wheye, 1988; Harrison, 1979), and clutches at high northern latitudes are often even larger than those at lower latitudes (Lack, 1968; Morton, 1974). The geographic differences in clutch size are found in birds of various taxonomic and biogeographic affiliations, habitats, and nesting habits. Even closely related species commonly, but not invariably, exhibit the difference in clutch size: *Turdus migratorius* (3–4) versus *T. falklandii* (2–3); *Colaptes auratus* (5–8) versus *C. pitius* (4–6); *Troglodytes aedon* (6–7) and *T. troglodytes* (5–6) versus *T. aedon* (4–7); *Tachycineta bicolor* (5–7) versus *T. leucopyga* (4–6); *Zonotrichia leucophrys* (3–5) versus *Z. capensis* (3–4); *Picoides pubescens* (4–5) versus *P. lignarius* (3–5). In addition, several genera not found in our censuses also show this pattern (e.g., *Mimus*, *Agelaius*, *Sturnella*, and *Cistothorus*).

Guild Structure

Big-tree users comprised equivalent proportions of the avifauna, by species (refer to part A in Table 11.2) and by abundance of individuals (part B, Table 11.2), in Chiloé and Alaska conifer stands, and a significantly lower proportion in Alaska deciduous stands. Species using covered nests were most frequent in Chiloé, and least frequent in Alaska deciduous forest (part A, Table 11.2). The relative abundance of covered-nesting species was lowest in Alaska deciduous forest and slightly higher in Chiloé than Alaska coniferous forest (part B, Table 11.2). Thus, although the diversity of covered nesters in Chiloé was much higher than in Alaska coniferous forest, their relative abundance was less different, reflecting the commonness of *Troglodytes troglodytes* in conifer forests and the relative rarity of some Chilean covered-nesting species.

Understory birds comprised a greater proportion of the avifauna, by both species richness (part A, Table 11.2) and abundances (part B, Table 11.2) in Alaska deciduous stands, followed by Alaska conifer stands. Chilote forests held the fewest species and individuals of this guild (parts A and B, Table 11.2). Leaf-gleaners were least speciose and least abundant in Chiloé (parts A and B, Table 11.2). In contrast, aerial feeders were most diverse and most abundant in Chiloé.

Table 11.2. Guild Structure of Forest Avifauna of Southeast Alaska and Chiloé

Part A) Average proportion of regular species in each guild in each region (range in parentheses). An * between two values in the same row marks the likely source of significant differences, i.e., no overlap or a Mann-Whitney U pairwise test between adjacent entries in the table would be signifficant.

Guild	Alaska				Chiloé	Kruskal-Wallis	p
	Deciduous (n = 3)		Coniferous (n = 6)		Broadleaf (n = 4)		
Big-Tree Users	11% (—)	*	17% (14%–20%)		18% (13%–26%)	6.54	0.038
Covered Nesters	17 (—)	*	23 (20–27)	*	47 (38–53)	10.53	0.005
Understory Birds	70 (67–72)	*	48 (36–60)	*	39 (32–44)	8.68	0.013
Aerial Feeders	6 (6–7)		5 (0–8)	*	12 (10–14)	8.03	0.018
Canopy Seed Predators	0 (—)		2 (0–7)	*	13 (6–16)	8.68	0.013
Leaf-Gleaners	39 (28–44)		27 (20–33)	*	17 (13–19)	8.93	0.012

Part B) Proportion of individuals in each guild in each region. * As in A, an * indicates marginal significance.

Guild	Alaska				Chiloé	Kruskal-Wallis	p
	Deciduous (n = 3)		Coniferous (n = 6)		Broadleaf (n = 4)		
Big-Tree Users	4% (3%–4%)	*	9% (6%–21%)		15% (6%–21%)	7.11	0.029
Covered Nesters	4 (<1–6)	*	23 (11–37)		40 (30–50)	9.51	0.009
Understory Birds	74 (69–78)	*	60 (42–75)	(*)	36 (26–50)	7.81	0.020
Aerial Feeders	9 (5–13)		7 (<1–14)	*	30 (22–38)	7.94	0.019
Canopy Seed Predators	1 (<1–3)		1 (0–3)	*	8 (5–11)	7.83	0.020
Leaf-Gleaners	32 (29–36)		28 (14–45)	(*)	12 (7–17)	5.27	0.072

The results for canopy seed predators, indicating a higher frequency of species and individuals in Chiloé than Alaska (parts A and B, Table 11.2), is at least partly an artifact. Many of the seed predators of the North American conifer forest are sporadic and even nomadic in occurrence and often do not breed during the season used by most birds for nesting (Benkman, personal communication). Therefore, they are not often recorded in spring "breeding-bird" censuses and are certainly underestimated in this study (especially *Loxia* species).

We compared guild structure of the Chilean bird community to that of Alaskan coastal forests and that of Canadian interior forests to the east of the coastal range of mountains near Juneau to determine if the bird communities of the two rainforests (Alaska coastal, Chiloé) are convergent in structure. The guild structure of Chilote forest birds is no more similar to that of Alaskan coastal forests than it is to that of Canadian forests. The proportionate frequency of species in Chiloé is most similar to Alaskan forests (specifically conifer forests) for three guilds (big-tree users, understory birds, leaf-gleaners), and most similar to Canadian forests of the remaining three guilds (covered nesters, aerial feeders, canopy seed predators), as shown in Table 11.3. This result suggests that guild structure of the Chilote forest birds is not particularly similar to that of Alaskan coastal forest birds.

Comparison of Guild Structure in North and South Temperate Forests

Patterns

Data from our censuses were combined with data extracted from published studies, and classified by location (northwest North America, midwest North America, south temperate) and forest type (broadleaf, conifer). The resulting six categories were subjected to two-way ANOVA, in a search for general patterns. Because we already know that there is some heterogeneity within the northwest region (i.e., between Alaskan coastal and Canadian interior forests), this exercise serves chiefly to look for patterns on a continental scale. Where direct comparisons of these results with those of the preceding section yield different results, that difference may reflect the difference in scale and perhaps point to interesting exceptions to generalizations.

Broadleaf forests consistently held more species of big-tree users than did conifer forests, although the ANOVA showed a significant interaction of location and forest type (see part A in Table 11.4). Location had no detectable effect on species diversity of big-tree users. Birds in this guild were less abundant in northwestern forests and marginally less abundant in conifer than in broadleaf forests (especially in midwest and south temperate; refer to part B in Table 11.4).

Table 11.3. Comparison of Guild Structures of Forest Birds in Coastal Southeast Alaska, Adjacent Interior Canada, and Chiloé

Guild						Kruskal-Wallis	*p*
Big-Tree Users	CC (16.5)	CD (15.2)	AD (12.3)	AC (8.0)	Ch (2.5)	12.86	0.012
Covered Nesters	CC (17.5)	Ch (14.3)	CD (9.3)	AC (6.3)	AD (4.5)	12.54	0.014
Understory Birds	AD (14.7)	CD (14.0)	CC (12.3)	AC (8.3)	Ch (4.3)	8.01	0.066
Aerial Feeders	CD (18.0)	Ch (14.5)	CC (10.0)	AD (5.5)	AC (5.3)	15.01	0.005
Canopy Seed Predators	CC (17.5)	Ch (14.3)	CD (9.3)	AC (6.5)	AD (4.5)	14.01	0.007
Leaf-Gleaners	CC (16.5)	CD (15.2)	AD (12.3)	AC (8.0)	Ch (2.5)	14.94	0.005

The frequency of species in each guild for each regional forest type is compared by Kruskal-Wallis test, ranked in descending order (average ranks in parentheses), and the position of Chiloé birds in this rank–order is determined by inspection. Codes are as follows: AD = Alaska deciduous; AC = Alaska coniferous; Ch = Chiloé; CD = Canada deciduous; CC = Canada coniferous. By region, Chiloé is closest to Alaska for three guilds, and to Canada for the other three guilds. By forest type, Chiloé is closest to northern conifer forests for five guilds and about equidistant between conifer and deciduous for one guild (aerial).

Diversity of species using covered nests was higher in broadleaf than in conifer forest, but their abundance was similar (parts A and B, Table 11.4). South temperate forests held more species of covered-nesters than the other forests, and this guild was least abundant in northwestern forests. Thus, midwestern forest birds of this guild had relatively fewer species but equal proportions of individuals, as compared to south temperate forests, and equal proportions of species but more individuals, as compared to northwestern forests. This indicates that there was a greater average density per species in the midwestern users of covered nests.

Understory species diversity showed no significant trends with location or forest type (part A, Table 11.4). Abundance was least in northwestern forests, and greater in conifer than in broadleaf forests (part B, Table 11.4). Aerial feeders were least speciose and least abundant in northwestern forests, and more diverse and more abundant in broadleaf than coniferous forests (parts A and B, Table 11.4).

Canopy seed predators were most diverse in south temperate forests, followed by northwestern and midwestern forests (part A, Table 11.4). Abundance patterns followed the same ranking (part B, Table 11.4). However, the value for northwestern forests is underestimated, as discussed earlier. Furthermore, the low values for midwestern forests reflect the extinction of two major canopy seed predators in the last century (*Ectopistes migratorius* and *Conuropsis carolinensis*). Therefore, this trend is spurious but serves to emphasize the point that canopy seed-eaters in the Chilote forests are a noteworthy component of the avifauna.

Leaf-gleaners were least speciose and least abundant in south temperate forests (parts A and B, Table 11.4). They were consistently (but not significantly) more diverse in conifer than broadleaf forests, and significantly more abundant in conifer forests.

Predictions About Comparative Guild Structure

First, the architectural form of most broadleaf trees produces many large branches, whereas that of conifers is commonly a set of relatively small branches off the main trunk. If the difference in branching pattern is associated with a parallel difference in availability of foraging surfaces and cavity nest-sites, then one might predict a greater diversity or abundance of big-tree users in broadleaf forests. The linear array of the six categories of forest (3 locations × 2 types) showed that, indeed, broadleaf forests tend to host more species of big-tree users than conifer forests (part A, Table 11.4), although the Tukey post-hoc test did not detect significant differences at the predicted point in the array. In contrast, corresponding differences in abundances were not apparent between broadleaf and conifer forests (part B, Table 11.4). Large broadleaf trees may offer more avenues for exploitation, but not necessarily more resources.

Table 11.4. Regional Comparison of Guild Structure of Forest Birds, for South Temperate Rainforest in Chile and Argentina, North Temperate Forests in the Midwest of North America, and Northwestern Forests of Alaska and Canada (Codes: S = south temperate; M = midwest; N = northwestern; B = broadleaf; C = conifer; D = deciduous; for comparison of broadleaf forests with conifer forests, B includes D. Samples sizes are N_D (19), N_C (17), M_D (20), M_C (3), S_B (14), S_C (2). Significant differences in pattern, and interactions were determined by two-way ANOVA (see text); an * indicates significant interaction. Data are shown when the direction of difference is consistent (even when the interaction in significant or the overall ANOVA is not significant); data are not shown when the interaction is significant and the direction of difference is not consistent. Nonsignificant differences in linear ranks, determined by Tukey post-hoc test, are underlined; the numbers accompanying the rankings are the average percents for each regional forest type.)

Part A) By species.

Guild	Interaction	Pattern		Rankings
		Leaf type	Location	
Big-Tree Users	*[a]	B > C	M = S = N	S_B M_D N_D M_C N_C S_C
		(25% > 15%)	(25% = 21% = 20%)	27 26 22 18 18 0%
Covered Nesters	no	B > C	S > M = N	S_B S_C M_C N_D M_C N_C
		(32% > 22%)	(47% > 28% = 23%)	50 29 29 25 24 22%
Understory Birds	*	—	—	S_C N_D N_C M_C S_B M_D
				55 53 48 39 30 27%
Aerial Feeders	no	B > C	S = M > N	M_D S_B S_C N_D M_C N_C
		(15% > 6%)	(16% = 15% > 8%)	19 15 12 11 7 6%
Canopy Seed Predators	no	C > B	S > N > M	S_C S_B N_C M_C N_D M_D
		(5% > 2%)	(11% > 3% > 0.4%)	12 11 5 2 2 0%
Leaf-Gleaners	no	C > B	M = N > S	M_C M_D N_C N_D S_C S_D
		(28% > 23%)[b]	(33% = 29% > 9%)	42 31 28 26 13 9%

[a] Interaction significant, but direction of difference consistent.
[b] ANOVA not significant, but direction of difference consistent.

Table 11.4.

Part B) By abundance.

Guild	Interaction	Pattern: Leaf type	Pattern: Location	Rankings
Big-Tree Users	*	B > C[a]	M = S > N	M_D S_B M_C N_C N_D S_C
		(14% > 7%)	18% = 14% > 7%	21 18 9 8 6 0.7%
Covered Nesters	no	B = C	S = M > N	S_B M_D M_C S_C N_C N_D
		(24% = 19%)	29% = 28% > 16%	31 30 21 19 19 15%
Understory Birds	*	C > B	—	S_C M_C N_C N_D M_D S_B
		(51% > 32%)[a]		66 52 47 41 27 26%
Aerial Feeders	no	B > C	S = M > N	S_B M_D S_C N_D M_C N_C
		(22% > 5%)	(29% = 19% > 10%)	32 22 18 16 4 3%
Canopy Seed Predators	no	C = B	S > N > M	S_B S_C M_C N_C N_D M_D
		(2% = 1%;)	(6% > 1% > 0.3%)	7 4 2 1 0.9 0%
Leaf-Gleaners	no	C > B	N = M > S	M_C N_C M_D N_D S_C S_B
		(37% > 26%)	(36% = 36% > 5%)	47 40 34 33 11 4%

[a] $p = 0.06$.
[b] Direction of difference consistent.

Second, covered nests might be thought to be adaptations to protect the eggs and young from weather, nest predators, or brood parasites (although covered nests do not necessarily prevent brood parasitism by cowbirds; Cruz & Wiley, 1989). If the advantage of a covered nest lies in a protective microlimate, then one might predict that covered nests would be most frequent in the northwest, where birds often nest early, when snow is still on the ground. Alternatively, one could argue that the rainy climates select for protected nests and predict that covered nests would be more frequent in northwestern and southern forests than in the midwest. On the other hand, if the advantage of a covered nest lies in protection from nest predators and brood parasites, one might predict them to be most common in the midwest, where cowbirds, snakes, and avian and mammalian predators abound, or least common in the south temperate forests, which host relatively few species of nest robbers. None of these patterns is seen in the array of six categories of forest, suggesting that neither factor, taken alone, accounts for the observed pattern.

Third, some of these forests are largely evergreen, and some are largely deciduous. One might argue that evergreenness is associated with long-lived leaves, which are allocated higher levels of chemical and morphological protection than deciduous leaves, and that insect herbivores are consequently less abundant on evergreen leaves. If this were true, and if the diversity and abundance of birds is associated with the size of the potential prey population, then leaf-gleaners should be most abundant or diverse in deciduous stands. This is not the case, however: Although southern forests have low levels of leaf gleaners, North American coniferous forests tend to have proportionately more abundant (though no more diverse) populations of leaf gleaners than deciduous forests (although the Tukey test does not detect significant differences at the predicted points in the array). Thus, the hypothesized relationship between evergreenness and insect abundance seems not to explain much of the observed pattern.

Fourth, conifer and broadleaf forests differ in leaf shape and in branching patterns of branchlets and twigs, and they might offer differing availability and diversity of noncavity nest-sites or foraging surfaces. One might predict that nest-sites and diversity of foraging surfaces would be greater in broadleaf forests than in conifers and that the birds using these forests would be more diverse and abundant there. However, there tend to be relatively more species of open-nesters and proportionately more leaf-gleaners in conifer forests (refer to Table 11.4).

Conclusions

Comparisons of Alaskan and Chilean Coastal Rainforests

Site diversity was somewhat higher in Chilean forest and Alaskan deciduous forest than in Alaskan conifer forests. Point-diversity was highest for

Alaskan deciduous forests, intermediate for Chiloé, and lowest for Alaskan coniferous forests. However, the average number of individual birds censused was markedly higher at each census point in Chilean forests, as suggested by Jaksic and Feinsinger (1991), and lowest for Alaskan coniferous forest. The forests of Chiloé, therefore, support an avifauna within the range of diversity of Alaskan coastal forests but substantially denser.

Covered nests are much better represented in the Chilean forest avifauna than in Alaska or the rest of North America. This difference is not apparently associated with adaptation to microclimate or probable risk of nest destruction by vertebrates (although measurements of rates of nest failure are needed to examine this more closely). Instead, there seems to be a taxonomic association: Psittacids, furnariids, and rhinocryptids typically nest in holes or build dome nests and constitute a notable proportion of the Chilean avifauna. Other covered-nesters occur in both regions: Troglodytids and *Tachycineta* always use covered nests, and one species of each occurs in the censused forests of each region. Picidae always nest in cavities, and occur in both regions, but are not common. However, a taxonomic association begs the questions of the origin of the habit. In addition, one reviewer suggested the possibility that long-distance migrants lack time to build covered nests, but the only Neotropical migrant in the Chiloé forests is *Elaenia albiceps*, which builds an open nest.

The frequency of covered-nesting species on Chiloé contrasts also with that of New Zealand, where the forest trees have many taxonomic similarities, the climate is similarly wet, and the latitude is comparable, but the phylogenetic history of many indigenous bird species is very different from that of Chile. About 20% to 25% of the native species of *Nothofagus* forest sites in New Zealand use covered nests (census data from Kikkawa, 1966; nest information from Anonymous, 1985; Falla, Sibson, & Turbott, 1981). The contrast between the frequency of covered-nesters in New Zealand and Chile seems to reinforce a taxonomic interpretation of the high frequency of covered-nesters in Chilean forest birds. However, a taxonomic correlation in Chile does not provide an explanation that is generalizable to New Zealand, because there is within-family variation in nest type for New Zealand birds (Anonymous, 1985; Falla, Sibson, & Turbott, 1981).

Ecological correlates of small clutches in the northern hemisphere may include high risk of nest predation, low adult survival, or the existence of multiple breeding opportunities during the season (Bennett & Harvey, 1988; Martin, 1993a, 1993b, 1995; Martin & Li, 1992; Saether, 1988). However, such correlations often seem to break down when comparing temperate and tropical regions (Karr, Nichols, Klimkiewicz, & Brawn, 1990), and preliminary evidence from Chile also suggest that at least some of these correlations may not hold there. Similarly, most native passerines in New Zealand produce small clutches (fewer than four eggs), and correlations of clutch sizes and number of broods per year with nesting habits are not clear (data from Anonymous, 1985; Falla, Sibson, & Turbott 1981; but see Yom-Tov, Christie, & Iglesias, 1994). Regional comparisons offer the opportunity to test the ecological hypotheses about the evolution of clutch size on a larger scale.

Some of the Chilean species may have small clutches for historical or phylogenetic reasons (as opposed to ecological reasons). Some belong to taxa that characteristically produce small clutches everywhere (e.g., pigeons and hummingbirds). Clutch size of others might be constrained if they were geologically recent immigrants to southern forests from the lowland tropics, where small clutches are common. Such an hypothesis leaves unexplained the relatively large clutch sizes of several Chilean species, unless they immigrated from other source faunas or did so earlier. The high level of endemism in the Chilean avifauna (Vuilleumier, 1985) suggests that there has been time for the evolution of significant morphological and behavioral differences in many taxa, and there is no apparent reason why clutch size should be more constrained than morphology and behavior.

The avifaunas of coastal rainforests in Chiloé and southeast Alaska differed somewhat in guild structure. Guild structure of Chilote forest avifaunas was not more similar to Alaskan coastal forest than to Canadian forest avifaunas in the same general region. Long-distance migrants are more often among the common birds in Alaska (and to a slightly lesser degree in Washington and Oregon; Carey, Hardt, Horton, & Biswell, 1991; Gilbert & Allwine, 1991; Manuwal, 1991) than in Chiloé and comprise a much larger proportion of the avifauna. Thus, biogeographical differences and biological interactions (or other factors) appear to override the climatic and topographic similarities in structuring the bird community in these forests.

Guild Structure of North and South Temperate Forests

Broad regional comparisons documented the high richness of covered nesters in south temperate forests. South temperate forests appeared to have relatively few species and birds that forage principally by leaf gleaning, and northwestern forests supported relatively few species and individuals of aerial feeders. Big-tree users were proportionately more speciose and often more abundant in broadleaf forests than in conifer forests, and covered nesters were more diverse in broadleaf forests. Aerial feeders were proportionately more diverse and abundant in broadleaf forests. In contrast, leaf-gleaners were relatively more diverse and abundant in conifer forests, and understory birds were relatively more abundant there.

We did not analyze another interesting guild—that of frugivores that disperse the seeds of fleshy fruits. Fleshy fruits and frugivory are common in both Chiloé and Alaska in late summer (our observations; Armesto,

Smith-Ramírez, Sabag, this volume). However, very few birds in coastal Alaska eat fruits during the breeding season, in contrast to several species in Chiloé. Perhaps the most interesting questions to ask of this guild concern the effect of fruits on the breeding ecology of Chilean birds and the comparisons of foraging ecology in the nonbreeding season in the two regions.

As is common with first efforts to build and understand comparisons, this study raises more questions than it answers. This preliminary study documents some contrasting patterns and identifies some related questions that require more detailed examination both of the patterns themselves and of the processes that form them.

Acknowledgments. We are grateful for the financial support of NSF-International Programs (Mary F. Willson), a visiting professorship for Willson at the Universidad de Chile funded by Fundación Andes, and Fondecyt (Juan J. Armesto). Cecilia Smith helped in many ways with the fieldwork. Jackie Canterbury kindly permitted the use of her data on mist-net capture rates from coniferous forest near Ketchikan, AK. Landowners of Chiloé graciously consented to let us use their forests for this study. Kathryn E. Sieving and anonymous reviewers provided helpful comments on the manuscript.

References

Alaback, P. 1991. Comparative ecology of temperate rainforests of the Americas along analogous climatic gradients. *Rev Chil Hist Nat*, 64, 399–412.

Anonymous. 1985. *Complete Book of New Zealand Birds*. Sydney: Readers Digest Services.

Bennett, P.M., and Harvey, P.H. 1988. How fecundity balances mortality in birds. *Nature*, 333, 216.

Bond, R.R. 1957. Ecological distribution of breeding birds in the upland forests of southern Wisconsin. *Ecol Monog*, 27, 351–384.

Carey, A.B., Hardt, M.M., Horton, S.P., and Biswell, B.L. 1991. Spring bird communities in the Oregon Coast Range. *U.S.D.A. Forest Service General Technical Report 285* (pp. 123–142).

Cody, M.L. 1970. Chilean bird distribution. *Ecology*, 51, 455–464.

Cruz, A., and Wiley, J.W. 1989. The decline of an adaptation in the absence of a presumed selection pressure. *Evolution*, 43, 55–62.

Ehrlich, P.R., Dobkin, D.S., and Wheye, D. 1988. *The Birder's Handbook*. New York: Simon and Schuster.

Falla, R.A., Sibson, R.B., and Turbott, E.G. 1981. *Collins Guide to the Birds of New Zealand*. Auckland: Collins.

Flack, J.A.D. 1976. Bird populations for aspen forests of western North America. *Ornithol Monog*, 19, 1–97.

Gilbert, F.F., and Allwine, R. 1991. Spring bird communities in the Oregon Cascade Range. In L.F. Ruggiero, K.B. Aubry, A.B. Carey, and M.H. Huff (eds.), *Wildlife and Vegetation of Unmanaged Douglas-Fir Forests* (pp. 145–158). Portland, OR: U.S.D.A., Forest Service, Pacific Northwest Research Station, Technical Report PNW-GTR-285.

Harrison, H.H. 1975. *A Field Guide to Bird's Nests*. Boston: Houghton Mifflin.

Harrison, H.H. 1979. *A Field Guide to Western Bird's Nests*. Boston: Houghton Mifflin.
Holmes, R.T., Bonney, Jr., R.E., and Pacala, S.W. 1979. Guild structure of the Hubbard Brook community: A multivariate approach. *Ecology*, 60, 512–520.
Holmes, R.T., and Recher, H.F. 1986. Determinants of guild structure in forest bird communities: An intercontinental comparison. *Condor*, 88, 427–439.
Jaksic, F.M., and Feinsinger, P. 1991. Bird assemblages in temperate forests of North and South America: A comparison of diversity, dynamics, guild structure, and resource use. *Rev Chil Hist Nat*, 64, 491–510.
James, F.C., and Rathbun, S. 1981. Rarefaction, relative abundance, and diversity of avian communities. *Auk*, 98, 785–800.
Johnson, A.W. 1967. The birds of Chile and adjacent regions of Argentina, Bolivia and Peru. Vol. 2. Buenos Aires: Platt Establecimientos Gráficos.
Karr, J.R., Nichols, J.D., Klimkiewicz, M.K., and Brawn, J.D. 1990. Survival rates of birds of tropical and temperate forests: Will the dogma survive? *Am Nat*, 136, 277–291.
Kessler, W.B., and Kogut, T.E. 1985. Habitat orientation of forest birds in southeastern Alaska. *NW Sci*, 59, 58–65.
Kikkawa, J. 1966. Population distribution of land birds in temperate rainforest of southern New Zealand. Transactions Royal Society New Zealand, *Zoology*, 17, 215–277.
Lack, D. 1968. Ecological adaptations for breeding in birds. London: Methuen.
Manuwal, D.A. 1991. Spring bird communities in the southern Washington Cascade Range. In L.F. Ruggiero, K.B. Aubry, A.B. Carey, and M.H. Huff (eds.), *Wildlife and Vegetation of Unmanaged Douglas-Fir Forests* (pp. 161–174). Portland, OR: U.S.D.A., Forest Service, Pacific Northwest Research Station, Technical Report PNW-GTR-285.
Martin, T.E. 1988a. On the advantages of being different: Nest predation and the coexistence of bird species. *Proceedings of the National Academy of Sciences* (USA), 85, 2196–2199.
Martin, T.E. 1988b. Processes organizing open-nesting bird assemblages: Competition or nest predation? *Evol Ecol*, 2, 37–50.
Martin, T.E. 1993a. Evolutionary determinants of clutch size in cavity-nesting birds: Nest predation or limited breeding opportunities? *Am Nat*, 142, 937–946.
Martin, T.E. 1993b. Nest predation among vegetation layers and habitat types: Revising the dogmas. *Am Nat*, 141, 897–913.
Martin, T.E. 1995. Avian life history evolution in relation to nest sites, nest predation, and food. *Ecol Monog*, 65, 101–127.
Martin, T.E., and Li, P. 1992. Life history traits of open- vs. cavity-nesting birds. *Ecology*, 73, 579–592.
Meserve, P.L., and Jaksic, F.M. 1991. Comparisons of terrestrial vertebrate assemblages in temperate rainforests of North and South America. *Rev Chil Hist Nat*, 64, 511–535.
Morton, M.L. 1974. Adaptive strategies of *Zonotrichia* breeding at high latitude or high altitude. *Proceedings of the International Ornithological Congress*, 16, 322–336.
Ralph, C.J. 1985. Habitat association patterns of forest and steppe birds on northern Patagonia, Argentina. *Condor*, 87, 471–483.
Root, R.B. 1967. The niche exploitation pattern of the blue-gray gnatchatcher. *Ecol Monog*, 37, 317–350.
Sabag, C. 1993. *El rol de las aves en la dispersión de semillas en un bosque templado secundário de Chiloé (42° S)*. Unpublished masters thesis, Facultad de Ciencias, Universidad de Chile.

Saether, B.-E. 1988. Pattern of covariation between life-history traits of European brids. *Nature*, 331, 616–617.
Sokal, R.R., and Rohlf, F.J. 1981. Biometry 2nd ed. San Francisco: Freeman.
Spindler, M.A., and Kessel, B. 1980. Avian populations and habitat use in interior Alaska taiga. *Syesis*, 13, 61–104.
Theberge, J.B. 1976. Bird populations in the Kluane Mountains, southwest Yukon, with special reference to vegetation and fire. *Can J Zool*, 54, 1346–1356.
Vuilleumier, F. 1972. Bird species diversity in Patagonia (temperate South America). *Am Nat*, 106, 266–271.
Vuilleumier, F. 1985. Forest birds of Patagonia: Ecological geography, speciation, endemism, and faunal history. *Ornithol Monog*, 36, 255–305.
Willson, M.F. 1974. Avian community organization and habitat structure. *Ecology*, 55, 1017–1029.
Willson, M.F. 1986. Avian frugivory and seed dispersal in eastern North America. *Curr Ornithol*, 3, 223–279.
Willson, M.F. 1991. Dispersal of seeds by frugivorous animals in temperate forests. *Rev Chil Hist Nat*, 63, 537–554.
Willson, M.F., and Comet, T.A. Bird communities of northern forests: Patterns of diversity and abundance. Unpublished manuscript.
Willson, M.F., De Santo, T.L., Sabag, C., and Armesto, J.J. 1994. Avian communities of fragmented south-temperate rainforests in Chile. *Conserv Biol*, 8, 508–520.
Yom-Tov, Y., Christie, M.I., and Iglesias, G.J. 1994. Clutch size in passerines of southern South America. *Condor*, 96, 170–177.

12. The Importance of Plant-Bird Mutualisms in the Temperate Rainforest of Southern South America

Juan J. Armesto, Cecilia Smith-Ramírez, and Carlos Sabag

In the Neotropics, numerous nectar- and fruit-eating birds engage in loose mutualisms with plants that provide them with food (Feinsinger, 1983, 1987; Snow, 1981). While feeding on the fruit pulp or flower nectar, the birds act as seed dispersers or pollinators. These relationships contribute significantly to species diversity in many tropical forests (Gentry, 1982; Stiles, 1985).

Most plant-disperser and plant-pollinator associations with birds can be classified as "diffuse" mutualisms; that is, one species of frugivore or nectarivore may feed on various plant species, while at the same time, the pollen or seeds of a given plant species may be transported by several bird species (Feinsinger, 1983; Herrera, 1982). Mutualistic relationships are thus structured around guilds, or groups of interdependent plants and birds (e.g., Herrera 1982, 1984; Jordano, 1987; Murray et al., 1987), rather than species pairs.

Howe (1984) has argued that in these diffuse mutualistic systems, plant species that flower or fruit during periods of food scarcity or bird species that are active during times of low pollinator or frugivore activity may play a critical role in the maintenance of the entire assemblage. These are called pivotal or keystone species (Gilbert, 1980; Howe, 1984). Knowledge about these pivotal or keystone species is of critical importance for understanding community dynamics and for evaluating the potential impact of biodiversity losses on ecosystem function (Chapin, Schulze, & Mooney, 1992).

In temperate forests, the geographical distribution and overall importance of plant–avian mutualistic interactions are still poorly understood (e.g., Armesto, Rozzi, Miranda, & Sabag, 1987; Herrera, 1984; Willson, 1986, for seed dispersal). In temperate forests of southern Chile, about 70% of the forest flora, including more than 90% of subcanopy and successional trees, bear fleshy fruits (Armesto & Rozzi, 1989), which are consumed by birds and other frugivorous vertebrates (Armesto et al., 1987; Willson, 1991). The proportion of fleshy-fruited plant species in southern Chile is higher than that reported for many North American temperate forests (Willson, 1986) and similar to some tropical forests (Gentry, 1982). Recent work in southern Chile (Smith-Ramírez, 1993) revealed that at least 14 plant species are pollinated by hummingbirds. In contrast, nectar-eating birds (Jaksic & Feinsinger, 1991) and hummingbird flowers seem to be rare in conifer or temperate deciduous forests of North America.

Because of the critical role of plant–animal mutualistic associations in the maintenance of local and regional diversity, we ask whether pivotal species occur among avian seed dispersers and pollinators in South American temperate rainforests. To answer this question, we review the information on bird population densities and diets and examine the flowering and fruiting patterns of their food plants. We examine the guild structure of these mutualistic assemblages and discuss how Chilean temperate forests differ from or resemble North American temperate and Neotropical forests with respect to keystone plant–avian relationships. Finally, we speculate on the susceptibility of temperate forest assemblages to habitat disruption.

Study Area

Evergreen temperate rain forests occur in southern Chile between 38° S and 55° S (Armesto et al., 1987; Arroyo et al., this volume). Our studies of plant–frugivore interactions have been conducted primarily in a 40-year-old, second-growth temperate rainforest, at 150 m elevation, in central Chiloé Island, Chile (42°30′ S). Phenological records and observations of hummingbird visits to plants have been conducted in a broader range of habitats, including second-growth forest sites as well as primary forests at higher elevations (150–400 m). Lowland forests (below 200 m) have been cleared to a large extent to open land for pastures and for timber production, leaving isolated patches of old-growth and secondary forest. Dominant trees in these secondary forests are *Drimys winteri* (Winteraceae), *Ovidia pillo-pillo* (Thymeliaceae), *Amomyrtus luma*

(Myrtaceae), and *Nothofagus dombeyi* (Fagaceae). Shrubs include *Berberis darwinii*, *B. buxifolia* (Berberidaceae), and *Gaultheria phillyreifolia* (Ericaceae). Primary forests are dominated by *Amomyrtus luma* and *Laurelia philippiana* (Monimiaceae) (Armesto & Figueroa, 1987). These rainforests are representative of the Valdivian rainforest type, which is widespread over most of south-central Chile, particularly on the coastal range (Veblen, Schlegel, & Oltremari, 1983). These forests grow in a maritime climate, characterized by low temperature fluctuations and high humidity year-round (Alaback, 1991; Arroyo, Armesto, Squeo, & Gutiérrez, 1993). Average annual rainfall exceeds 2000 mm; and average monthly maximum (February) and minimum (August) temperatures at sea level are 20.7°C and 3°C respectively. More details on the vegetation, climate, and soils of the study area are given by Holdgate (1961), Villagrán (1985), and Armesto and Rozzi (1989).

Methods

We surveyed bird populations by mist-net captures and sight/sound censuses during the fruiting/flowering seasons of 1988, 1990, and 1991. Bird censuses were conducted in the first 3 hours after sunrise, along a fixed-strip transect 800 m long and 15 m wide. Avian densities reported here are the average of five transects surveyed on five consecutive days during January each year. Black mist nets, 12 m long and 2 m wide, with a mesh size of 25 mm, were set at four different locations in the second-growth forest during daylight hours for a period of 5 to 10 days every summer. Mist nets are known to provide biased estimates of bird densities (Karr, 1981) and, therefore, they were used only as a complement of transect surveys and for collecting samples of birds' gut contents.

Consumption of fruit by birds was determined by examining feces obtained from mist-net captured birds placed in paper sacks for 2 to 10 min (cf. Herrera, 1984). Voided seeds were identified to species later in the laboratory. Over a 3-year period, we examined the gut contents of a total of 351 birds. For one species of bird, the ground-dweller *Scelorchilus rubecula*, data on fruit consumption are those reported by Correa et al. (1990).

At monthly intervals (bimonthly during the winter of 1988) during 1987 and 1988, we recorded the phenology of 61 species of trees, shrubs, and woody vines and epiphytes occurring in both old-growth and secondary forests between 200 m and 500 m elevation. Flower and ripe fruit stages were distinguished (Smith-Ramírez & Armesto, 1994). An average of 7 individuals (range 4–14) per species was considered.

Hummingbird visitation was assessed by recording visits to all the flowering plants bearing red (or pink) tubular corollas in the study area in February, June, and November of 1989 (Smith-Ramírez, 1993).

Results

The Birds

Field surveys indicate that 23 species of birds occur in secondary forests in the study area, as shown in Table 12.1. We classified as frugivores those species (with a reasonable sample size) in which more than 50% of the individuals had evidence of fruit consumption in their gut contents and voided intact seeds. Only four bird species fitted in this category (Table 12.1). Other species consumed only insects or the seeds were destroyed by passing through the bird's gut. All the individuals of the tyrannid *Elaenia albiceps* (white-crested Elaenia) captured in mist nets had intact seeds or fruit remains in their feces. Seeds identified in feces of this species belong to 10 species of plants (see Table 12.2), which represent 43.5% of the fleshy fruited species in the study area. Because we collected feces only during 5 to 10 days in January each summer, a brief period of the fruiting season, this percentage may underestimate the actual breadth of the frugivorous diet of *Elaenia*. We suspect that additional species will be included in the diet of this bird when surveys are extended to the entire fruiting season (December–March). This may also be true for *Turdus falcklandii*, an important frugivore that is resident in the area. The latter species and *Pyrope pyrope* are underrepresented in mist-net captures because of their larger size (Table 12.1).

The frugivore *Elaenia albiceps* represented 55% of all mist-net captures and 85% of all frugivore captures over the 4 years of study (Table 12.1). This species was dominant every year during the peak of the fruiting season, as shown in Figure 12.1. *Elaenia albiceps* emigrates from the area during fall (late March), and reappears in spring (late September), soon becoming the most abundant bird species, as shown in Figure 12.2. *Turdus falcklandii* was the most important frugivore present during winter.

Table 12.1. Avifauna of Second-Growth Forests of Chiloé

	Mist-Net Captures		Bird Counts	
	N	%	*N*	%
Frugivores[1]				
Elaenia albiceps	227	55.1	276	42.9
Turdus falcklandii	17	4.1	72	11.2
Pyrope pyrope	4	1.0	8	1.2
Scelorchylus rebecula	—	—	17	2.6
Nectarivores				
Sephanoides galeritus	45	10.9	38	5.9
Other birds (18 species)	119	28.9	233	36.2

Note: [1] More than 50% of individuals eat fruit during summer.
[2] *N* = number of individuals captured or counted over three summers; % = relative abundance.

Table 12.2. Major Frugivores and Their Food Plants in the Forest of Chiloé

Plant Species	FRUGIVORES				
	Elaenia albiceps	*Scelorchilos rubecula*	*Turdus falcklandii*	*Pyrope pyrope*	*Pseudalopex fulvipes*
Amomyrtus luma	X	X			X
Berberis buxifolia	X				X
Berberis darwini	X				
Drimys winteri	X	X	X	X	X
Fuchsia magellanica	X				
Gaultheria phillyreifolia	X				
Myrceugenia ovata	X				
Nertera spp.	X				X
Ovidia pillo-pillo	X	X			X
Pernettya mucronata	X	X			X
Relbunium hypocarpium	X				

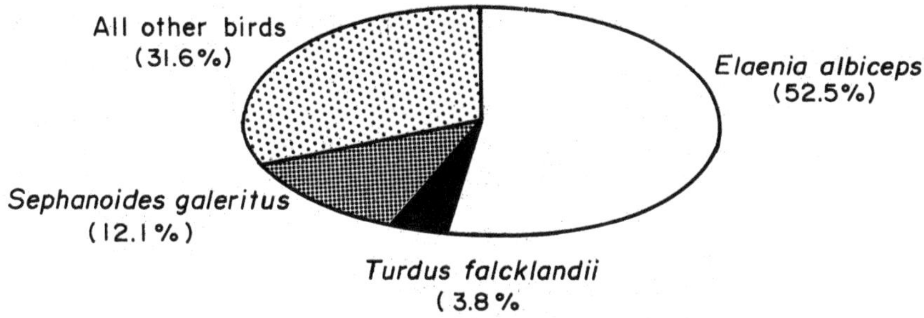

Figure 12.1. Proportion of the avifauna represented by the hummingbird *Sephanoides galeritus* and the frugivores *Elaenia albiceps* and *Turdus falcklandii*. Based on 472 mist-net captures during three consecutive summers in a second-growth forest in Chiloé, southern Chile.

Only one hummingbird species, *Sephanoides galeritus* (fire-crowned hummingbird), was recorded in mist nets and visual counts in the 3 years of study (Table 12.1). *Sephanoides* was the second most common bird captured in mist nets, accounting for nearly 10% of all captures every year (see Figure 12.1). The relative density of *Sephanoides* in the avifauna of southern forests increased during fall and early winter, when other birds decreased in abundance (Figure 12.2). These data agree with that reported in other areas of Chiloé (Willson, De Santo, Sabag, & Armesto, this volume).

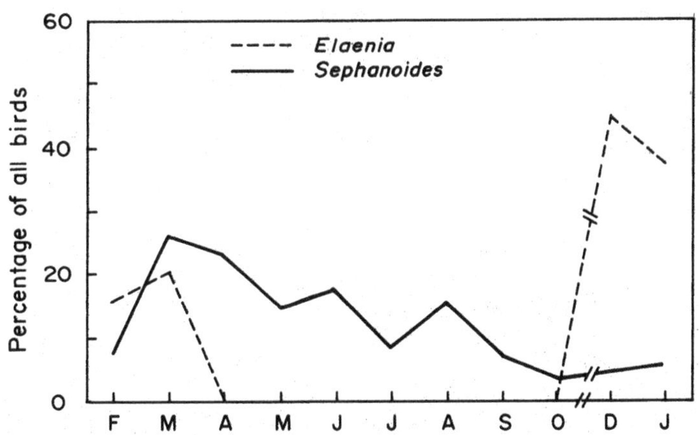

Figure 12.2. Seasonal changes in the representation of the hummingbird *Sephanoides galeritus* and the migrant frugivore *Elaenia albiceps* in the avifauna of a forest dominated by Myrtaceae, in the delta of Río Gol-Gol, 40°30′ S, southern Chile. Based on bird counts reported by García (1982).

Hummingbird Flowers

Fourteen plant species that grow in the study area were visited by the hummingbird *Sephanoides galeritus*, as shown in Table 12.3. This assemblage of hummingbird-visited plants includes six species of vines and epiphytes, two hemiparasites, four shrubs, and two tree species. As a group, they represent 23% of all the flowering plant species listed for the study area by Smith-Ramírez & Armesto (1994). Colors of flowers are predominantly red or pink, except in the case of the bromeliad *Fascicularia bicolor*, where the base of the leaves becomes bright red during the flowering season. Visits of *Sephanoides galeritus* to all these species have been recorded at various times during the year (Smith-Ramírez, 1993). In addition, pollen of *Embothium coccineum*, *Fuchsia magellanica*, *Notanthera heterophylla*, *Tristerix corymbosus*, *Philesia magellanica* and three species of Gesneriaceae have been recovered from the bodies of 23 hummingbirds caught in mist nets in the study area (Smith-Ramírez, 1993). Other studies (Riveros, 1991) show that the flowers of most

Table 12.3. Species Visited by Hummingbirds in the Rainforest of Chiloé

Species	Life Form	Flower Color	Breeding System[1]	Pollinator Dependence[2] %
Astherantera ovata	Epiphyte	Red	SI	0
Campsidium valdivianum	Vine	Red	?	
Crinodendrum hookerianum	Shrub	Red	?	
Desfontainia spinosa	Shrub	Red/yellow	SI	0
Embothrium coccineum	Tree	Red	PI	0
Escallonia rosea	Shrub	Pink	SC	0
Fascicularia bicolor	Epiphyte	Light blue[3]	?	
Fuchsia megellanica	Shrub	Red/violet	SC	10
Mitraria coccinea	Epiphyte	Red	SC	0
Notanthera heterophylla	Hemiparasite	Pink Pink	?	
Philesia magellanica	Shrub, vine		?	
Rhaphithamnus spinosus	Tree	Violet	PI	5
Sarmienta repens	Epiphyte	Red	?	
Tristerix corymbosus	Hemiparasite	Red	?	

[1] From data in Riveros (1991). *SI* = totally self-incompatible; SC = totally self-compatible; *PI* = partly self-incompatible.
[2] Percentage of flowers that form fruit in the absence of pollinator.
[3] Basal leaves surrounding the inflorescence are bright red.

of these species are seldom visited by insects. In a few cases where insect visits have been reported, flower morphology should prevent insect pollination (Smith-Ramírez, 1993). In addition, insect visitors of hummingbird flowers are known to be generalists, visiting the flowers of many other plant species (Riveros, 1991), and hence it is unlikely that they can be effective pollinators. Insect behavior contrasts sharply with the visiting pattern of the hummingbird, which tends to concentrate on one species at a time. It has also been noted that *Sephanoides* carries pollen of morphologically distinct flowers on different parts of its body, which may reduce the chances of interspecific pollen transfer (Smith-Ramírez, personal observation). This interesting observation needs to be quantified in future work.

Flowering Phenology

In the temperate rainforest region of Chile, hummingbirds are active all year long, as can be observed in Figure 12.2, and, not surprisingly, flowers visited by hummingbirds are found in bloom throughout the year. Due to this prolonged flowering season the phenological pattern of ornithophilous (bird-pollinated) species is statistically different (Fisher's exact test, $p < 0.05$) from that of all other species in the rain forest, as shown in Figure 12.3. The main differences are that plant species visited by hummingbirds have a less pronounced flowering peak, and their flowering extends into the winter months. Consequently, ornithophilous species had a longer flowering period on the average (5.2 ± 2.5 months, mean $\pm$ 1 SD) than nonornithophilous species (2.8 ± 1.0 month, Wilcoxon test, $p = 0.002$). This difference in the length of the flowering season

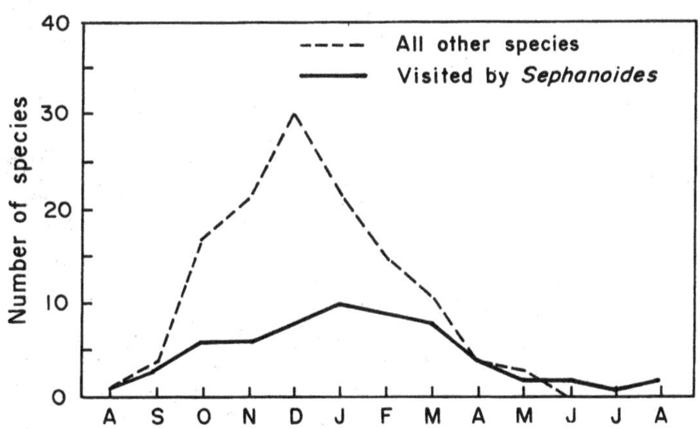

Figure 12.3. Flowering phenology of plants visited by the hummingbird *Sephanoides galeritus* compared with the flowering pattern of all other species in a rainforest of Chiloé Island, southern Chile.

is due mainly to the extended blooming of the hemiparasite *Tristerix corymbosus*.

Fruiting Phenology

Phenological records indicate that most fleshy-fruited species in Chiloé bear ripe fruits between February and April, as shown in Figure 12.4. This tendency for the number of species in fruit coincides with the seasonal trend in the production of ripe fruit for all the species in the study area (Smith-Ramírez & Armesto, 1994). At the same time, this period of maximum fruit production in the area coincides with the period of the year when *Elaenia albiceps*, the major frugivore, is the most abundant bird in the temperate forest (Figure 12.4). Nevertheless, many fleshy-fruited species extend their ripening season year-round, with a low in early spring (September–October). Fruit-eating habits of birds during fall and winter need to be studied to have a better picture of the dispersal process in southern forests.

Conclusions

Guild Structure and Pivotal Species

Bird species seem to play fundamental roles with regard to pollination and seed dispersal of plants in the rain forest of Chiloé. Although the

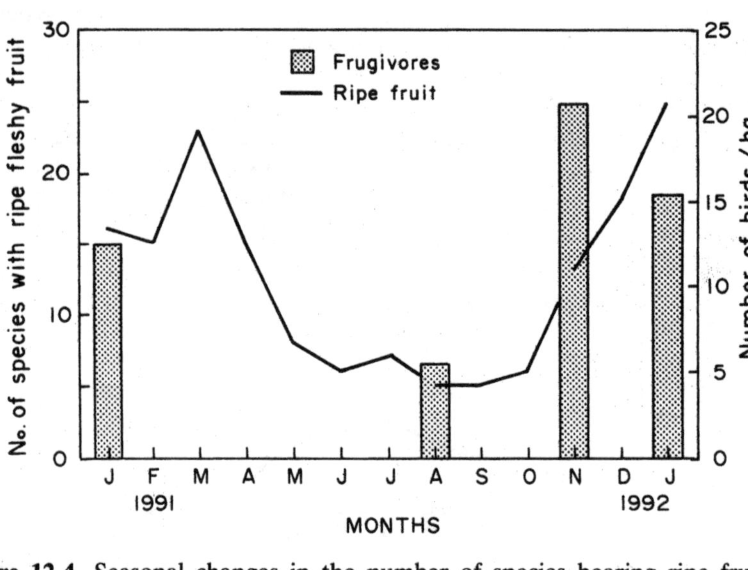

Figure 12.4. Seasonal changes in the number of species bearing ripe fruit and numerical abundance of avian frugivores in a second-growth forest of Chiloé Island, southern Chile.

reported data come mainly from second-growth forests, the same trees and birds are found in primary forest in the study area (Armesto, unpublished data). Further observations in primary forests are necessary to understand how bird behavior and phenology patterns may differ from those found in secondary forest patches.

The presence of abundant populations of *Sephanoides galeritus* throughout the year and of *Elaenia albiceps* during summer may be critical for the reproductive success of many forest species in the temperate forest of Chiloé. Because of the effectiveness of *Sephanoides galeritus* as compared to other, less frequent, nonspecific visitors of flowers with large, red, tubular corollas, this bird species appears as the main pollinator of a guild of 14 plant species found in the canopy of old-growth and secondary forest stands. Pollinator dependence of the plant species in the guild is attested by Rivero's (1991) data on breeding systems and seed yield. Six out of the seven plant species for which the breeding system has been studied are entirely dependent on the pollinator to produce fruit, because they are either genetically or morphologically self-incompatible, as shown in Table 12.3.

We have also shown that during the period of peak abundance of ripe fruit of forest species in Chiloé, the migrant passerine *Elaenia albiceps* accounts for more than 50% of all fruit-eating birds in the forest and that this frugivorous species disperses a major proportion of the seeds that fall into open second-growth areas (Table 12.2). A recent study by Sabag (1993) shows that *Elaenia* feeds on fruits of a large proportion of the fleshy-fruited plants of second-growth and primary forests. Although three other resident bird species are components of the frugivorous guild in the forest of Chiloé, their abundance is low compared to this migrant species (see Table 12.1 and Figure 12.2), and they seem to consume fruits of fewer species (Table 12.2). We believe, however, that the importance of these latter frugivores as seed dispersers may increase as further studies become available (see Willson et al., this volume). Another member of the frugivore guild in the forest is the Chiloé fox (*Pseudalopex fulvipes*), which consumes large numbers of berries, particularly in summer (Armesto et al., 1987; Medel et al., 1990), but it is unknown how effective it may be as compared to birds in disseminating seeds to suitable germination sites.

For most tree species in old-growth forest, seed germination or seedling establishment is restricted to tree-fall gaps (Armesto & Fuentes, 1988; Veblen, 1985). Adequate seed dispersal becomes increasingly important for forest regeneration as open areas become larger and more distant from seed sources as a consequence of human disturbance. A vast seed shadow, such as that generated by the dispersal of seeds by birds, is important in undisturbed forest stands, because it increases the probability that a seed will reach a canopy gap or a site where a canopy gap may soon form (Hubbell & Foster, 1986).

We propose that *Sephanoides* and *Elaenia* are pivotal species for the maintenance of regional diversity in the temperate rainforest of Chile and that a reduction in their populations as consequence of increasing habitat destruction (e.g., replacement of native forests by conifer plantations, logging of native forest) could affect the regeneration and colonizing ability of forest plants, resulting in a decline in these populations in other areas of the temperate rainforest. This hypothesis is supported by two lines of evidence: 1) the rich plant species assemblages that depend on these two bird species for their reproduction; and 2) the apparent lack of alternative species that could substitute *Sephanoides* and *Elaenia* in their roles as major seed disperser and pollinator respectively.

From the point of view of the birds, another species that may also be pivotal (sensu Howe, 1984) within the guild of hummingbird-pollinated plants is the hemiparasite *Tristerix corymbosus*. This is the only species providing a nectar source for hummingbirds during winter in Chiloé. Although hummingbirds are reported to migrate locally (Vuilleumier, 1985), we do not know what fraction of the summer population comes from immigrants versus overwinter residents. The survival of the latter may depend of the extended flowering of *Tristerix*. For the resident fruit-eating birds (e.g., *Turdus falcklandii*), winter-fruiting species, such as *Pernettya mucronata, Luma apiculata*, and *Myrceugenia ovata*, may be an important food source yet to be documented.

Geographical Comparisons

In general, South American temperate forests are characterized by an impoverished avifauna (Vuilleumier, 1985), which is reflected in the low bird species richness as compared to other forests (Willson et al., this volume). This low species richness can be atrributed to avifaunal extinctions during glacial times and to a lack of postglacial recolonization due to the extreme isolation of southern temperate forests from other forested rigions in South America.

Only one species of hummingbird is found in the forest of Chiloé, which feeds on at least 14 species of plants. This is one of the largest assemblages of plant species dependent on a single hummingbird species that we have found in a survey of guilds of hummingbird-pollinated plants in various forest ecosystems, as shown in Table 12.4. Furthermore, unlike in many tropical systems, there is no "interchangeable alternative" hummingbird for *Sephanoides* were its population to decline. This emphasizes the importance of *Sephanoides* for the guild of plants with red, tubular flowers. Recently, we have identified other nectarivorus birds, all passerines, feeding on flowers of *Embothrium coccineum* (Proteaceae). Although these species may have some role in pollination (Smith-Ramírez & Armesto, unpublished manuscript), this does not decrease the relevance of *Sephanoides* as the main pollinator in this forest.

Table 12.4. Assemblages (Guilds) of Hummingbird-Pollinated Plants in Various Forest Ecosystems

No. of Plant Species	No. of Hummingbird Species	Plants/ Hummingbird Ratio	Forest Type	Reference
19	1[1]	19	Cloud forest, Costa Rica	Murray et al. (1987)
17	1[2]	17	Cloud forest, Costa Rica	Murray et al. (1987)
14	1	14	Temperate rainforest, Chile	Smith-Ramírez (1993)
13	3	4.3	Dry shrubland and forest Puerto Rico	Kodric-Brown, Brown, Byers, and Gori (1984).
23	6	3.8	Deciduoustropical forest Jalisco, Mexico	Arizmendi and Ornelas (1990)
9	4	2.5	Chaparral and conifer forest Arizona, USA	Brown and Kodric-Brown (1979)
13	16	0.8	Tropical forest, Trinidad and Tobago	Feinsinger, Wolfe, and Swarm (1982)
10	14	0.7	Secondary forest, Costa Rica	Feinsinger (1978)
5	9	0.6	Tropical forest, Trinidad Island	Snow and Snow (1972)

[1] One long-billed hummingbird accounts for 97% of the visits.
[2] One short-billed hummingbird accounts for 68.2% of the visits.

Nectarivory seems to be infrequent in the avifaunas of North American temperate forests as well (i.e., 3 or fewer species of hummingbirds are reported for temperate areas in North America; Jaksic & Feinsinger, 1991); however, eastern deciduous or conifer-dominated northwestern forests of North America seem to be very poor in hummingbird-pollinated flowers as compared to southern Chile. Since we found no studies that evaluated the importance of hummingbirds in the pollination of temperate forest plants in North America, we utilized flower descriptions in published local floras. In a survey of the flora of Mount Rainier National Park, Washington (Franklin et al., 1988), only one species had ornithophilous flowers (0.6% of the flora). In the flora of Glacier National Park, Montana (Lesica, 1985), eight species had ornithophilous flowers (0.8% of the flora). In fact, the percentage of the flora pollinated by hummingbirds in Chiloé may be quite high even when compared to some tropical forests. Stiles (1985) reports that 2.4% of the woody flora (15 species) are pollinated by hummingbirds in Guanacaste, Costa Rica; 3.1% (29 species) in Barro Colorado, Panama; and 2.9% (39 species) in La Selva, Costa Rica.

Similarly, a survey of avian frugivore assemblages and their food plants in various forest ecosystems of the Northern and Southern Hemispheres (see Table 12.5) illustrates that the fleshy-fruited plants from Chiloé depend on a species-poor guild of frugivores. Among these species, *Elaenia* appears to be critical for the dispersal of summer fruiting species, while *Turdus* and *Scelorchilus* may contribute to the dispersal of winter fruiting species.

In conclusion, geographical comparisons reveal that an important proportion of the flora of Chilean temerate forests depend on nectar and fruit-eating birds, yet these rich plant guilds must rely on few species of avian frugivores or nectarivores.

Table 12.5. Comparison of Avian Frugivorous Guilds in Several Temperate Forest Ecosystems

	No. of Species Avian Frugivores	No. of Plant Species with Fleshy Fruits	Reference
Florida, 29° N	22	41	Skeate (1987)
Kansas, 39° N	30	16	Stapanian (1982)
Illinois, 40° N	11	8	Malmborg and Willson (1988)
Spain, 37°59′ N	5–14	7–11	Herrera (1984)
France, 43°40′ N	20	65	Debussche and Isenmann (1989)
Australia, 37° S	18	23	French (1991)
Chile, 42°30′ S	4	23	Sabag (1993)

Susceptibility to Disturbance

Although its geographical range extends into the subtropical woodland of Paraguay and northern Argentina, it is likely that the population numbers of the migrant frugivore *Elaenia albiceps* are regulated in the temperate forests of southern South America, where this species breeds (Vuilleumier, 1985). Summer populations in Chiloé consist largely of juveniles (Sabag, 1993). Recent expansion of the timber and wood-chip industry in Chile has resulted in a rapid decrease in the area of natural forest and an increase in the area planted with exotic species, particularly *Pinus radiata* and *Eucalyptus* spp. (INFOR, 1991). Forest fragmentation and reduction in area have a negative impact on the species richness and population density of breeding birds in North America (Bolger, Alberts, & Soulé, 1991; Lynch, 1987). A similar decline is occurring in the temperate forest of southern Chile (Willson et al., 1994). Long-term censuses of bird populations in forest fragments are necessary to assess these trends.

We predict that a negative-feedback dynamics driven by human disturbance to forests may result in a serious drop in bird abundances. Habitat destruction will lead to a rapid decline in bird populations due to the diminishing food resources, which in turn may cause further reduction in plant populations. The reductions in bird populations are likely to affect entire guilds of plants, eventually leading to a loss of those species that fail to reproduce or disperse their seed to suitable sites for establishment.

Fruit-eating birds, particularly *Elaenia albiceps*, are likely to be rare or absent from areas planted with exotic conifers, which are entirely devoid of fleshy fruits. In these areas, which currently occupy more than 1.5 million hectares between 35° S and 39° S (Armesto, Smith-Ramirez, León, & Arroyo, 1992), the recovery of native forest will be severely impaired in the future by low seed inputs to the soil seed bank.

Open areas covered by grassland and lacking shrubs or trees are also barren territory for frugivores. Guevara, Purata, and Maarel (1986) have shown that isolated fruiting trees left standing in open areas may help initiate the recolonization process, because they attract birds carrying seeds from surrounding forests. We suggest that leaving some fleshy-fruited trees standing should be a policy in clearing processes where eventual restoration is desired. Also patches of intact forest must remain in the regional landscape in order to provide suitable habitat for frugivorous birds and sources of seed.

On the other hand, plantations of introduced *Eucalyptus* species, abundant in south-central Chile, attract large numbers of hummingbirds during their flowering season (Smith-Ramírez personal observation). Although this phenomenon may compensate for the lack of suitable native habitat in some areas, these introduced trees may attract the birds away from their native food plants, negatively affecting the seed crop of the latter. This hypothesis can be tested by examining the reproductive success of hummingbird-pollinated species whose flowering period overlaps with that of *Eucalyptus*. Since many hummingbird-pollinated plants are canopy dwellers, it is possible that a decline of these populations cannot be readily detected. Studies of the dynamics of these canopy species could be very useful in detecting future trends.

It is apparent that future expansions of the grasslands at the expense of the native forest area or plans for large-scale substitution of native forests by exotic tree plantations should have negative effects on the species that integrate the mutualistic guilds described here. Because of its isolation from other forested areas and its species-poor avifauna, the temperate rainforest of the Pacific coast of South America must be perceived as an islandlike ecosystem, where species losses are unlikely to be compensated by immigration from contiguous forest habitats or by increases in the populations of other ecologically similar species.

Acknowledgments. Data collection in Chiloé has been funded by Fondecyt, Grants 860-88 and 1135-92, and by the Universidad de Chile. The paper was written under the sponsorship of Grant 1135-92 from Fondecyt, Chile. Peter Feinsinger, Douglas J. Levey, Steward T.A. Pickett, and Mary F. Willson provided critical and helpful reviews of earlier versions.

References

Alaback, P.B. 1991. Comparative ecology of temperate rainforests of the Americas along analogous climatic gradients. *Rev Chil Hist Nat*, 64, 399–412.

Arizmendi, M. del C., and Ornelas, J.F. 1990. Hummingbirds and their floral resources in a tropical dry forest in Mexico. *Biotropica*, 22, 172–180.

Armesto, J.J., and Figueroa, J. 1987. Stand structure and dynamics in the temperate rainforests of Chiloé archipelago, Chile. *J Biogeog*, 14, 367–376.

Armesto, J.J., and Fuentes, E.R. 1988. Tree species regeneration in a mid-elevation, temperate rain forest in Isla de Chiloé, Chile. *Vegetatio*, 74, 55–78.

Armesto, J.J., and Rozzi, R. 1989. Seed dispersal syndromes in the rain forest of Chiloé: Evidence for the importance of biotic dispersal in a temperate rain forest. *J Biogeog*, 16, 219–226.

Armesto, J.J., Rozzi, R., Miranda, P., and Sabag, C. 1987. Plant/frugivore interactions in South American temperate forest. *Rev Chil Hist Nat*, 60, 321–336.

Armesto, J.J., Smith-Ramírez, C., León P., and Arroyo, M.K. 1992. Biodiversidad y conservación del bosque templado en Chile. *Amb Des* (Chile), 8, 19–24.

Arroyo, M.T.K., Armesto, J.J., Squeo, F., and Gutiérrez, J. 1993. Global change: Flora and vegetation of Chile. In H.A. Mooney, E.R. Fuentes, and B.I. Kronberg (eds.), *Earth System Responses to Global Change: Contrasts Between North and South America* (pp. 239–263). San Diego: Academic Press.

Bolger, D.T., Alberts, A.C., and Soulé, M.E. 1991. Occurrence patterns of birds species in habitat fragments, sampling, extinction, and nested species subsets. *Am Nat*, 137, 155–166.

Brown, J.H., and Kodric-Bown, A. 1979. Convergence, competition and mimicry in a temperate community of hummingbird-pollinated flowers. *Ecology*, 60, 1022–1035.
Chapin III, F.S., Schulze, E.D., and Mooney, H.A. 1992. Biodiversity and ecosystem processes. *Tr Ecol Evol*, 7, 107–108.
Correa, A., Armesto, J.J., Schlatter, R.P., Rozzi, R., and Torres-Mura, J.C. 1990. La dieta del Chucao (*Scelorchilus rubecula*), un Passeriforme terricola del bosque de austro Sudamérica. *Rev Chil Hist Nat*, 63, 197–202.
Debussche, M., and Isenmann, P. 1989. Fleshy fruit characters and the choice of bird and mammal seed dispersers in a Mediterranean region. *Oikos*, 56, 327–338.
Feinsinger, P. 1978. Ecological interactions between plants and hummingbirds in a successional tropical community. *Ecol Monog*, 48, 269–287.
Feinsinger, P. 1983. Coevolution and Pollination. In D.J. Futuyma and M. Slatkin (eds.), *Coevolution* (pp. 282–310). Sunderland MA: Sinauer Associates.
Feinsinger, P. 1987. Approaches to nectarivore-plant interactions in the New World. *Rev Chil Hist Nat*, 60, 285–319.
Feinsinger, P., Beach, J.H., Linhart, Y.B., Busby, W.H., and Murray, K.G. 1987. Disturbance, pollinator predictibility, and pollination success among Costa Rican could forest plants. *Ecology*, 68, 1294–1305.
Feinsinger, P., Wolfe, J.A., and Swarm, L.A. 1982. Island ecology: Reduced hummingbird diversity and the pollination biology of plants, Trinidad and Tobago, West Indies. *Ecology*, 63, 494–506.
Franklin, J.F., Moir, W.H., Hemstrom, M.A., Miles, A., Greene, S.E., and Smith, B.G. 1988. *Forest Communities of Mount Rainier National Park*. Atlanta: USDA, National Park Service.
French, K. 1991. Characteristics and abundances of vertebrate-dispersed fruit in temperate wet sclerophyll forests in southeastern Australia. *Aus J Ecol*, 16, 1–13.
García, J.R. 1982. *Comunidad avifaunística del delta rio Gol-Gol una necesidad de conservación*. Tesis, Instituto de Zoología, Universidad Austral de Chile, Valdivia.
Gentry, A. 1982. Patterns of neotropical plant species diversity. *Evol Biol*, 15, 1–84.
Gilbert, L.E. 1980. Food web organization and conservation of neotropical diversity. In: M.E. Soulé and B.A. Wilcox (eds.), *Conservation Biology* (pp. 11–33). Sunderland, Ma: Sinauer Associates.
Guevara, S., Purata, S.E., Maarel, E.V. 1986. The role of remnant forest trees in tropical secondary succession. *Vegetatio*, 66, 77–84.
Herrera, C.M. 1982. Seasonal variation in the quality of fruits and diffuse coevolution between plants and avian dispersers. *Ecology*, 63, 773–785.
Herrera, C.M. 1984. A study of avian frugivores, bird-dispersed plants and their interaction in mediterranean scrublands. *Ecol Monog*, 54, 1–23.
Holdgate, M.W. 1961. Vegetation and soils in the south Chilean island. *J Ecol*, 49, 359–380.
Howe, H.F. 1984. Implications of seed dispersal by animals for tropical reserve management *Biol Cons*, 30, 261–281.
Hubbell, S.P., and Foster, R. 1986. Biological, chance and history and the structure of tropical rain forest tree comunities. In J. Diamond and T.J. Case (eds.), *Community Ecology* (pp. 314–330). New York: Harper & Row.
INFOR. 1991. Estadísticas forestales 1990. *Boletín Estadístico No. 21*, Instituto Forestal, Santiago.

Jaksic, F.M., Feinsinger, P. 1991. Bird assemblages in temperate forests of North and South America: A comparison of diversity, dynamics, guild structure, and resource use. *Rev Chil Hist Nat*, 64, 491–510.

Jordano, P. 1987. Patterns of mutualistic interactions in pollination and seed dispersal: Connectance, dependence, asymmetries, and coevolution. *Am Nat*, 129, 657–677.

Karr, J.R. 1981. Surveying birds with mist nets. *St Av Biol*, 6, 62–67.

Kodric-Brown, A., Brown, J.H., Byers, G.S., and Gori, D.F. 1984. Organization of a tropical island community of hummingbirds and flowers. *Ecology*, 65, 1358–1368.

Lesica, P. 1985. Checklist of the vascular plants of Glacier National Park, Montana, U.S.A. *Monograph No. 4, Montana Academy of Sciences.*

Lynch, J.F. 1987. Responses of breeding bird communities to forest fragmentation. In: D.A. Saunders, G.W. Arnold, A.A. Burbidge, and A.J.M. Hopkins (eds.), *Nature Conservation: The Role of Remnants of Native Vegetation* (pp. 123–140). London: Surrey Beatty and Sons, in association with CSIRO and CALM.

Malmborg, P.K., and Willson, M.F. 1988. Foraging ecology of avian frugivoures and some consequences for seed dispersal in an Illinois woodlot. *Condor*, 90, 173–186.

Medel, R.G., Jiménez, J.E., Jaksic, F.M., Yañez, J.L., and Armesto, J.J. 1990. Discovery of a continental population of the rare Darwin's fox, *Dusicyon fulvipes* (Martin, 1837) in Chile. *Biol Cons*, 51, 71–77.

Murray, K.G., Feinsinger, P., Busby, W.H., Linhart, Y.W., Beach, J.H., and Kinsman, S. 1987. Evaluation of character displacement among plants in two tropical pollination guilds. *Ecology*, 68, 1283–1293.

Riveros, M. 1991. *Aspectos sobre la biología reproductiva en dos comunidades del sur de Chile, 40° S.* Tesis de doctor, Facultad de Ciencias, Universidad de Chile, Santiago.

Sabag, C. 1993. *El rol de las aves en la dispersión de semillas en un bosque templado secundario de Chiloé (42° S).* Tesis de magister, Facultad de Ciencias, Universidad de Chile, Santiago.

Skeate, S.T. 1987. Interactions between birds and fruit in a northern Florida hammock community. *Ecology*, 68, 297–309.

Smith-Ramírez, C. 1993. Los picaflores y su recurso floral en el bosque templado de la isla de Chiloé, Chile. *Rev Chil Hist Nat*, 66, 65–73.

Smith-Ramírez, C., and Armesto, J.J. 1994. Flowering and fruiting patterns in the rain forest of Chiloé, Chile: ecological and climatic constraints. *J Ecol*, 82, 353–365.

Snow, D.W. 1981. Tropical frugivorous birds and their food plants: A world survey. *Biotropica*, 13, 1–14.

Snow, B.K., and Snow, D.W. 1972. Feeding niches of hummingbirds in a Trinidad valley. *J An Ecol*, 41, 471–485.

Stapanian, M.A. 1982. Evolution of fruiting strategies among fleshy-fruited plant species of eastern Kansas. *Ecology*, 63, 1422–1431.

Stiles, F.G. 1985. Conservation of forest birds in Costa Rica: Problems and perspectives. In A.W. Diamond and T.E. Lovejoy (eds.), *Conservation of Tropical Forest Birds* (pp. 141–168). Cambridge England: ICBP (International Couincil for Bird Preservation) Technical Publication No. 4.

Veblen, T.T. 1985. Forest development in tree-fall gaps in the temperate rain forest of Chiloé. *Nat Geog Res*, 1, 162–183.

Veblen, T.T., Schlegel, F.M., and Oltremari, J.V. 1983. Temperate broad-leaved evergreen forests of South America. In: J.D. Ovington (ed.) *Ecosystems of the World.* Vol. 10: *Temperate Broad-Leaved Evergreen Forests* (pp. 5–32). Amsterdam: Elsevier.

Villagrán, C. 1985. Análisis palinológico de los cambios vegetacionales durante el Tardiglacial y Postglacîal en Chiloé, Chile. *Rev Chil His Nat*, 58, 57–69.
Vuilleumier, F. 1985. Forest birds of Patagonia: ecological, geography, speciation, endemism, and faunal history. *Ornithol Monog*, 36, 255–302.
Willson, M.F. 1986. Avian frugivory and seed dispersal in easthern North American. *Cur Ornithol*, 3, 223–279.
Willson, M.F. 1991. Dispersal of seeds by frugivorous animals in temperate forests. *Rev Chil His Nat*, 64, 537–554.
Willson, M.F., De Santo, T., Sabag, C., and Armesto, J.J. 1994. Avian communities of fragmented south-temperate rainforests in Chile. *Conser Biol*, 8, 508–520.

13. The Temperate Rainforest Lakes of Chile and Canada: Comparative Ecology and Sensitivity to Anthropocentric Change

Doris Soto and John G. Stockner

Temperate rainforests are among the most environmentally sensitive and biologically diverse ecosystems in the world. They are currently under heavy exploitive pressures and the subject of international controversy. Along with their roles as providers of unique and diverse species habitats and as major players in global carbon cycling, they also serve as collection basins for many large lakes of ecological and economic importance. The catchments of Chilean Araucanian lakes (39° to 42° S) and south coastal British Columbian (BC) lakes (49° to 51° N) (Figure 13.1) are in rainforest biogeoclimatic zones and share common properties, for example, climate, age and glacial origin, morphometry, and the oligotrophic state (i.e., $<10\,\mathrm{ug \cdot L^{-1}}$ average total phosphorus (TP) content) (Vollenweider, 1968; Campos, 1984; Soto & Zuñiga, 1991; Stockner, 1981, 1987).

This paper has three objectives: (1) to provide a brief description of the major temperate rainforest lake systems; (2) to compare and contrast Chilean lakes with BC lakes of similar latitude and size; and (3) to compare their sensitivity and potential for change due to anthropogenic activities, both short and longer term. It is possible that limnological factors (e.g., temperature, nutrient ratios, food web structure) may serve as sensitive barometers of future change induced by human activities on both regional (catchments) and global (climate) scales.

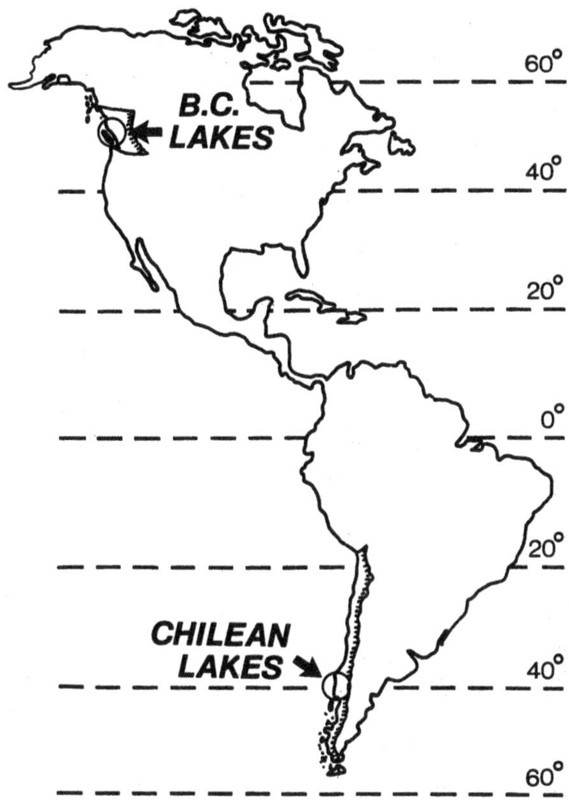

Figure 13.1. Location of study lakes in Chile (Araucanian region) and British Columbia, Canada (south coastal).

Data Sources and Methods

This comparative study focuses on 12 Araucanian lakes (hereafter called "Chilean" lakes) and on 8 south coastal BC lakes (hereafter called "BC" lakes) (Great Central, Henderson, Hobiton, Kennedy, Long, Nimpkish, Sproat, and Woss). The review of Chilean lakes is largely based on previous publications (Campos, 1984; Campos, Arenas, Jara, Gonser, & Prins, 1984; Campos, Steffen, Aguero, Parra, Zuñiga, 1988; Campos, Steffen, Aguero, Parra, Zuñiga, 1990; Montecino, 1991; Soto & Zuñiga, 1991; Soto, Campos, Steffen, Parra, & Zuñiga, 1993); and unpublished data from Llanquihue and Yelcho lakes (D. Soto). Data from BC lakes are from Stockner (1981, 1987), Stockner and Costella (1980), Stockner and Shortreed (1985), Shortreed and Stockner (1986, 1990). Some BC lakes were experimentally fertilized for enhancement of sockeye salmon (*Oncorhynchus nerka*) production (Stockner, 1987), but data used in this study are from "untreated" stations and/or lakes.

Samples were analyzed using standard limnological methods that are described in greater detail elsewhere (Soto et al., 1993; Stockner & Shortreed, 1985). Although there were some minor differences in laboratory protocol in chemical and biological analyses reported, they were not of sufficient importance to account for any of the major differences discussed here. Water samples were taken with a Van-Dorn bottle at discrete depths within the epilimnion and hypolimnion. Nutrient samples were analyzed within 24 h of collection and were kept cold prior to analysis. Total nitrogen (TN) was obtained by summing values of all inorganic and organic nitrogen analyses.

Chlorophyll *a* concentrations were obtained by extracting filtered samples in 90% acetone and measuring absorbance with a Shimadzu UV-150-02 spectrophotometer (Chilean lakes) or fluorescence with a Turner Model 111 fluorometer (BC lakes) (Strickland & Parsons, 1968). Primary productivity was estimated using the ^{14}C sodium bicarbonate technique and in situ light and dark bottle incubations over midday at 5 to 10 discrete depths within the euphotic zone (Strickland, 1960). Production values were integrated over the euphotic zone and over the year based on measurements from a 12 to 14 month sampling period in Chilean lakes and from a 7 to 9 month sampling period in BC lakes.

Zooplankton from Chilean lakes were collected with a Ruttner net (100-um mesh) using 0 to 100 m vertical tows and from BC lakes by a SCOR-type net (100-um mesh) and 0 to 25 m or 0 to 50 m vertical tows (Rankin, Ashton, & Kennedy, 1984). Zooplankton biomass from Chilean lakes was estimated using length weight regressions of Dumont, Van de Velde, Dumont (1975) and Culver, Bouchele, Bean, and Fletcher (1985). Plankton samples from BC lakes were split and one half dried at 90°C for 24 h and weighted to determine zooplankton dry weights and biomass.

Results and Discussion

Climate, Hydrology

Chile is a long (17° to 56° S) and slender country lying between the Andes Range to the east and the south Pacific Ocean to the west. Most lakes lie within three regions: Araucanian, Patagonian, and Torres del Paine (38° to 52° S), whose major hydrographic basins run from east to west. The Province of British Columbia (48° to 60° N), Canada, lies along the north Pacific Ocean and extends eastward for >1000 km, a region of variable climatic and biogeographic diversity (Farley, 1979). The greatest number of lakes (but not the largest) lie in coastal regions, but there are numerous large, deep lakes on the vast interior regions east of the coast mountains (Stockner, 1987, Figure 1).

The Chilean Araucanian study lakes include several large and deep lakes that have large catchments and one major outlet flowing to the

Pacific Ocean (Thomasson, 1963). Annual precipitation averages from 1500 to 3000 mm, and most of the rivers and streams show marked seasonal hygrographs with high flows in winter and low flows in summer. A similar pattern occurs in BC coastal lakes and rivers which experience annual precipitation >2000 mm with most occurring during the wet, winter period (November to March). The BC lakes are generally fast-flushing, <3.5 y (Stockner, 1981), while the Chilean lakes show a longer water residence time, >10 y (Table 13.1).

Catchment Basin

The soils on the Chilean lake catchments are predominately volcanic and the original Valdivian rainforests are a diverse assemblage of nonconiferous evergreens represented by various combinations of *Nothofagus dombeyi*, *N. nitida*, *Aextoxicon*, *Laurelia*, *Drimys*, *Weinmannia*, *Podocarpus*, *Fitzroya*, and so on (Alaback, 1991). Human disturbance has altered from 5 to 60% of this original rainforest vegetation during the last 200 years (Campos, 1984; Campos et al., 1988). The catchments now

Table 13.1. Some salient limnological variables from Chilean (Araucanian) and Canadian (British Columbia south coastal) rainforest lakes. (Values are annual means)

	Chilean (n = 12)	British Columbian (n = 8)
Area (km^2)	209.3	17.9
Mean depth (m)	139.4	66.2
Volume (km^3)	31.5	3.2
Residence (yr)	10.4	3.0
Secchi (m)	13.2	7.8
Compensation Depth (m)	36.5	10.9
Temp. summer (max. °C)	22.5	15.9
winter (max. °C)	10.0	5.4
Growing Season (mo)	8	7
Total P ($ug \cdot L^{-1}$)	4.8	2.6
Nitrate ($ug \cdot L^{-1}$)	18.2	14.2
Tot. N ($ug \cdot L^{-1}$)	78.6	220.0
Na ($mg \cdot L^{-1}$)	3.2	1.2
Ca ($mg \cdot L^{-1}$)	3.6	4.4
Mg ($mg \cdot L^{-1}$)	1.3	0.8
TDS ($mg \cdot L^{-1}$)	63.0	32.9[1]
pH	7.4	6.9
CHL a ($ug \cdot L^{-1}$)	0.7	1.3
Prim. prod. ($gC*m^{-2}*yr^{-1}$)	89.4	32.4
Zooplankton biomass ($ug \cdot L^{-1}$)	18.8	10.2
Bacteria range (no. $mL^{-1} \times 10^6$)	0.08–1.3	0.06–1.2
Auto. picoplankton range (no. $mL^{-1} \times 10^3$)	11–132	9–60

[1] Untreated lakes (n = 5) (G. Central, Hobiton, Kennedy (2 yrs), Long lakes)

show varying degrees of agricultural and recreational development and human settlement.

The catchment basins of BC lakes lie within the large western hemlock (*Tsuga heterophylla*) biogeoclimatic zone (Farley, 1979) that experiences a moist, cool Mediterranean climate with wet winters and dry summers. Soils can be characterized as shallow, coarse, textured, and highly organic with pockets of glacial till. These soils tend to be strongly adsorptive of phosphorus ions but less so with nitrogen (Feller, 1977). The predominant forest vegetation is western hemlock, Douglas fir (*Pseudotsuga menziesii*), western red cedar (*Thuja plicata*), and Sitka spruce (*Picea sitchensis*). There is human settlement around only a few lakes (Sproat, Great Central lakes), but the most significant disturbance has been the removal of much of the old growth rainforest from most lake basins since the turn of the century (e.g., Great Central, Nimpkish, Woss, Sproat, Henderson lakes). Because of the steep, rocky terrain and the nature of the catchment soils, little land is suitable for agricultural development and the limited accessibility (roads) has prevented significant recreational or urban development.

Physicochemical

The Chilean lakes are warm-monomictic with stable stratification occurring from October-April. In some lakes the epilimnion can attain depths >40 m (Campos et al., 1988). In winter during periods of deep mixing water temperatures rarely fall below 10°C. BC lakes are also warm-monomictic with stable stratification from April to mid-October, but the average depth of the epilimnion seldom is >20 m (Stockner & Costella, 1980). During deep mixing periods in winter, water temperatures range from 4° to 6°C.

Average total phosphorus (TP) concentrations are higher in Chilean lakes than BC lakes—3.8 and 2.4 $ug \cdot L^{-1}$, respectively. Total nitrogen (TN) values in Chilean lakes are about threefold lower than in BC lakes and average nitrate concentrations in both systems are very low, <20 $ug \cdot L^{-1}$. The major ion composition of Chilean lakes suggests a greater marine influence than BC lakes with a predominance of sodium, chloride, and magnesium ions, and a slightly alkaline average pH value (Table 13.1). The waters of BC lakes are soft and of low alkalinity, with an average total dissolved solids (TDS) value (32.9 $mg \cdot L^{-1}$) that is considerably lower than the average Chilean lake value (63 $mg \cdot L^{-1}$).

Biological

Picoplankton

Picoplankton (size range 0.2 to 2.0 um) have not been well studied in Chilean lakes and values reported here come only from Lake Riñihue (S. Wolf, personal communication), one of the most productive of the

Araucanian lakes. More information on bacterioplankton and autotrophic picoplankton densities in Chilean lakes is needed before meaningful comparisons can be made. Nonetheless, it is of interest to note that bacteria and autotrophic picoplankton (APP) show similar density ranges in both lake systems although APP, as is expected, reach greater densities in the more productive Chilean lakes (Table 13.1).

Phytoplankton

The Chilean lakes are dominated by the diatoms *Aulacoseira* (formerly *Melosira*) and *Rhizosolenia* and the chrysophyceaen *Dinobryon* spp. during much of the year (Campos et al., 1988), with a predominant spring peak in microphytoplankton biomass and production (Campos, 1984; Campos et al., 1988). *Cyclotella*, *Rhizosolenia*, *Aulacoseira*, and *Fragilaria* are the most common diatoms in BC lakes, but their abundance is very low compared to APP, and chrysophyte, and chryptomonad nanoflagellates (Stockner, 1987). Large spring peaks of microphytoplankton biomass and/or production do not commonly occur in coastal lakes, but at fertilized stations within lakes sometimes they do take place (Stockner & Costella, 1980, Stockner, 1987).

Zooplankton

Zooplankton communities of Chilean lakes are dominated by Boeckellid calanoid copepods (*Boeckella gracilipes*, *B. michaelseni*, and *B. meteoris*). Also present, but less common, are the small cladoceran *Neobosmina chilensis*, some cyclopoid copepods and rotifers (Soto and Zuñiga, 1991). Large copepods or cladocerans are rare, notably daphnids that are common in most lakes of the northern hemisphere. Most of the Boeckellids are herbivorous and regulate the abundance of certain large flagellates, for example *Dinobryon*, *Peridinium*.

The small bodied cladocerans *Bosmina* and *Eubosmina*, cyclopoid copepods (*Diacyclops*), and rotifers are the major zooplankton groups in coastal BC lakes (Rankin & Ashton, 1980). Because of low food and high fish-grazing pressure, larger daphnids or calanoid copepods are seldom abundant in these lakes.

Fishes

Chilean lakes have a predominance of planktonic and benthic feeding fishes (*Galaxias*, *Basilichthys*, *Aplochiton*, *Trichomycterus*) and a low abundance of piscivorous fish (Figure 13.2). Percas (*Percichthys trucha*) were common until the introduction of rainbow trout (*Oncorhynchus mykiss*) in the early 1900s. Both trout and percas in Chilean lakes have alternative benthic prey available in the littoral zone, for example, insects, snails, freshwater crabs, crayfish, which reduces their predation on pelagic zooplanktivores (Soto & Zuñiga, 1991).

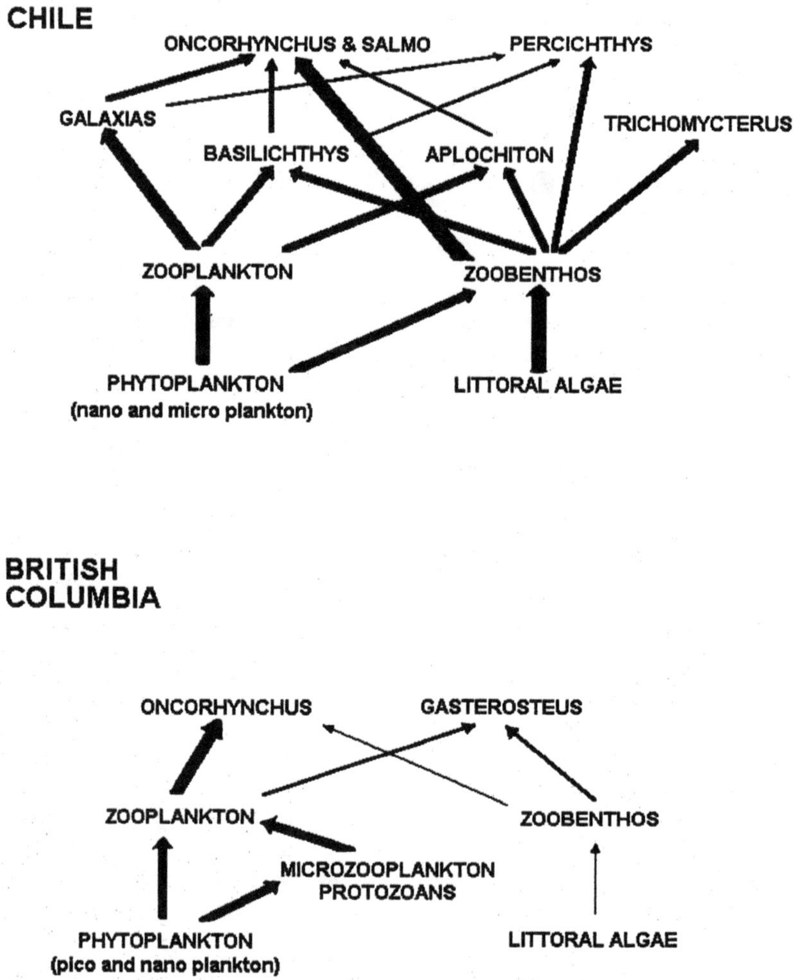

Figure 13.2. Schematic of food web structure of Chilean Araucanian and BC coastal lakes. Thickness of arrows denotes the relative strength of interactions.

The major zooplanktivore in BC lakes is juvenile sockeye salmon (*Oncorhynchus nerka*), but in a some lakes (e.g., Long, Great Central lakes) pelagic sticklebacks (*Gasterosteus aculeatus*) can attain comparable densities and compete with sockeye for zooplankton prey (O'Neill & Hyatt, 1987) (Figure 13.2). The abundance of juvenile salmon is highly variable (500 to 5000 $\text{fish}\cdot\text{ha}^{-1}$) and dependent on annual adult escapement levels (Hyatt & Stockner, 1985). The major piscivores are cutthroat trout (*Salmo clarkii*), dolly varden (*Salvelinus malma*), and rainbow trout (*O. mykiss*), but their abundance is generally very low (Simpson, Wo, & Miki, 1981).

Lake System Comparisons

There are many similarities between Chilean and BC lakes. The lakes and their catchments are of similar glacial origin and age, and they lie within temperate rainforest biogeoclimatic zones. They are strongly influenced by a cool, moist, maritime climate, and both experience high annual precipitation during winter months. BC lakes are considerably cooler than Chilean lakes, but both have similar ranges between their winter minimum and summer maximum temperatures (12°C) and both have 7 to 8 month growing seasons (periods of stratification). Chilean lakes are clearer, with a higher dissolved ion content, but a lower dissolved organic content, than BC lakes. This suggests major differences in catchment soil characteristics and nutrient (ion) retention capacities (Feller, 1977). Apart from sharing some similar phytoplankton genera, all endemic zooplankton and fishes in Chilean lakes belong to genera and/or families from other southern hemisphere continents (e.g., Australia, New Zealand) (Campos, 1977; Pezzani, 1977). The pelagic food webs of both lake systems are configured by size selective planktivores that reduce the average size of macrozooplankton to between 0.6 to 1.5 mm (Soto & Zuñiga, 1991; O'Neill & Hyatt, 1987). Although both systems have low ambient nutrient levels, Chilean lakes have higher TP levels and lower N levels than BC lakes. One of the more important differences between systems is that Chilean lakes have higher rates (>2X) of annual and daily primary production with lower phytoplankton biomass (CHL) levels. Using annual production (aP) and average CHL (B) values yields an aP/B ratio that is almost 5-fold higher biomass turnover rate in Chilean lakes. Both warmer winter minimum and summer maximum lake temperatures would partially explain the faster turnover of organic carbon and higher metabolic rates seen in Chilean lakes.

Resilience

Most temperate lakes show a significant CHL response to large increases in nutrient income (TP, TN), due largely to increases in the abundance of large microphytoplankton, such as diatoms, dinoflagellates, and colonial cyanobacteria (Dillon & Rigler, 1974; Smith, 1982; Vollenweider, 1968). This has been the case in BC lakes, where significant CHL peaks occurred when some lakes were fertilized with N and P to increase salmon production (Stockner, 1981, 1987; Hardy, Shortreed, & Stockner, 1986). However in some Chilean lakes (e.g., Lake Llanquihue), increasing TP inputs over the past decade have not appreciably increased average CHL content (D. Soto, unpublished data) (Figure 13.3). What are some of the factors that may account for this apparent resilience by Chilean lakes to increasing TP income?

1. Nitrogen limitation. The Chilean lakes are more consistently N-limited and apparently respond only to N inputs rather than P (Soto et

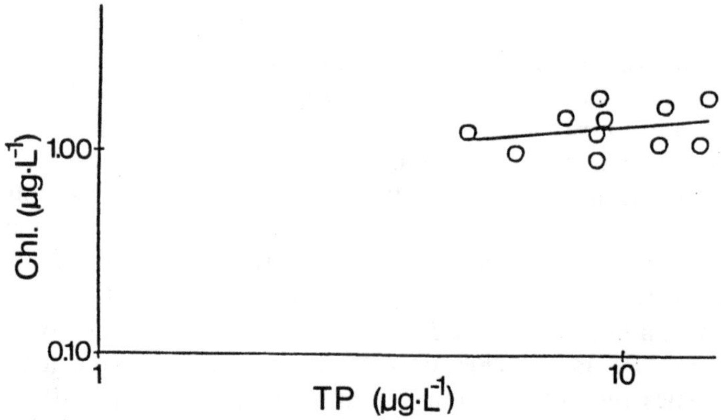

Figure 13.3. Annual average total phosphorus (TP) vs chlorophyll (CHL) relationship in several sites with different TP loadings in Lake Llanquihue, Chile.

al., 1993). Such severe N-limitation would slow the onset of eutrophication in these lakes. BC lakes commonly show periods of colimitation when both P and N levels are so low as to be growth limiting (Suttle, Cochlan, & Stockner, 1991). The TP : CHL relationship for N-limited Chilean lakes has a lower functional slope value (1.05) than colimited BC lakes (1.21) and a much lower value than P-limited north temperate lakes (Figure 13.4, Table 13.2). Nitrogen-fixing colonial cyanobacteria which can occur in more productive meso- or eutrophic N-limited lakes rarely occur in Chilean or BC lakes. This is because of very low (subthreshold) ambient concentrations of P that limit the growth of larger colonial cyanobacteria (Spencer & King, 1985; Stockner & Shortreed, 1988).

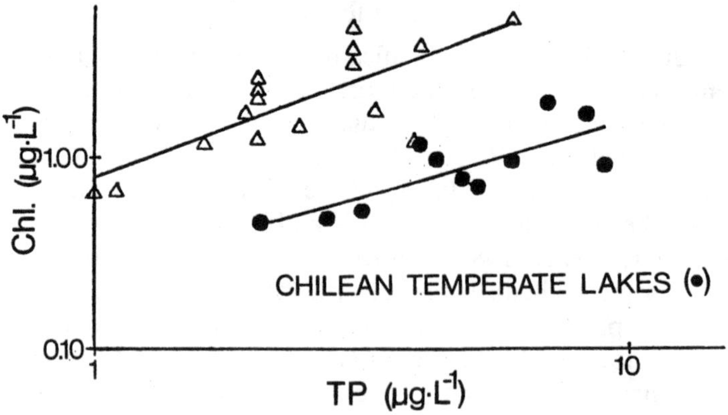

Figure 13.4. Annual average total phosphorus (TP) vs chlorophyll (CHL) relationships in Chilean and British Columbia[1] lakes. ([1]from Stockner & Shortreed, 1985).

Table 13.2. Relation between average chlorophyll (CHL) and average total phosphorus (TP) concentrations for Chilean lakes British Columbian and other north temperate lake systems[1]

Lakes	a	b	r^2	p	$^2b/r^2$
Lake Washington, USA[3]	−0.90	1.35	0.94	≪0.01	1.39
Alberta, Canada[4]	−1.19	1.62	0.73	<0.01	1.90
BC coastal[5]	−0.09	0.92	0.58	<0.01	1.21
Nova Scotia, Canada[6]	−0.58	0.89	0.56	<0.01	1.19
Yukon Territories, Canada[7]	−0.39	0.70	0.35	<0.01	1.18
Chilean Araucanian	−0.59	0.79	0.57	<0.01	1.05

[1] Log Chl = a + b * log TP
[2] Bivariate analysis
[3] Edmondson and Lehman, 1981
[4] Prepas and Trew, 1983
[5] Stockner and Shortreed, 1985
[6] Kerekes, 1975
[7] Shortreed and Stockner, 1986

2. Food web structure. There are some differences in the food web structure of these lakes (Figure 13.2). Because of a scarcity of available nutrients and very low primary production rates, the plankton communities of BC lakes throughout much of the growing season tend to be dominated by autotrophic picoplankton (APP), small nanoflagellates, ciliates, and microzooplankton (Stockner & Porter, 1988). Large cladocerans (e.g., *Daphnia*) are not common because of limitation by food (bottom-up) and predation by juvenile sockeye (top-down). Microphytoplankton abundance is also low and so is the rate of export (sedimentation) of primary production to benthic communities (Jackson, Stockner, & Harrison, 1990). In BC lakes, the majority of energy and material flows are through pelagic food webs and not littoral or benthic food chains (Stockner, 1987) (Figure 13.2). Microbial food webs also process a significant fraction (50 to 70%) of the primary production, and account for much of the energy flow and nutrient regeneration in the pelagic community (Stockner & Shortreed, 1989; Weisse & Stockner, 1994). Because microbial food webs tend to dissipate energy and have high respiratory losses, fish production in BC coastal lakes is very low (Stockner, 1987; Stockner & Shortreed, 1989).

As noted previously, Chilean lakes are considerably warmer and have higher rates of primary production and biomass turnover than BC lakes and can sustain a more pronounced spring microphytoplankton increases, for example, diatoms (Campos, 1984; Campos et al., 1988; D. Soto, unpublished data). As is typical for these more productive oligotrophic lakes, a significant fraction of spring phytoplankton production usually sinks from the euphotic zone and supports a more diverse and productive benthic and littoral communities (Scavia & Fahnenstiel, 1988; Soto &

Zuñiga, 1991). More productive oligotrophic and mesotrophic lakes have more complex food webs but shorter pathways that tend to be more efficient and provide more direct carbon flows to fish (Stockner, 1987) (Figure 13.2). Microbial food webs in these lakes are more ephemeral and tend to be less important in material and energy transfers (Weisse & Stockner, 1994). It is possible that the food web structure of Chilean lakes, which in contrast to BC lakes includes a littoral/benthic component, may partially explain why the biomass turnover rates (aP:B) are considerably faster and appear to support a greater abundance and variety of fishes.

Future Changes

Short Term

With increased disturbance of the catchments comes the possibility of higher nutrient loads and accelerated lake production (eutrophication) in both Chilean and BC lakes. Both agricultural development and rainforest harvesting with reforestation using monocultures of introduced conifers (exotics) are presently a serious problem for Chilean lakes. The endemic nonconiferous evergreen rainforests have extensive root systems and low leaf export that in combination with volcanic soils provide and efficient nutrient retention system on the catchment (Perez, Armesto, & Ruthsatz, 1991). The continued loss of the native forests, with their buffering effect, will increase the possibility of soil erosion and nutrient loss, especially nitrogen, with the likelihood of rapid lake eutrophication. Logging of the BC coastal rainforests has not had a major effect on the primary productivity of BC lakes or rivers. This is largely because the soil structure is highly adsorptive of phosphorus and little increase in TP has been seen following logging of the catchments (Feller, 1977).

The introduction of exotic Pacific coho salmon (*Oncorhynchus kisutch*) and Atlantic salmon (*Salmo salar*) to Chilean lakes could alter food web structure and affect lake metabolism. Over the past few years escaped salmon have become common in Chilean lakes and there is concern that their planktivory as juveniles and piscivory as adults could affect the abundance of percas, the only endemic piscivore in the lakes (Soto & Zuñiga, 1991). In some BC lakes, where net-pen rearing of Atlantic salmon fry is being done on an experimental basis, there is now concern about the possible impacts of the introduction of Atlantic salmon, another piscivore (exotic), on the fish communities and food-web structure.

Long Term

The lakes of Chile and British Columbia are both likely to be affected by global climate change. There are three aspects to consider: rising water temperatures (e.g., epilimnion), increased UVB, and increased winter precipitation. We expect that higher rainfall and elevated surface and groundwater temperatures will influence both lake systems. They will experience more rapid winter flushing, a longer period of stratification (growing season), and a higher seasonal epilimnetic temperature, factors that would tend to lower annual rates of primary production. Such conditions would also affect food-web structure, especially in BC lakes where a warmer epilimnion and sustained thermal barrier will restrict access of juvenile sockeye to their zooplankton forage. This could reduce freshwater production of juvenile sockeye and increase the abundance of more thermal tolerate species, e.g., rainbow and cutthroat trout. It has been proposed that global warming will reduce annual primary production in some large interior BC lakes because of lower nutrient inputs during the longer, drier and warmer growing season that is expected (Henderson, Stockner, & Levy, 1992).

In both regions, increasing air temperatures will have the most immediate and major impact on lakes receiving glacial input. More rapid ice melt will alter hydrologic, thermal, optical, and chemical properties of the glacial rivers that will increase turbidity and lower temperatures of glacial lakes in both regions.

Increased UVB will likely affect lake phytoplankton community structure and production most severely in Chilean lakes where compensation depths are 3 times higher than in BC lakes. However the magnitude of the impact on both species composition and primary production is still unclear and awaits further research in fresh water systems. Extrapolation of results from UVB work on Antarctic marine phytoplankton would suggest a small reduction (<15%) in daily production and some shifts in species composition from larger microphytoplankton (diatoms, dinoflagellates) toward smaller cyanobacterial or chlorophyte dominance (Smith, 1989). There is also evidence to suggest that zooplankton may be affected by UVB by restricting access to there epilimnetic forage base in oligotrophic lakes (Williamson, Zagarese, Schulze, Hargreaves, & Seva, 1994).

Acid rain and atmospheric nutrient deposition has already affected lakes of North America to a greater extent than in South America due to greater industrial development. But neither Chilean nor BC lakes have been seriously affected by these perturbations and should not be affected in future under the influence of global warming due to their geographic location along the Pacific Ocean. It appears that Chilean lakes and their catchment basins are far more sensitive to both floral and faunal species losses than BC lakes because of their low potential for species replacement (e.g., geographic isolation).

Acknowledgments. We thank H. Campos and C. Jara for providing unpublished information and for constructive discussions; F. Jara, E. MacIsaac, K. Shortreed, and T. Weisse for insightful comments and review of the manuscript. Earl MacIsaac and Bruce Nidle helped with figure preparation. We also thank the organizers of the second AMIGO meeting for providing the interdisciplinary environment which encouraged some of the ideas for this paper.

References

Alaback, P.B. 1991. Comparison of the temperate rain forests of the Americas. *Rev. Chil His Nat*, 64, 399–412.

Campos, H. 1977. Osteichtyes. In S. Hurlbert (ed.), *Biota Acutatica de Sud America Austral* (pp. 330–334). San Diego: San Diego State University Foundation.

Campos, H. 1984. Limnological study of Araucanian Lakes (Chile) *Verh Intern verein Limnol*, 22, 1319–1327.

Campos, H., Arenas, J., Jara, C., Gonser, T., and Prins, R. 1984. Macrozoobentos y fauna ictica de las aguas limnéticas de Chiloé y Aisén continentales. *Revista Medio Ambiente*, 52–64.

Campos, H., Steffen, W., Aguero, G., Parra, O., and Zuñiga, L. 1988. Limnological study of lake Llanquihue (Chile). Morphometry, physics, chemistry, plankton and primary productivity. *Archiv fur Hydrobiol*, Supplement, 81(1), 37–67.

Campos, H., Steffen, W., Aguero, G., Parra, O., and Zuñiga, L. 1990. Limnological study of lake Todos los Santos (Chile). Morphometry, physics, chemistry, plankton and primary productivity. *Archiv fur Hydrobiol*, 117, 453–484.

Culver, D.A., Bouchele, M.M., Bean, D.J., and Fletcher, J.W. 1985. Biomass of freshwater crustacean zooplankton from length-weight regressions. *Can J Fish Aquat Sci*, 42, 1380–1390.

Dillon, P., and Rigler, F. 1974. The phosphorus-chlorophyll relationship in lakes. *Limnol Oceanogr*, 19, 767–773.

Dumont, H.J., Van de Velde, I., and Dumont, S. 1975. The dry weight estimate of biomass in a selection of Cladocera, Copepods, and Rotifera from the plankton, periphyton, and benthos of continental waters. *Oecologia*, 19, 75–95.

Edmondson, W.T., and Lehman, J.T. 1981. The effect of changes in the nutrient income on the condition of Lake Washington. *Limnol Oceanogr*, 26, 1–29.

Farley, A.L. 1979. *Atlas of British Columbia*. Vancouver: University of British Columbia Press.

Feller, M.C. 1977. Nutrient movement through western hemlock/western red cedar ecosystems in SW British Columbia. *Ecology*, 58, 1269–1283.

Hardy, F.J., Shortreed, K.S., and Stockner, J.G. 1986. Bacterioplankton, phytoplankton, and zooplankton communities in a British Columbia coastal lake before and after nutrient reduction. *Can J Fish Aquat Sci*, 43, 1504–1514.

Henderson, M., Stockner, J.G., and Levy, D.A. 1992. Probable consequences of climate change on freshwater aspects of the production of Adams River sockeye salmon (*Oncorhynchus nerka*). *Geo Journal*, 28, 51–59.

Hyatt, K.D., and Stockner, J.G. 1985. Responses of sockeye salmon (*Oncorhynchus nerka*) to fertilization of British Columbia lakes. *Can J Fish Aquat Sci*, 42, 320–331.

Jackson, L.J., Stockner, J.G., and Harrison, P.J. 1990. Contribution of *Rhizosolenia erinensis* and *Cyclotella* spp. to the deep chlorophyll maximum in Sproat Lake, British Columbia, Canada. *Can J Fish Aquat Sci*, 47, 128–136.

Kerekes, J.J. 1975. Phosphorus supply in undisturbed lakes in Kejimkujik National Park, Nova Scotia. *Verh Intern verein Limnol*, 19, 1–18.

Montecino, V. 1991. Primary productivity in South American temperate lakes and reservoirs. *Rev Chil Hist Nat*, 64, 555–567.

O'Neill, S.M., and Hyatt, K.D. 1987. An experimental study of competition for food between sockeye salmon (*Oncorhynchus nerka*) and three sticklebacks (*Gasterosteus aculeatus*) in a British Columbia coastal lake. In H.D. Smith, L. Margolis, and C.C. Wood (eds.), Sockeye salmon (*Oncorhynchus nerka*) population biology and future management. *Can Spec Publ Fish Aquat Sci*, 96, 143–160.

Perez, C., Armesto, J.J., and Ruthsatz, B. 1991. Descomposición de hojas, biomasa de raices y características de los suelos en bosques mixtos de coníferas y especies laurifolias en el Parque Nacional de Chiloe. *Rev Chil Hist Nat*, 64, 479–490.

Pezzani, S. 1977. Copepoda. In S. Hurlbert (ed.), *Biota Acuatica de Sud America Austral* (pp. 139–144). San Diego: San Diego State University Foundation.

Prepas, E.E., and Drew, D.O. 1983. Evaluation of the phosphorus—chlorophyll relationship for lakes off the Precambrian Shield in western Canada. *Can J Fish Aquat Sci*, 40, 27–35.

Rankin, D.P., and Ashton, H.J. 1980. Crustacean zooplankton abundance and species composition in 13 sockeye salmon (*Oncorhynchus nerka*) nursery lakes in British Columbia. *Can Tech Rep Fish Aquat Sci*, No. 1957.

Rankin, D.P., Ashton, H.J., and Kennedy, O.D. 1984. Zooplankton abundance in lakes sampled by the 1978 lake enrichment program. Part 2: Central and North Coast, Skeena and Nass Systems. *Can Data Rep Fish Aquat Sci*, No. 459.

Scavia, D., and Fahnenstiel, G.L. 1988. From picoplankton to fish: Complex interactions in the Great Lakes. In R. Carpenter (ed.), *Complex Interactions in Lake Communities* (pp. 68–85). New York: Springer-Verlag.

Shortreed, K.S., and Stockner, J.G. 1986. Trophic status of 19 sub-arctic lakes in the Yukon Territories. *Can J Fish Aquat Sci*, 43, 797–805.

Shortreed, K.S., and Stockner, J.G. 1990. Effect of nutrient additions on lower trophic levels of an oligotrophic lake with a seasonal deep chlorophyll maximum. *Can J Fish Aquat Sci*, 47, 262–273.

Simpson, K.L., Wo, L. Hop, and Miki, I. 1981. Fish surveys of 15 sockeye salmon (*Oncorhynchus nerka*) nursery lakes in British Columbia. *Can Tech Rep Fish Aquat Sci*, 1022.

Smith, V. 1982. The nitrogen and phosphorus dependence of algal biomass in lakes: An empirical and theoretical analysis. *Limnol Oceanogr*, 27, 1101–1111.

Smith, R.C. 1989. Ozone, middle ultraviolet radiation and the aquatic environment. *Photochem Photobiol*, 50, 459–468.

Soto, D., and Zuñiga, L. 1991. Zooplankton assemblages of Chilean temperate lakes: A comparison with North American counterparts. *Rev Chil Hist Nat*, 64, 569–581.

Soto, D., Campos, H., Steffen, W., Parra, O., and Zuñiga, L. 1993. A case of potentially N-limited pristine lakes and ponds in Chilean Patagonia. *Archiv Hydrobiol*, 99, 181–197.

Spencer, G.N., and King, D.L. 1985. Interaction between light, NH_4, and CO_2 in buoyancy regulation of *Anabaena flos-aquae* (Cyanophyceae). *J Phycol*, 22, 194–199.

Stockner, J.G. 1981. Whole lake fertilization for the enhancement of sockeye salmon (*Oncorhynchus nerka*) in British Columbia, Canada. *Verh Internat verein Limnol*, 21, 293–299.

Stockner, J.G. 1987. Lake fertilization: The enrichment cycle and lake sockeye salmon (*Oncorhynchus nerka*) production. In H.D. Smith, L. Margolis, and C.C. Wood (eds.), Sockeye salmon (*Oncorhynchus nerka*) population biology and future management, (pp. 198–215). *Can Spec Publ Fish Aquat Sci*, No. 96.

Stockner, J.G., and Costella, A.C. 1980. The paleolimnology of eight sockeye salmon (*Oncorhynchus nerka*) nursery lakes in British Columbia, Canada. *Can Tech Rep Fish Aquat Sci*, No. 979.

Stockner, J.G., and Porter, K.G. 1988. Microbial food webs in fresh water planktonic ecosystems. In S.R. Carpenter (ed.), *Complex Interactions in Lake Communities*, (pp. 71–84). New York: Springer-Verlag.

Stockner, J.G., and Shortreed, K.S. 1985. Whole-lake fertilization experiments in coastal British Columbia Lakes: empirical relationships between nutrient inputs and phytoplankton biomass and production. *Can J Fish Aquat Sci*, 42, 649–658.

Stockner, J.G., and Shortreed, K.S. 1988. Response of *Anabaena* and *Synechococcus* to manipulation of nitrogen: phosphorus ratios in a lake fertilization experiment. *Limnol Oceanogr*, 33(6, 1), 1348–1361.

Stockner, J.G., and Shortreed, K.S. 1989. Algal picoplankton production and contribution to food webs in oligotrophic British Columbia Lakes. *Hydrobiologia*, 173, 151–166.

Strickland, J.D.H. 1960. Measuring the production of marine phytoplankton. *Bull Fish Res Board Can*, No. 122.

Strickland, J.D.H., and Parsons, T.R. 1968. A practical handbook of seawater analysis. *Bull Fish Res Board Can*, No. 167.

Thomasson, K. 1963. Araucanian lakes. *Acta Phytogeographica Sue*, 47, 1–139.

Vollenweider, R.A. 1968. Scientific fundamentals of the eutrophication of lakes and flowing waters, with particular reference to Nitrogen and Phosphorus as factors in eutrophication. Paris, *Rep Organism Economic Cooperation and Development*, DAS/CSI/68.27.

Weisse, T., and Stockner, J.G. 1994. Eutrophication: the role of microbial food webs. *Mem Instit Ital Idrobiol* (in press).

Williamson, C.E., Zagarese, H.E., Schulze, P.C., Hargreaves, B.R., and Seva, J. 1994. The impact of short-term exposure to UV-B radiation on zooplankton communities in north temperate lakes. *J Plankton Res*, 16, 205–218.

3. Forest System Responses to Human Activities

14. Implications of Patch Dynamics for Forested Ecosystems in the Pacific Northwest

G.A. Bradshaw, Frederick J. Swanson, and Maria R. Fiorella

It is not possible to discuss ecosytem processes and functions without explicit reference to the relevant scales of analysis. Although the concepts of heterogeneity and patchiness have been used for nearly twenty years (e.g., Pickett, 1991; Wiens, 1976) terms such as "pattern", "hierarchical", and "scale" appear with increasing frequency in journals from disciplines as disparate as geophysics, ecology, and oceanography (e.g., Grant, Swanson, & Wolman, 1990; Moloney, Morin, & Levin, 1991; Powell, 1989). Two current approaches to understanding patterns in the ecosystem are patch dynamics and hierarchy theory (O'Neill, DeAngelis, Wade, & Allen, 1986; Pickett and White, 1985; Steele et al., 1992; Wiens, 1976). The concepts of patch dynamics and hierarchy theory focus on relationships and phenomena across spatial and temporal scales. Hierarchy theory views the system as organized into levels by its intrinsic structure, where flux between levels is regulated by regionalized variables. On the other hand, patch dynamics has focused attention on the heterogeneous nature of the landscape, where the landscape evolves as a mosaic of patches. In ecosystems, this heterogeneity or variability is expressed as patterns in the distribution of plants, animals, and abiotic processes. More recently, these concepts have been incorporated into conservation design and forest and stream management plans (e.g., U.S.D.A., 1993; Swanson et al., 1994).

Concurrent with this conceptual shift has been the dramatic ascendence of technology used in the natural sciences. Computer accessibility and the development of accurate and innovative measurement tools have broadened the scope of field and laboratory experimentation previously hindered by feasibility constraints. For instance, satellite imagery and GIS maps provide media by which spatially explicit information from several disciplines (e.g., geomorphology, vegetation, stream hydrology, and fish population biology) can be integrated (Michener et al., 1994; Goodchild, 1993). In conjunction with these conceptual and technological advances, there has been increased concern regarding human impacts on the environment. The result has been increased conflict over natural resource management. Recognition of the cumulative effects of pollution on global climate has underscored the connectivity of local environments to events at distances and scales outside everyday perception; the range of human impact on ecosystems extends well beyond artificial borders. As a result of all three factors (i.e., conceptual, technological, and social), research and management interests have been directed to spatial and temporal aspects of natural ecosystems. The manipulation of forested landscapes, and the consequences thereof, are now consciously considered in conservation and management designs. Increasingly, management concerns are shifting to the "forest" rather than the "trees."

The temperate forests of the Pacific Northwest (PNW), U.S.A., represent a prime example of a setting in which scientific and socioeconomic concerns have converged. The Northwest landscapes form a focal point for many current political and socioeconomic debates in the U.S. environmental arena (U.S.D.A., 1993). The region has undergone a significant change in its disturbance regime due to intensive and extensive land-use practices (e.g., timber harvesting, grazing, fire suppression, and mining) (see Harris, 1984; Spies, Ripple, & Bradshaw, 1994). Before the onset of intensive timber harvesting activity, which began roughly 100 years ago and has continued intensely on public lands the past forty years, the PNW lands west of the Cascade Range crest were characterized by a fairly continuous forest cover. At present, the public and private forests of Washington, Oregon, and even Alaska are punctuated with clearcuts and managed plantations (Figure 14.1; Spies, Ripple, & Bradshaw, 1994). Widespread declines and shifts in the habitat and abundances of late successional animal species have forced a reevaluation of forest management practices regarding the effects on landscape fragmentation and population viability (Harris, 1984). Management and conservation designs developed for this region serve as a template for analogous plans in others North American regions and countries.

Ironically, land-use practices in these forests have provided landscape ecologists with a regional-scale fragmentation experiment. A strong pattern outside the natural range of variability has been superimposed on

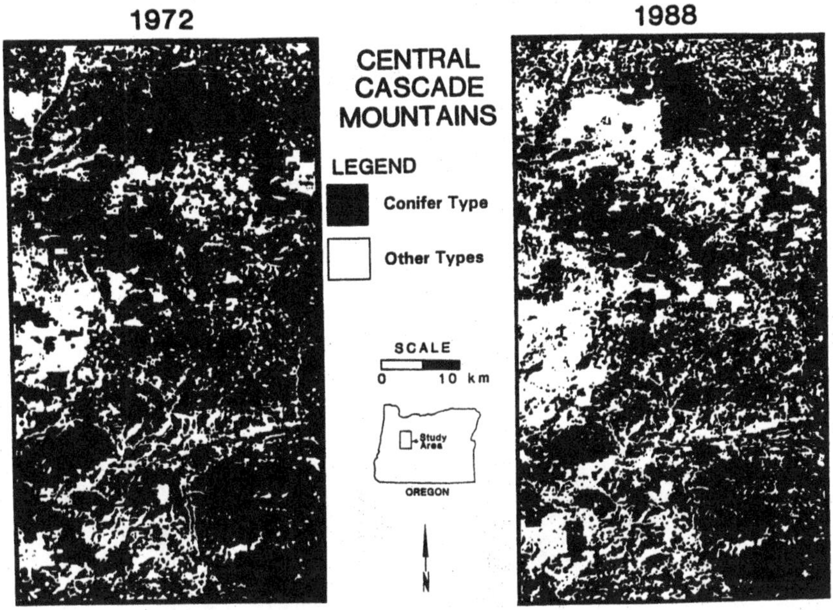

Figure 14.1. Classified satellite images depicting changes in landscape patterns that have resulted from anthopogenic disturbances created during the years between from 1972 and 1988 (Spies et al., 1994). The area covered lies west of the Cascade Range, central Oregon, U.S.A. spanning an area of 3,500 square kilometers.

the region where the same spatially and temporally varying processes are at work. We revisit the topic of landscape heterogeniety for the case of the Pacific Northwest forests with the intent of highlighting the interactions of patterns and processes across spatial and temporal scales in context of historical patterns of fragmentation. More specifically, we seek to (1) describe dominant sources and types of factors responsible for the generation of spatial structure and landscape heterogenity in PNW temperate forests; and (2) examine possible consequences of increased spatial heterogeneity of forested landscapes introduced by land conversion. It is our intent to emphasize the need to not only document the fragmentation of the landscape but understand the changing relationships between the landscape patches and matrix. Because a contol landscape at the landscape and regional scales is lacking, modeling has taken a more predominant role to elucidate large-scale processes and interactions. A simple heuristic model is employed to illustrate potential effects of local patterns (e.g., land conversion at the local scale) on the overall landscape composition over time (e.g., forested or nonforested landscape composition).

Natural Sources of Landscape Heterogeneity

Spatial Heterogeneity

The most prominent pattern observed at the regional scale within the Pacific Northwest is the distinct longitudinal anisotropy manifested as broad north- to south-running vegetation (Figure 14.2). These vegetation patterns reflect interactions between strong west to east temperature and precipitation gradients modulated by long-term climatic variation (Franklin & Dyrness, 1973; Franklin, Swanson, Harman et al., 1991; Swanson, Franklin, & Sedell, 1990). Although precipitation generally increases as one moves northward, local variability in moisture patterns occurs as a

Figure 14.2. AVHRR image (resolution 1 km × 1 km) of the Pacific Northwest region covering Oregon and Washington states.

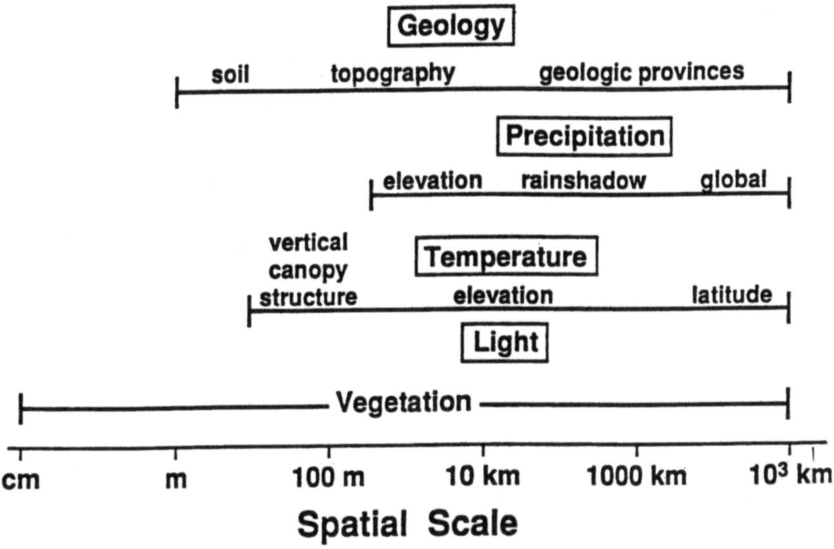

Figure 14.3. Abiotic sources of heterogeneity responsible for creating patterns in vegetation. Vegetation is viewed as a response variable to a suite of abiotic factors across scales.

result of storm patterns and local topographic contraints. A similar longitudinal temperature gradient exists, but is modulated in the coastal areas by the buffering warmer ocean current (Alaback & McClellean, 1993). However, closer examination of the broad-scale vegetation pattern reveals a less homogeneous distribution that reflects local climatic conditions coincident with topographic (i.e., aspect, slope) and structural (e.g., stand and within-gap) gradients. The interaction of abiotic and biotic processes (e.g., competition and predation) further embroider the multiscale landscape mosaic at several scales by controlling vegetation density (i.e., spatial arrangement), composition, and vertical structuring (e.g., multilayer canopies) within conifer stands (Figure 14.3). Examination of the spatial distribution of a given species can be viewed hierarchically. We illustrate this perspective by examining the spatial distribution of western hemlock.

In general, the regional distribution of many vegetation species reflect such a regional temperature and moisture "envelope" (Franklin & Dyrness, 1973). As a shade tolerant species, western hemlock (*Tsuga heterophylla*) ranges west of the Cascade crest with its northerly and southerly limits correlated with temperature and moisture extrema, respectively. Successful hemlock seedling regeneration is a function of several factors operating at various scales. Among other biotic variables, seed availability for forest regeneration is a function of seed production, species composition, seed tree proximity, and timing of release. To a certain extent, species production and timing of release are autogenically determined; however, the amount and timing of seed release are also related to regional climatic dynamics (J. Means, personal communication). Because abiotic factors (e.g., solar radiation, temperature, and moisture availability) are also modulated at several scales, the probability of seedling survivorship may depend on their values and interaction within a hierarchical framework (e.g., subgap, gap, stand, and landscape levels (Figure 14.4). Thus, heterogeneity of hemlock distribution within the region reflects the intergation of abiotic processes at several scales.

Experimental and field studies indicate that the relative position of a western hemlock seedling within a gap, gap size, slope, and aspect contribute to moisture and insolation variability (Spies et al., 1992). At the landscape scale, western hemlock occurrence is generally negatively correlated with southern aspects (Figure 14.5). This topographic signature becomes more apparent with decreasing latitude. A second scale of pattern is correlated with light and temperature gradients within gaps. A north to south zonation of light and moisture has been identified within Douglas

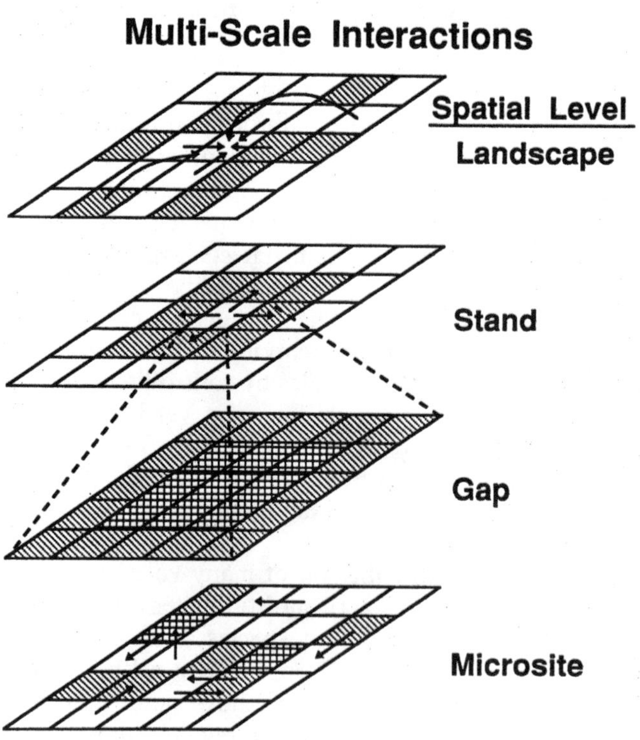

Figure 14.4. Patterns are created by a connecting hierarchy of relationships in the spatial and temporal domains. This figure illustrates how processes such as seedling regeneration may be affected by patterns at several scales.

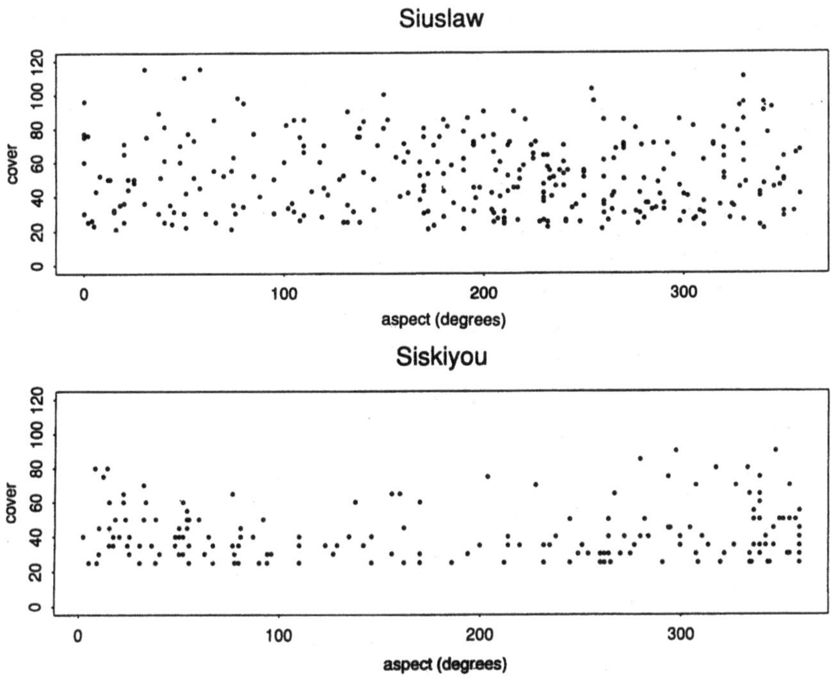

Figure 14.5. Percent cover hemlock (*Tsuga heterophylla*) plotted as a function of aspect (degrees) for forests at two latitudes in western Oregon. The data are derived from the Siuslaw (northwest) and Siskyou (south central) National Forest inventories. The cover-aspect pattern is more pronounced in the southern forest (i.e., Siskyou National Forest).

fir gaps on flat slopes (Figure 14.6; Gray, 1995). This within-gap gradient will vary with gap size and adjacent tree height. The interception of incoming solar radiation and precipitation by canopy trees creates a zonation of moisture, light, and radiation orthogonal to the north–south azimuth within the gap. Accordingly, seedling survivorship varies differentially along the gradient. However, environmental zonation is not consistent among all gaps. Because the amount of incoming solar radiation reaching the forest floor is a function of the geometry and adjacent tree size, the zonation created by direct light is predicted only in moderate to large gaps (>25 meters in diameter) and on slopes less than 50% (M. Easter, personal communication). A difference in hemlock seedling survivorship is expected when northern and southern aspects are compared under these conditions; lower moisture stress and leaf temperatures on northern slopes increase the likelihood of seedling survivability. On steep, southern slopes, increased exposure can generally exceed hemlock seedling tolerance and successful establishment is retarded. As a result, environmental differentiation effected by gap size is overridden by large-scale

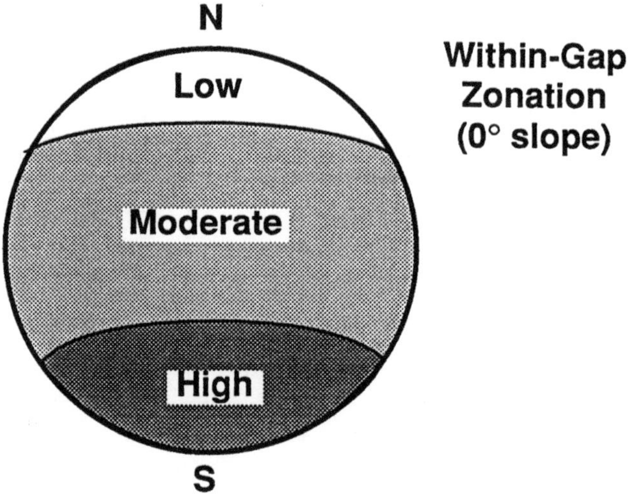

Figure 14.6. Seedling establishment zonation within a gap at zero slope. High, moderate, and low refer to relative levels of predicted seedling survivability.

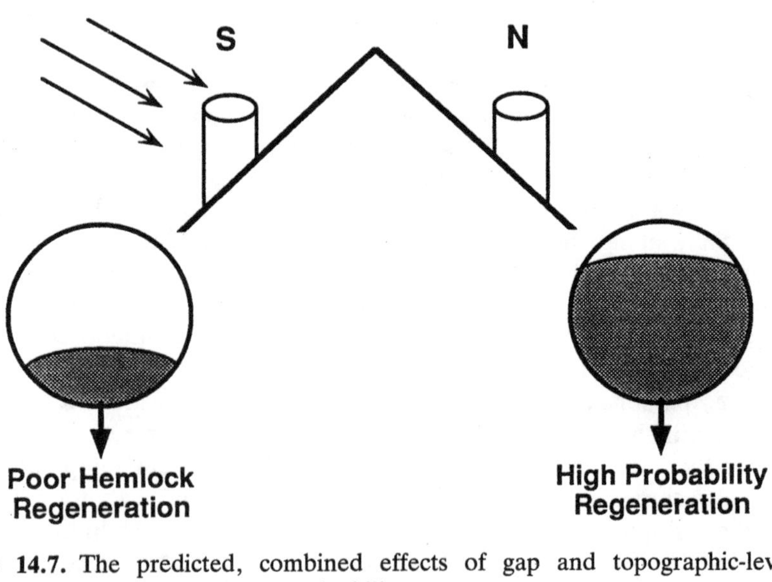

Figure 14.7. The predicted, combined effects of gap and topographic-level constraints on hemlock seedling survivability.

topographic factors when a threshold slope grade is exceeded (Figure 14.7).

These relationships may be viewed within the landscape context. A set of stand condition maps were created based on Thematic Mapper satellite imagery, classified according to age and structure. Companion elevational maps were developed for the Douglas fir dominant forest in the H.J. Andrews Experimental Forest located in central Oregon (Fiorella & Ripple, 1992). Old-growth Douglas-fir dominant stands characterized on average by moderate-to-large gaps were overlaid on slope and aspect maps derived from the topographic GIS coverage. The intersection of these maps predicts potential sites where the probability of hemlock seedling establishment is high (Figure 14.8). Thus, the distribution of hemlock seedling survival at the landscape scale may be viewed as controlled by both within-gap landscape interactions. Because the spatial distribution of seedlings predicted in the landscape depends on stand age, there is an implicit consideration of time where the timing and frequency

Figure 14.8. GIS map derived from the intersection of a satellite image-derived old-growth, slope and aspect maps depicting predicted distributions of high survivability sites of western hemlock (dark patches).

of disturbances characteristic of the area direct the distribution of stands of varying successional stages.

Temporal Heterogeneity

Disturbances such as wildfire, windthrow, flooding, and insect outbreaks also play an important role in imposing pattern on the landscape by redistributing resources (Morrison & Swanson, 1990; Spies & Franklin, 1991). Within-stand heterogeneity of resources (e.g., downed wood, canopy gap structure) reflects the integration and coupling of disturbance patterns (e.g., mortality due to wildfire, root pathogens) at several temporal and spatial scales. In a study comparing a set of old growth, mature, and young Douglas-fir dominant stands in the western Cascades, Oregon, hemlock seedling distribution showed a gradual increase with stand age and percent woody debris cover (Bradshaw, 1991). More interestingly,

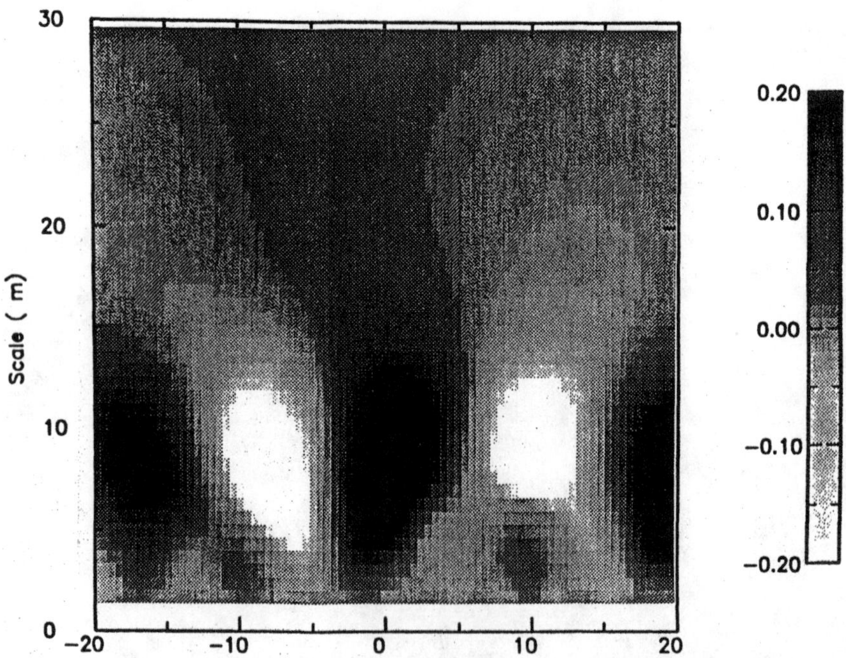

Figure 14.9. Wavelet cross-covariance calculated between hemlock seedling and downed log distributions. The *x*-axis refers to the lag at which correlation between the two variables occurs; the *y*-axis refers to the scale at which correlation occurs (meters). Downed logs and hemlock seedling distribution are correlated on average at approximately nine meters in this mature Douglas-fir stand.

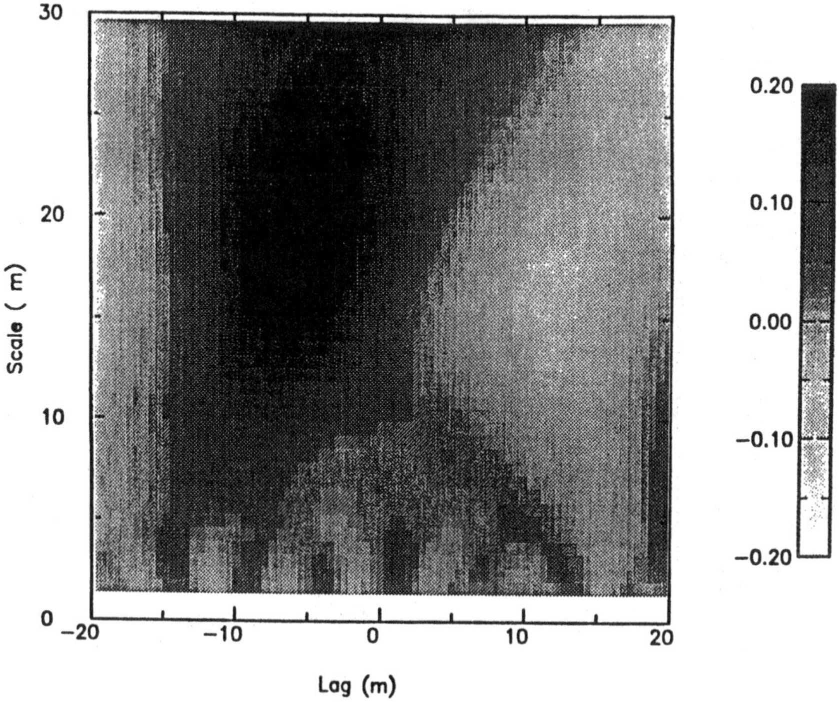

Figure 14.10. Wavelet cross-covariance calculated between hemlock seedling and downed log distributions. The x-axis refers to the lag at which correlation between the two variables occurs; the y-axis refers to the scale at which correlation occurs (meters). Downed logs and hemlock seedling distribution are correlated on average at 3 m and 18 m in this old growth Douglas-fir stand.

analyses quantifying the spatial distribution of hemlock seedling revealed a relationship between canopy and ground-level patterns. Hemlock seedling patches were correlated with patches of woody debris at different scales or patch sizes according to stand age and condition. In a mature stand characterized by high abundances of downed wood, hemlock seedling and wood occured coincidently in patches averaging 9 m (Figure 14.9).

In contrast, an old growth stand showed wood-hemlock correlation in patches at two different scales: 3 m and 18 m (Figure 14.10). The spatial distribution of hemlock seedlings and canopy gaps of the overstory were also correlated but at scales of 5 m and 25 m (Figure 14.11). These results suggest a coupled relationship between the distribution of hemlock seedlings in the understory and a composite pattern of patches related to the distribution of abiotic factors. The distribution of abiotic factors is in turn linked temporally by stand age and disturbance history.

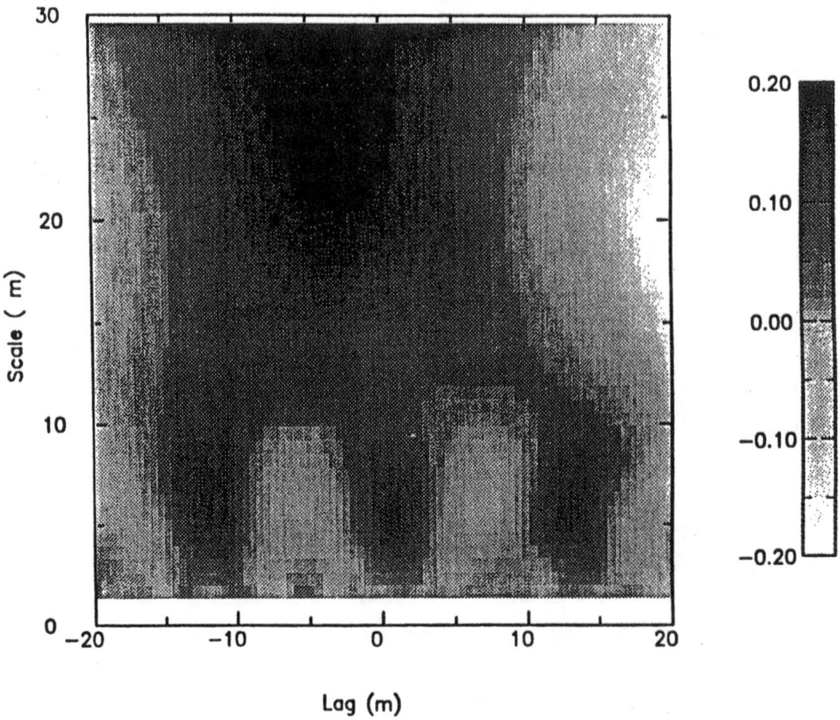

Figure 14.11. Wavelet cross-covariance calculated between hemlock seedling and downed log distributions. The *x*-axis refers to the lag at which correlation between the two variables occurs; the *y*-axis refers to the scale at which correlation occurs (meters). Overstory canopy gap and hemlock seedling distribution are correlated on average at 5 m and 25 m in this old growth Douglas-fir stand described in Figure 14.10.

From the geomorphic perspective, landforms processes influence ecosystem heterogeneity in several ways, by (1) modulating resource quality and availability (e.g., elevational modulation of temperature, nutrient distribution); (b) controlling flow of energy and material through the landscape (e.g., water movement); and (c) affecting the timing, frequency, and distribution of disturbances (e.g., wildfire, erosion, and wind; Grant et al., 1990; Swanson, Kratz, Caine, & Woodmansee, 1988). The timing distribution, and spatial arrangement of land movement and disturbance in general relates to the "range of natural variability," that is, a set of conditions or states that an ecosystem may have occupied prior to influence of European settlers (Swanson et al., 1994). Successional patterns and forest stand mosaics vary with disturbance intensity, location, and extent reflecting local and regional controls of climate, geology, vegetation dynamics (Delcourt, Delcourt, & Webb, 1983). It is instructive to view these processes from two perspectives of the temporal domain—frequency and duration. This distinction falls out naturally when viewed from a time series perspective; pattern can be described in terms of the peak structure (i.e., frequency) and in terms of the length or width of the peak (i.e., duration). Frequency and duration of disturbance will affect the type and rate of recovery in the ecosytem.

The frequency and duration characterizing a disturbance event may not always be commensurate or dependent on the same variables. Duration of a wildfire is primarily a function of several factors, including fuel load, topography, and present climate conditions (i.e., precipitation and wind; Agee and Huff, 1987; Morrison and Swanson, 1990). Although fire frequency is also a function of the same factors, it is constrained by frequency of initiation events, such as lightning strikes (i.e., elevation and combustible snags). Similarly, the frequency and rate of propagation of bark beetle infestation in PNW forests is strongly influenced by the order of disturbance events; large concentrations of downed wood or stressed live trees resulting from windthrow events and drought conditions, respectively, can initiate epidemic outbreaks. Frequency of infestation events (approximately every 20 to 30 yrs) is related to the probablities of these two types of disturbances and the order in which they occur (Bradshaw & Garman, 1994). In contrast, the duration of an infestation event is more dependent on the overall landscape composition and connectivity (i.e., an outbreak will only be sustained if there is available appropriate material). However, although infestation initiation may require a forest stand which has sustained physiological stress (e.g., drought), the spread can be sufficiently severe as to attack apparently healthy, live trees (i.e., a threshold condition may be breached) (Bedard, 1950). Duration of an outbreak may be expected to be dependent on the area and perhaps spatial pattern of the stand and landscape composition.

Modeling Anthropogenic Effects on Landscape Heterogeneity

In the previous section, landscape heterogeneity in the Pacific Northwest temperate forests was described in terms of pattern-generating processes within a spatial and temporal context. However, a glance at most regional satellite imagery reveals that the most striking source of landscape heterogeneity is anthropogenically derived patterns of historic land-use and land-conversion patterns. With the intensive and extensive resource extraction sustained over the recent decades, the temperate forests are markedly fragmented (Franklin and Forman, 1987). The implementation of a national policy of fire suppression has also contributed to the alteration of landscape dynamics by favoring species composition to fire intolerants and increasing the probability of insect outbreak (Borchers, 1994). In contrast to natural disturbance patterns, such human activities as timber harvesting and road construction are geometric in shape and occupy large areas of the landscape for sustained periods of time simoultaneously (Ripple, Bradshaw, & Spies, 1991). To examine the effects of such increased heterogenity and loss of connectivity in the landscape, we employ a simple heuristic model. A binary state (i.e., patch-matrix) model was constructed to observe how landscape rates and composition changes as fine-scale spatial pattern varies under different landscape conditions. Applications of the model and approach are discussed for two cases—disturbance propagation (e.g., bark beetle infestation) and seedling regeneration to which we refer for purposes of the present discussion as "dissipative" and "restorative" processes, respectively. Both seedling regeneration and bark beetle infestation are important agents sculpting landscape dynamics in the PNW (Franklin, Cromack, Denison et al., 1988; Furniss & Carolin, 1980; Harmon, Ferrell, & Franklin,

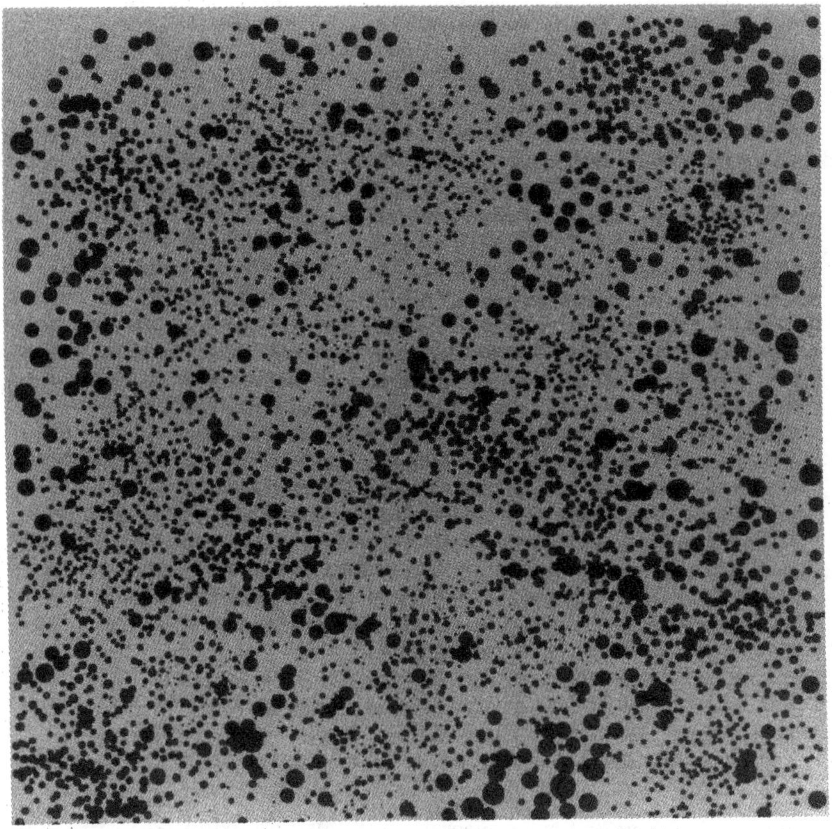

Figure 14.12. An example of the generated landscapes used in the model: small patch size.

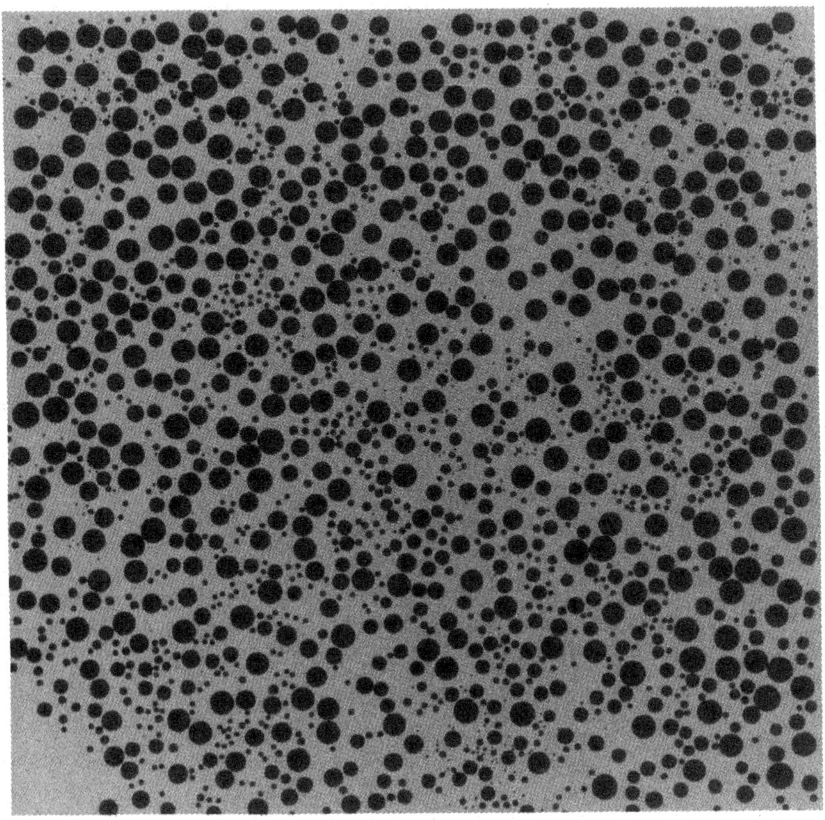

Figure 14.13. An example of the generated landscapes used in the model: large patch size.

1990). The term dissipative is used to connote a process which drives the forested system to an unforested state or one which deviates significantly from the present successional trajectory; restorative is used to denote processes which drive or sustain the system in its forested state. Although seedling regeneration initiated in canopy gaps may entail species replacement (e.g., replacement of standing Douglas fir to hemlock), it is nonetheless forested, and hence the process is considered "restorative" in the present context.

The model is constructed to test the hypothesis that local interactions between fine-scale features (e.g., tree aggregates, disturbance loci, gaps) and the matrix (e.g., intact forest) can alter the rate and direction of landscape change significantly. The landscape is defined as a set of randomly generated patches of homogeneous composition dispersed within a homogenous forest matrix (Figures 14.12 and 14.13). Two elements of spatial pattern are varied—mean patch size and patch-matrix contrast, as defined below. The initial landscapes have the same ratio of patch-to-matrix areas. Patterns resulting from patch-matrix contrast are examined by using a nearest-neighbor algorithm moved across the landscape of a given patch-size configuration. Depending on the cell type encountered (either patch or matrix) and the interaction algorithm applied, local change in landscape composition will occur. Patches are first-order approximations of harvested stands in a forest matrix; conversely, patches may be intact old growth stands within a matrix of young or harvested land (Figure 14.14).

Variability in patch size is achieved by specifying a mean patch radius and generating a normally distributed patch landscape. The patch-matrix contrast is varied in two ways: (1) radius of influence and (2) matrix susceptibility. A small radius of influence implies that only cells located in the immediate neighborhood of the patch will change; a large radius of influence implies that a larger zone of influence is exterted by the patch on the matrix. A difference in radii of influence is intended to correspond to rapid local change. Matrix susceptibility is defined as the tendency for the matrix to become patch. High susceptibility corresponds to an asymmetry in the direction of change between patch and matrix (i.e., low resistance of the matrix to resist patch-induced change) (Figure 14.15).

Results

The greatest difference exerted by these three parameters is associated with matrix susceptibility values. Under conditions of low susceptibility, the landscape changes over to matrix (Figure 14.16). Conversely, under conditions of high susceptibility, all four variants asymptotically approach a landscape composed of patch composition (Figure 14.17). Within each category, the effects of patch size and radius of influence have a lesser impact, but trends are visible. In the low susceptibility case (Figure 14.6b), the presence of a small radius of influence for patches of both sizes causes the landscape to converge to the forest matrix more rapidly in comparison to conditions where the radius of influence is high. As indicated by the case of small patches, total patch area increases. However, this trend reverses and the landscape eventually converges to forest matrix. In contrast (Figure 14.6a), patch size and radii of influence under conditions of high susceptibility appear to have a less noticeable effect on landscape change behavior; the rate and direction of change are dominated by the condition of high-matrix susceptibility and convergance is much more rapid. The experiment illustrates that local interactions and spatial relationships can effect vastly different landscape development and composition despite differences in local patterns. This experiment also illustrates the concept of a threshold; namely, the direction of landscape

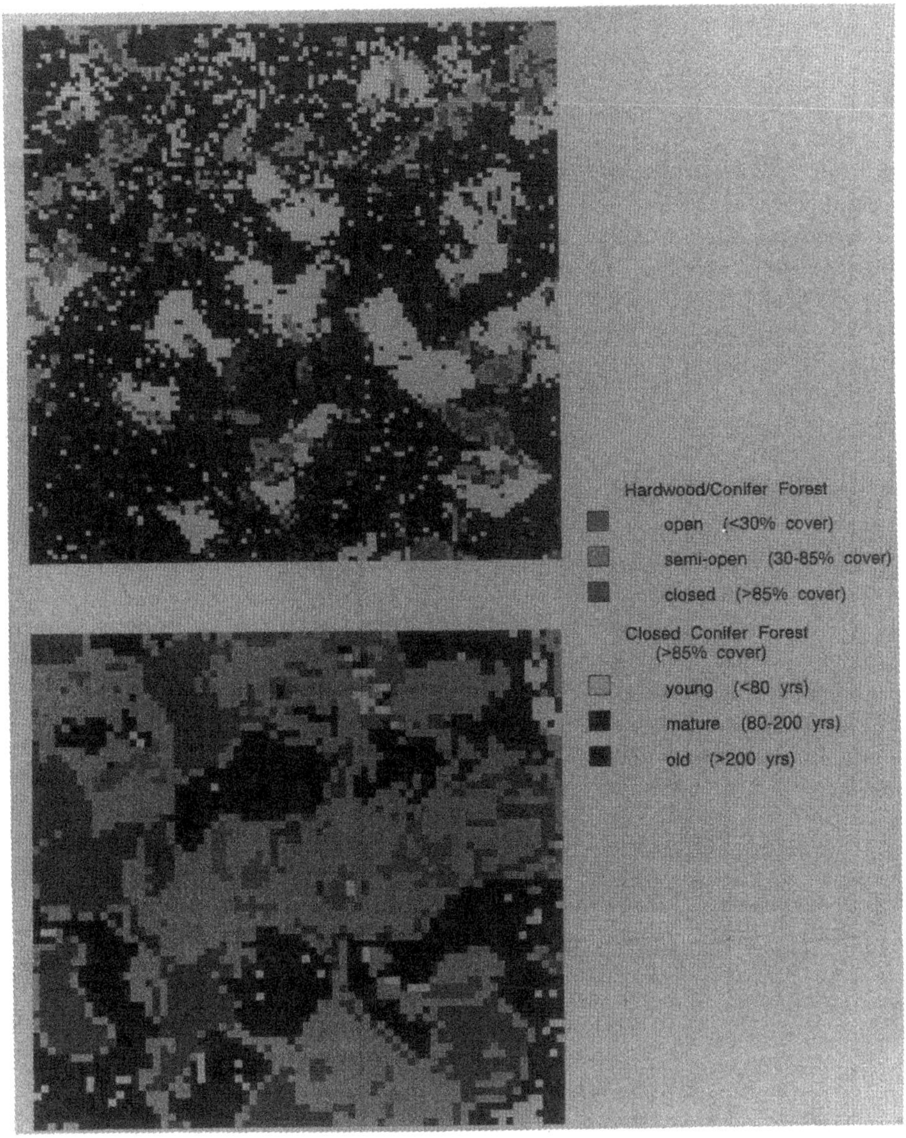

Figure 14.14. Classified thematic mapper satellite imagery depicting patch-matrix landscape for two representative forests in the Western Cascades, Oregon.

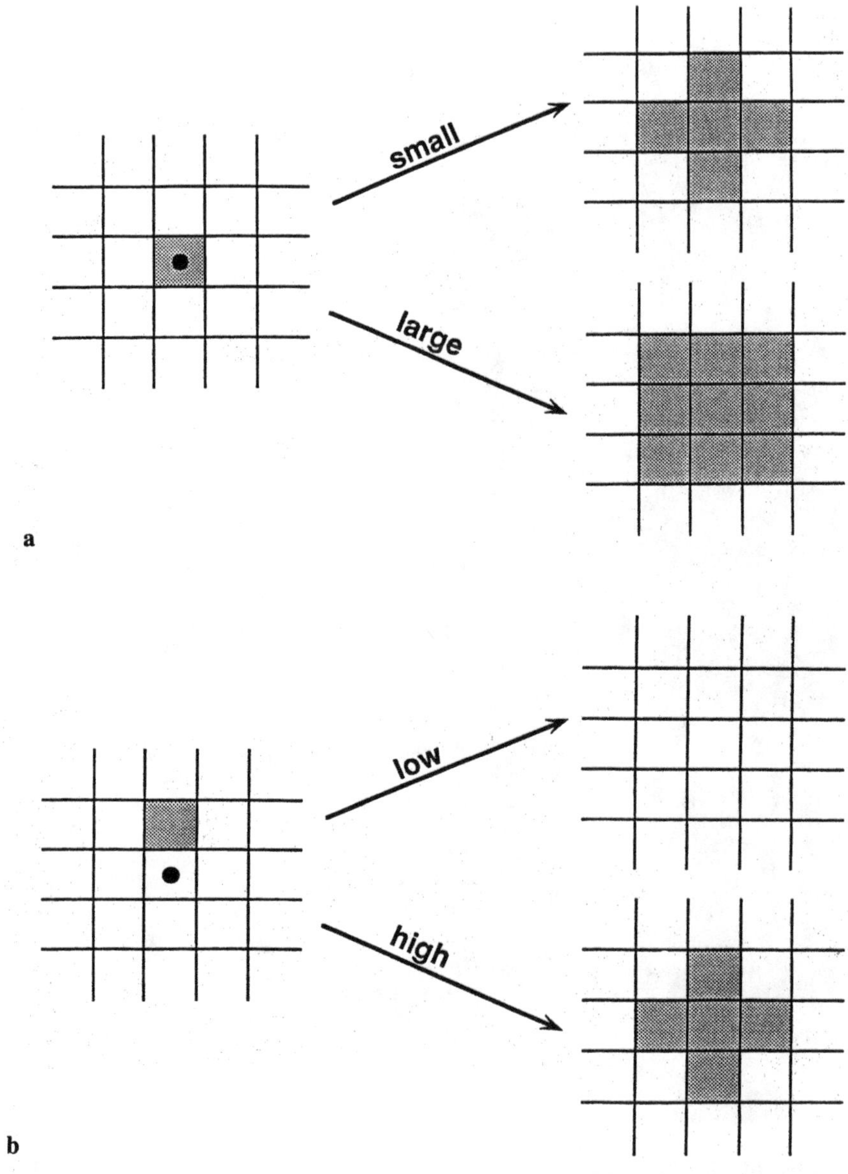

Figure 14.15. Rules describing two levels of patch-matrix interactions: (a) radius of influence and (b) matrix susceptibility. Two levels were considered in each case.

change may alter dramatically relative to initial perturbation magnitude (i.e., incremental change of spatial parameter).

As discussed, the forests of the PNW are significantly more fragmented prior to the settlement of Europeans. The present landscape is largely

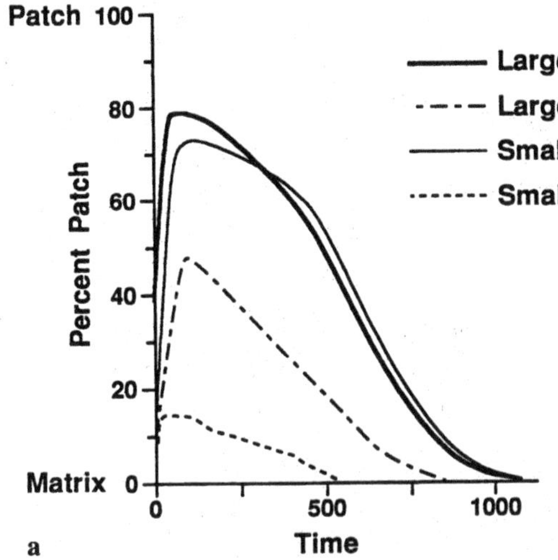

a

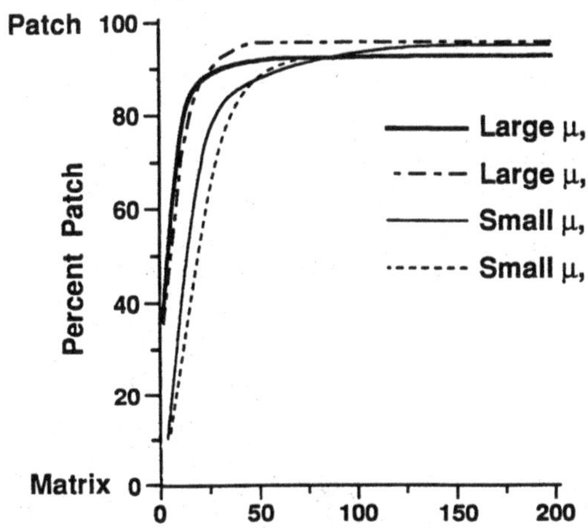

b

Figure 14.16. Resultant dynamics for the case of (a) high matrix susceptibility, and (b) low matrix susceptibility for low and high mean patch size radii and low and high conditions of small and large radii of influence (R). μ is the average patch size.

composed of patches within a matrix of varying composition and successional age. Each patch, as referred in the model, may be regarded as a potential disturbance locus (e.g., initiation site for insect outbreak) or refugia of faunal populations, respectively. Disturbance sites represent patches of downed wood remaining from previous wildfire or windthrow events from which insect outbreak may initiate. In such a disturbance setting, matrix susceptibility can be viewed as the tendency for Douglas fir trees to resist bark beetle infestation. In a heterogeneous, vital old growth stand, matrix susceptibility would be expected to be relatively low disturbance spread where would eventually die out despite initial patch size and potential rate of spread. Thus, the high matrix-patch contrast would act to insulate and isolate patches and retard disturbance spread. On the other hand, a landscape characterized by high susceptibility might correspond to aggregates of stressed trees (e.g., densely grouped young conifers). The weakened forest matrix would be less effective at resisting an outbreak. According to the model, propagation would occur locally at a faster rate. Levels of matrix heterogeneity, landscape connectivity, and dispersal capabilities of the insect would interact to determine the rate of spread.

In a second scenario, we consider conifer regeneration in a patchy landscape. In this case, the old growth and mature stands constitute the patches; the matrix is composed of young conifer and deciduous stands. The potential for conifer seedling establishment would be relatively high if the surrounding matrix was open and resource availability was not blocked by the presence of dense alder stands. In this case, matrix susceptibility would be considered high. Conversely, the presence of thick alder stands in the matrix would render seedling establishment more difficult and retard conifer regeneration. A similar scenario would be applicable for the dispersal patterns of faunal communities in the patch-matrix landscape.

Conclusions

Terrestrial ecology is now at the juncture where field and experimental analyses may be combined with the concepts of patch dynamics and hierarchy theory to develop a coherent spatially explicit, across-scale understanding of forest dynamics. Understanding what controls the rate and direction of local interactions as a function across spatial scales will be critical to developing appropriate management and conservation plans. We have engaged in a simple experiment to evaluate the effects of two elements of spatial pattern (i.e., patch-matrix contrast and patch size) on the rate and direction of landscape change. Two processes characteristic of Pacific Northwest forests, disturbance and regeneration, were discussed within the model context.

The model results suggest that anthopogenically introduced heterogeneity at a local level can dominate overall landscape change. Deviations in local patch-matrix interactions and pattern can effect significantly different rates of landscape change and the outcome of landscape composition. A differential in rates of change has several implications. If the rate of change effected by local patch-matrix interactions is slow relative to larger-scale phenomena (e.g., physiological response to global climate change), there is greater probability that the original successional trajectory will be altered. This phenomenon is illustrated with the topography-gap decoupling which may occur in the case of hemlock seedling regeneration and also in the case of understory vegetation response (Bradshaw, 1991; Bradshaw and Spies, 1992).

Landscape and regional model development should identify and evaluate critical components of pattern and their potential interactions over time. Depending on the particular landscape and ecological process, certain elements are more likely to exert influence than others (e.g., patch size versus patch–patch connectivity). Once again, while seeking for general principles, accomodation for the idiosyncracies of each ecosystem is necessary to fully capture system dynamics. It remains to determine to what extent local interactions and patterns influence variability observed within the region. Within the context of global change, several questions emerge. For example, as one moves latitudinally in the Northwest, the dominance of given landscape processes and patterns change accordingly. Will the effects of fine-scale change (e.g., stand-level land conversion) be uniform throughout the region? Will a change in regional climate multiply the effects of disturbance or will a threshold be reached and result in negative feedback shifting to a different dynamic regime? Will rates of change be constant throughout the region? Clearly, the construction of simple spatially and temporally explicit models in conjunction with large-scale field studies will contribute to our understanding.

Acknowledgments. The authors would like to thank Barbara Marks for programming assistance, William J. Ripple (ERSAL) for provision of satellite imagery, and Mark Harmon, and Mark Easter for helpful comments and discussion. Research has been conducted in part under the auspices of LTER site studies, H.J. Andrews Experimental Forest (NSF BSR 90 11663).

References

Agee, J.K., and Huff, M.H. 1987. Fuel succession in a western hemlock/Douglas-fir forest. *Can J For Res*, 17, 697–704.

Alaback, P.B., and McClellan, M.H. 1993. Effects of global warming on managed coastal ecosystem of western North America. In H.A. Moonpy, E.R. Fuentes, and B.I. Kronberg (eds.), *Earth System Responses to Global Change: Contrasts Between North and South America* (pp. 299–303). New York: Academic Press.

Bedard, W.D. 1950. The Douglas-fir beetle. U.S.D.A. Circular No. 918.

Borchers, J.G. 1994. A hierarchical context for sustaining ecosystem health. In R.G. Jaindl and T.M. Quigley (eds.), *Search for a Solution: Sustaining the Land, People, and the Economy of the Blue Mountains*. American Forests Publications.

Bradshaw, G.A. 1991. *Hierarchical Analysis of Patterns and Processes of Douglas-Fir Forests*. Unpublished doctoral dissertation, Oregon State University.

Bradshaw, G.A., and Garman, S. 1994. Detecting fine-scale disturbance in forested ecosystems as measured by large-scale landscape patterns. In W.K. Michener, J.W. Brunt, and S.G. Stafford (eds.), *Environmental Information Management and Analysis: Ecosystem to Global Scales* (pp. 533–550). London: Taylor and Francis.

Bradshaw, G.A., and Spies, T.A. 1992. Characterizing canopy gap structure in forests using wavelet analysis. *J Ecol*, 80, 205–215.

Delcourt, H.R., Delcourt, P.A., and Webb III, T. 1983. Dynamic plant ecology: The spectrum of vegetational change in space and time. *Quat Sci Rev*, 1, 153–175.

Fiorella, M., and Ripple, W. 1993. Determining successional stage of temperate coniferous forests with landsort satellite data. *Programmetric Engineering and Remote Lensing*, 59(2), 239–246.

Franklin, J.F., and Dyrness, C.T. 1973. *Natural Vegetation of Oregon and Washington*. U.S.D.A., Forest Service, General Technical Report PNW-8.

Franklin, J.F., Cromack, Jr., K., Denison, W., Mckee, A., Maser, C., Sedell, J., Swanson, F., and Juday, G. 1981. *Ecological Characteristics of Old-Growth Douglas-Fir forests*. Portland, OR: U.S.D.A., Forest Service, PNW Forest and Range Experiment Station. General Technichal Report PNW-118.

Franklin, J.F., Swanson, F.J., Harmon, M.E., et al. 1991. Effects of global climate change on forests in northwestern North America. In R. Peters and T. Lovejoy (eds.), *Climate Change and Biological Diversity*.

Franklin, J.F., and Forman, R.T.T. 1987. Creating landscape patterns by cutting: ecological consequences and principles. *Land Ecol*, 1, 5–18.

Furniss, R.L., and Carolin, V.M. 1980. *Western Forest Insects*. Washington, DC: U.S.D.A., Forest Service, Miscellaneous Publication No. 1339.

Goodchild, M., Parks, B.O. and Steyaert, L.T. 1993. *Environmental Modelling with GIS*. New York: Oxford University Press.

Grant, G.E., Swanson, F.J., and Wolman, M.G. 1990. Pattern and origin of stepped-bed morphology in high-gradient streams, Western Cascades, Oregon. *Geol Soc Am Bull*, 102, 340–352.

Gray, A.N. 1995. *Tree Seedling Establishment on Heterogeneous Microsites in Douglas-Fir Forest Canopy Gaps*. Unpublished doctoral dissertation. Oregon State University.

Harmon, M.E., Ferrell, W.K., and Franklin, J.F. 1990. Effects on carbon storage of conversion of old-growth forests to young forests. *Science*, 247, 699–702.

Harris, L.D. 1984. The fragmented forest: Island biogeographical theory and the preservation of biotic diversity Chicago: University of Chicago Press.

Kolasa, J., and Pickett, S.T.A. (eds.), 1991. *Ecological Heterogeneity*. New York: Springer-Verlag.

Michener, W.K., Brunt, J.W., and Stafford, S.G. 1994. *Environmental Information Management and Analysis: Ecosystem to Global Scales*. Bristol, PA: Taylor & Francis.

Moloney, K.A., Morin, A., and Levin, S.A. 1991. Interpreting ecological patterns generated through simple stochastic processes. *Land Ecol*, 5(3), 163–174.

Morrison, P.H., and Swanson, F.J. 1990. *Fire History and Pattern in a Cascade Range Landscape*. Portland, OR: U.S.D.A., Forest Service, Pacific Northwest Research Station, General Technical Report PNW-GTR-254.

O'Neill, R.V., DeAngelis, D.L., Wade, J.B., and Allen, T.F.H. 1986. *A Hierarchical Concept of Ecosystems. Princeton, NJ: Princeton University Press.*

Pickett, S.T.A., and White, P.S. 1985. *The Ecology of Natural Disturbance and Patch Dynamics*. Orlando, FL: Academic Press.

Powell, T.M. 1989. Physical and biological scales of variability in lakes, estuaries, and coastal ocean. In J. Roughgarden, R.M. May, and S.A. Levin (eds.), *Perspectives in Ecological Theory*. Princeton, NJ: Princeton University Press.
Ripple, W.J., Bradshaw, G.A., and Spies, T.A. Measuring landscape patterns in the Cascade Range of Oregon, USA. *Biol Conserv*, 57, 73–88.
Spies, T.A., Ripple, W.J., and Bradshaw, G.A. 1994. Dynamics and pattern of a managed coniferous forest landscape in Oregon. *Ecol Appl*, 4(3), 555–568.
Spies, T.A., Easter, M.J., and Gray, A.N. 1992. Light, temperature, and initial vegetation responses to experimental gaps in coniferous forests. *Bull Ecol Soc Am*, 77th Annual ESA Meeting, August 9–13, p. 352.
Spies, T.A., and Franklin, J.F. 1991. The structure of natural, young, mature, and old-growth forests in Washington and Oregon. In L.F. Ruggiero, K.B. Aubry, A.B. Carey, and M.H. Huff (eds.), *Wildlife and Vegetation of Unmanaged Douglas-Fir Forests* (pp. 91–109). Portland OR: U.S.D.A., Forest Service, Pacific Northwest Research Station, General Technical Report PNW-GTR-285.
Spies, T.A., and Franklin, J.F. 1989. Gap characteristics and vegetation response in coniferous forests of the Pacific Northwest. *Ecology*, 70(3), 543–545.
Spies, T.A., and Franklin, J.F. 1988. Coarse woody debris in Douglas-fir forests of western Oregon and Washington. *Ecology*, 69(6), 1689–1702.
Spies, T.A., Franklin, J.F., and Klopsch, M. 1990. Canopy gaps in Douglas-fir forests of the Cascade Mountains. *Can J For Res*, 20, 649–658.
Spies, T.A., Ripple, W.J., and Bradshaw, G.A. Dynamics and pattern of a managed coniferous forest landscape. *Ecol Appl*, 4(3), 555–568.
Swanson, F.J. 1981. Fire and geomorphic processes. In *Proceedings, Fire Regimes and Ecosystems Conference: December 11–15, 1979*. Washington, DC: U.S.D.A., Forest Service, General Technical Report WO-26.
Swanson, F.J., Franklin, J.F., and Sedell, J.R. 1990. Landscape patterns, disturbance and management in the Pacific Northwest, USA. In Zoenfeldt and Forman (eds.), *Changing Landscapes: An Ecological Perspective* (pp. 191–213). New York: Springer-Verlag.
Swanson, F.J., Kratz, T.K., Caine, N., and Woodmansee, R.G. 1988. Landform effects on ecosystem patterns and processes: Geomorphic features of the earth's surface regulate the distribution of organisms and processes. *BioScience*, 38(2), 92–98.
U.S.D.A., Forest Service. 1993. *Forest Ecosystem Management: An Ecological, Economic, and Social Assessment. A Report of the Forest Ecosystem Management Assessment Team (FEMAT)*. Washington, DC: U.S.D.A., Forest Service.
Wiens, J.A. 1976. Population responses to patchy environments. *Ann Rev Ecol Sys*, 7, 81–120.

15. Assessing and Responding to the Effects of Climate Change on Forest Ecosystems

David L. Spittlehouse

Management of our forest resources requires that we make decisions now that are based on future unknown climatic conditions. The Intergovernmental Panel on Climate Change (Houghton, Callander, & Varney, 1992) concluded that a 1.5°C to 4.5°C warming of the earth's climate and accompanying changes in precipitation and weather patterns are likely to occur by the end of the next century. However, there is much disagreement over the validity of the climate predictions, the reliability of the impact assessments, the economic costs, and our ability to develop responses (Reifsnyder, 1989; Schneider, 1990, 1994; Ausubel, 1991; Clark, 1991; Cline, 1992; Michaels & Stooksbury, 1992; Waterstone 1993). This debate over climate change, which is confusing to forest managers and the public, does little to help managers make decisions now that will have repercussions in 100 years. The uncertainty may, in fact, discourage forest managers from even considering the issue, particularly since the planning horizon for many forest management operations is often 10 years or less (Woodman, 1987).

Neither climate nor ecosystems are static. Organisms, including humans, have responded to climate changes in the past. The concerns are that human-induced changes may occur at a greater rate than in the past and that the changes may be detrimental to our welfare. Ausubel (1991) suggests that the technologies and social structures that buffer us from climatic stresses may make the changes irrelevant. Immediate concerns for meeting public demands for food and services and economic development rank well ahead of concern for distant, unknown climates (Watson, Bach, & Goklany, 1992). However, technologies are a buffer only as long as energy supplies last and climate fluctuations are small (Baker, Ruschy, & Skaggs, 1993). Even if human society is somewhat insulated from change, forest ecosystems are not. Forest managers need to assess how the ecosystems may change and develop management responses (Sandenburgh, Taylor, & Hoffman, 1987; Pollard, 1991). Management has to take place at a much smaller scale than the resolution of the current global climate models. Most assessments have been at the regional scale, and many of these analyses used inappropriate models (Clark, 1991; Malanson, 1993; Pastor, & Post, 1993). I will present a framework and examples for assessing forest sensitivity to climate change and developing response for managed forests at the site level. I make no assumptions about which, if any, of the global climate model scenarios is possible. The methodology could also aid managing forests for current climatic variability.

Framework for Responding to the Threat of Climate Change

Like Waterstone (1993), I believe that it is unlikely that international action will be sufficient to halt the equivalent doubling of the atmospheric carbon dioxide concentration. Forestry may be required to play a role in reducing the rate of increase, though the size of the gain is debatable (Pollard, 1991; Dixon, Brown, Houghton et al., 1994). I will concentrate on the need to maintain forest ecosystems healthy enough to tolerate climate changes, as recommended in Wall (1992). We need to consider all parts of the ecosystem, not just the species important for timber. Furthermore, forestry operations depend on the weather (e.g., warm winters can restrict winter logging). Responses can be determined from existing data, expressed in terms of the sensitivity of ecosystems to climate/weather variability. We must recognize that any climate change may not be gradual, and the rate of change will probably be more important than changes in the mean (Schneider, 1990; Pollard, 1991; Clark, 1991).

Development of a response plan requires a framework for the analysis. First there must be the identification of the issue of concern and the degree of change in the resource that would be considered a serious problem. Second is the development of scenarios, that is, determination of the sensitivity of ecosystems to changes in climate variables, and how this compares with potential future climate changes. The third step is the development of management responses. This includes activities to be taken at the time changes occur as well as activities needed now to facilitate future response. Benefit/cost analysis should be done here. The final step involves a monitoring program to determine if changes are taking place, and when the thresholds for concern and intervention have been reached. This will also aid in refining the response plan.

Management responses will be a mix of adaptation, mitigation, and protection measures. Prescriptions must be flexible and safe-fail, that is, they do not compromise the health of the forest if the climate does not change as predicted (Pollard, 1991). Responses will depend on whether we are dealing with managed or unmanaged ecosystems. Even where we are managing, it is likely that management will not be directed toward most organisms in the ecosystem. Assessment will improve our ability to select areas to target limited resources (Sandenburg, Taylor, & Hoffman, 1987; Woodman, 1987).

Adaptation is a response appropriate to managed ecosystems. An example is changing the ecotypes or species planted after disturbance (fire, harvesting) under a changed climate. The human adaptation response would be changing our expectations of, and demands on, the forest resource, for example, accepting a reduction in tree growth. Leaving migration corridors and reserve areas would be an example of mitigation. This may be the only way to address unmanaged ecosystems and wildlife within a managed landscape. Protection would involve responses to enhanced threats where current forests are still sustainable, such as, improving integrated pest management and fire portection programs. Many responses could fit under more than one definition, depending on how, when, or where they are applied. For example, development of drought tolerant stock would mitigate timber losses by adapting the trees to the new climate.

Species and ecosystem processes have different sensitivities to climate. Consequently, responses to climate change need to be at the species level (Leverenz & Lev, 1987; Clark, 1991; Smith, Shugart, Bonan, & Smith, 1992; Pastor & Post, 1993). Species responses will result from an interplay of competition, dispersal, and climatic factors (Malanson, 1993). Established trees tolerate large interannual variations in weather conditions (Reifsnyder, 1989), and future rapid, large-scale devastation of plant communities is unlikely. Populations of plants and animals at the edge of their ranges are likely to be the most sensitive to climate change and may act as early warning systems. Animal populations may move more rapidly than plant populations in response to a change, though this requires that appropriate habitats are available in other areas. Fragmentation of habitat will make it diffcult for even resilient populations to move. Response may be easier with managed rather than unmanaged ecosystems, since intervention is part of management (Woodman, 1987; Peters, 1990; Graham, Twiner, & Dale, 1990; Pollard, 1991; Wall, 1992). Any change is usually assumed to be negative. However, in some situations there will be an improvement over existing conditions. For example, warmer temperatures may result in increased productivity of the vegetation in presently cool areas. The increased carbon dioxide concentration may increase water use efficiency and growth of plants (Eamus & Jarvis, 1989; Graham et al., 1990; Ausubel, 1991; Michaels & Stooksbury, 1992). Franklin et al. (1992) and Veblen and Alaback (1995) suggest that disturbance regimes, such as those produced by fire or forest harvesting and renewal, interact with climate changes creating opportunities for adaptation by reducing the inertia of established forests. Disturbances can be viewed as events that speed adjustment of the vegetation to current environmental conditions, though they may be highly disruptive in the short to mid term.

Application of the Response Framework

Any analysis requires some form of a model. The problem is what to include and how to include it. The level of detail and scale depend on the type of planning being considered (Malanson, 1993). For example, Running and Namani (1991) assessed regional links between changes in water use and potential tree growth. Detailed simulations of ecosystem succession at the site level have been made for unmanaged ecosystems (Smith et al., 1992). The approaches I use here were chosen in recognition of the limited data available and the need to present information in a form suitable for use by forest managers. I will be assessing the effect of climate changes on a dominant tree species, Douglas-fir (*Pseudotsuga menziesii* (Mirb.) Franco), in coastal British Columbia, by considering the sensitivity of reforestation and growth of Douglas-fir to changes in rainfall and temperature. The approach is similar to that used by LeBlanc and Foster (1992) for oak in the midwestern United States.

Douglas-fir is an important commercial tree species occupying mid to low elevation, relatively dry sites over a wide latitudinal range in the Pacific Northwest. It is a seasonal rainforest (low summer rainfall) where disturbance is an important part of the life cycle, and biodiversity and biomass are high (Meidinger & Pojar, 1991; Boyle, 1995; Veblen & Alaback 1995). Paleoclimatological data indicate that the shift to modern forest composition began about 6000 years ago in the Pacific Northwest in response to a climatic cooling and increased precipitation (Brubaker, 1992). Global climate models predict a warmer climate with drier summers for the Pacific Northwest (Boer, McFarlane, & Lazare, 1992), which means that the range of Douglas-fir would move northward and upward in elevation (Leverenz & Lev, 1987). Woodman (1987) suggested that in this situation no overall change would be required in the management of Douglas-fir. I would disagree with this latter assessment, because responses will be required to manage sites where Douglas-fir ecosystems are threatened. Also there would be many challenges in introducing or enhancing Douglas-fir within any "new" range, because there are many factors besides climate that control tree growth.

Reforestation

Issue of Concern

Changes in climate, particularly rainfall, could affect the establishment of Douglas-fir seedlings after a disturbance such as harvesting. In situations with low summer rainfall, significant numbers of recently planted seedlings die through dessication. Seedlings are susceptible to variations in weather conditions since they exploit only the top 0.2 m or less of the soil profile. They require 30 to 60 days of adequate soil water (wetter than −0.1 MPa) after planting to develop their root system to allow them to withstand summer droughts (Spittlehouse & Childs, 1990).

Scenarios

Soil water availability is regulated by rainfall, evaporative demand, competing vegetation, and the ability of the soil to store water. Seedlings are planted in April, and snow melt and April rainfall result in a wet soil at planting time. Rainfall in May and June is critical for maintaining adequate soil water for establishment. The site considered is at an elevation of 350 m, on a 15° SE-facing slope with a stony loamy sand (50°28′ N, 123°00′ W). A physically based, layered, soil–water balance model and daily solar radiation, air temperature, and rainfall data were used to determine how soil water availability was affected by variations in May and June rainfall (Spittlehouse & Childs, 1990). These data were used to define three categories of suitbility for seedling establishment (soil water status) in terms of May and June rainfall. They are: adequate (no soil drying—May plus June rainfall >60 mm), marginal (some soil drying—May plus June rainfall between 30 and 60 mm), and too dry (severe soil drying—May plus June rainfall <30 mm). The relative variation in evaporative demand is much smaller than the changes in rainfall and is not considered. Each year of a monthly rainfall record for 1908 to 1989 was then categorized using the above criteria. Rainfall in May averaged 33 ± 20 mm (ranged from 2 to 93 mm), and 32 ± 21 mm (range from 0 to 103 mm) in June. Figure 15.1 shows the soil water status for each year, and the percentage of years in each category is presented in Table 15.1. About 50% of the years in the marginal category are at the drier end of the range. In this case, the timing of the rainfall and control of the competing vegetation is critical for successful establishment.

The sensitivity to changes in rainfall was determined by reevaluating the rainfall record after increasing or decreasing rainfall by 20%. The decrease in the mean rainfall is similar to changes predicted by global climate models (e.g., Boer et al., 1992). Average May rainfall was 40 mm

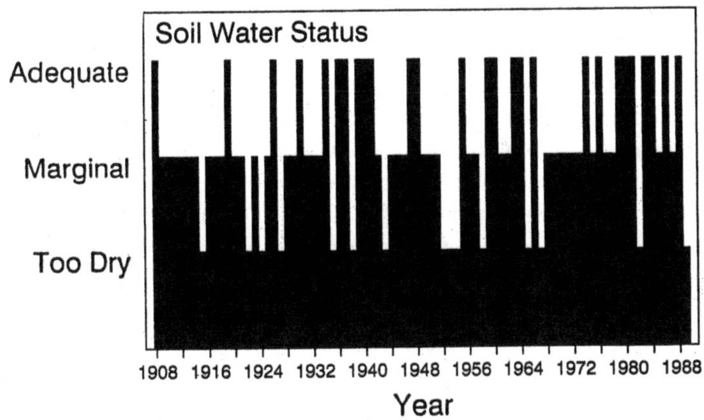

Figure 15.1. Interannual variation in the soil water status for establishment of Douglas fir seedlings. The height of the bar indicates the soil water status. The percentage in each category is given in Table 15.1

and June rainfall was 38 mm for the increased rainfall case, and 26 and 25 mm, respectively, for the reduced rainfall case. These changes could be considered as only a change in rainfall, or a surrogate for a change in evaporative demand plus a change in rainfall equivalent to a ±20% change in water availability. A 20% decrease in rainfall made conditions much worse (Table 15.1), particularly since about 65% of the marginal cases were close to the drier end of the range. Consecutive dry years may exacerbate conditions. Figure 15.1 shows only one such period of two or more years long occurred under the present climate. This changed to 4 times with a 20% reduction in rain. A 20% increase in rainfall gave a substantial improvement in establishment conditions.

A change in rainfall may be accompanied by changes in air temperature, humidity and cloudiness (Boer et al., 1992). This will effect evaporative

Table 15.1. Influence of soil water status on the likelihood of successful establishment after spring planting. Percentage of years from 1908–1989 in the establishment categories described in the text are shown for actual rainfall, and for a 20% increase and decrease in rainfall

	Percentage of years (1908–1989)		
	Too Dry	Marginal	Adequate
Actual	18	49	33
20% decrease	31	56	13
20% increase	13	37	50

demand. However, stomata tend to close as the vapor pressure deficit increases, thereby offsetting effects of temperature increases. Increasing the atmospheric carbon dioxide concentration could result in a reduction in stomatal conductance and increases in leaf area (Eamus & Jarvis, 1989). Rosenberg, McKenney, and Martin (1989) and de Bruin and Jacobs (1993) show that under a realistic range of conditions, transpiration rates will change by less than ±10%. Thus, it is probable that changes in rainfall will be the dominant factor in changing soil water status.

Droughts during July and August can also affect reforestation success. Spittlehouse and Childs (1990) found a summer drying pattern similar to that shown in Figure 15.1. Dry summers following dry springs can result in severe moisture stress even in established trees. This situation occurs 10% of the time at present, 16% with a 20% decrease in rainfall, and 6% a 20% increase. An increase in air temperature could result in heat stress since maximum temperatures at seedling height are already in the high 30s (°C) during the summer. The effect of changes in winter air temperature on chilling and budburst is important for growth in subsequent years and is considered later.

This analysis is species and site specific. However, the simulations could also be used to assess effects on natural regeneration of Douglas-fir, other plant species, and of soil texture. Fire risk could also be considered. Precipitation patterns and soils vary across the area where Douglas-fir occurs, but conditions described here are not atypical. Analyses could readily be done for other sites.

Management Implications and Response

Figure 15.1 and Table 15.1 indicate that these stony, sandy sites are extremely sensitive to variations in water availability. Any climate change that results in decreased rainfall in late spring and early summer is a serious threat to the establishment of Douglas-fir. This may be made worse if a warmer climate causes snow melt to occur earlier than at present. Table 15.1 indicates a relatively high chance of failure under present conditions. Consequently, strategies to deal with future changes are of use now.

A number of adaptation and mitigation options are available. Planting time could be change, though any move to earlier planting to avoid the droughts or to take advantage of warmer conditions needs to consider the potential for increased frost risk (McCreary, Lavender, & Hermann, 1990). A similar analysis would show whether there is less risk to fall planting. Second, the large size of the marginal class indicates that it is necessary to ensure good control of any vegetation that might compete for soil moisture. A third option is to consider alternates to clearcut harvesting, such as shelterwoods or partial cuts, to reduce the evaporative demand on the seedling and reduce soil surface temperature (Spittlehouse

& Childs, 1990). Fourth, increasing planting density or planting a mix in genetic sources of seedlings could be used on the expectation that within-population genetic variability will mean that some of the seedlings are more drought tolerant than others and will survive. Ledig and Kitzmiller (1992) suggest that since future change is unknown, it is prudent now to plant seedlings from families that do well under a range of climates. It is not feasible to plant alternate species now, in the expectation that they will be suited to future climates, when we do not know what these climatic conditions are or when they will occur. A fifth option is research to develop drought tolerant planting stock. This could including re-designing the seedling, e.g., larger root mass, or selection of drought-tolerant genotypes (Farnum, 1992).

Monitoring

The response strategy is not complete without a commitment to monitor what is happening. Whatever the chosen responses, there has to be a follow-up program in place. This acts both as an early warning system for future change as well as determining how and when actions should be taken.

Tree Growth

Issue of concern

Changes in growing season rainfall and air temperature will affect the annual growth of established Douglas-fir trees. Interannual variation in rainfall significantly affect the growth of Douglas-fir in coastal British Columbia (Spittlehouse 1985; Giles, Black, & Spittlehouse, 1985; Robertson, Jozsa, & Spittlehouse, 1990). From the timber management viewpoint, changes in the rate of tree growth will affect the time to harvest and the amount of wood available for harvest. Nontimber resources, such as the development of understory vegetation and animal habitat, will also be affected.

Scenarios

A forest water balance model was used to simulate April to October water use at two locations. Precipitation is adequate at other times of the year. The simulations for the first location (49°51′ N, 125°14′ W) were from 1965 to 1980 for a 0.75-m-deep gravelly sandy loam on an 8% NE-facing slope, 150 m elevation. Diameter growth varied annually by up to ±20%, and 80% of this was explained by differences in water use linked to variation in summer rainfall. A lack of rain in May and June had a greater effect than in August (Spittlehouse, 1985).

In the second analysis, I applied the water balance model to the seven sites described in Giles et al. (1985). These sites are on a SW-facing 10 to

35% slope, gravelly sandy loam 0.4 to 1.1 m deep, 190 to 300 m elevation (48°50′ N, 124°08′ W). Simulations were done using daily solar radiation, air temperature and rainfall data for the April to October period of 1951 to 1983, and were then repeated for a 20% reduction in mean rainfall. Transpiration deficit was correlated with the site index (the projected height of the trees at 100 years). Transpiration deficit is the difference between what the forest would transpire to meet the atmospheric demand if the soil had adequate water, and the actual transpiration. For the unadjusted rainfall, the regression between mean transpiration deficit (TD mm) and site index (SI m) was $SI = 54.9 - (0.5 \pm 0.07 * TD)$, se = ± 4.2 m, $R^2 = 0.910$. The average transpiration deficit for 1951 to 1983 for the reduced rainfall case was then used in this equation to predict the expected site index for the seven sites. The relationships are presented in Figure 15.2 in terms of site moisture class, which vary from very xeric (dry, rocky site at the top of the slope) to hygric (a water table present for much of the year) (Meidinger & Pojar, 1991). All sites are predicted to have reduced growth, but the dry sites are the most sensitive to the reduction in rainfall. Smith et al. (1992) noted different responses across site types.

Care must be taken to ensure that the model is not applied outside of conditions for which it was calibrated. Some of the years at the xeric and very xeric site are now drier than any experienced in the calibration range of the model. The frequency of extremely dry years increased, and the

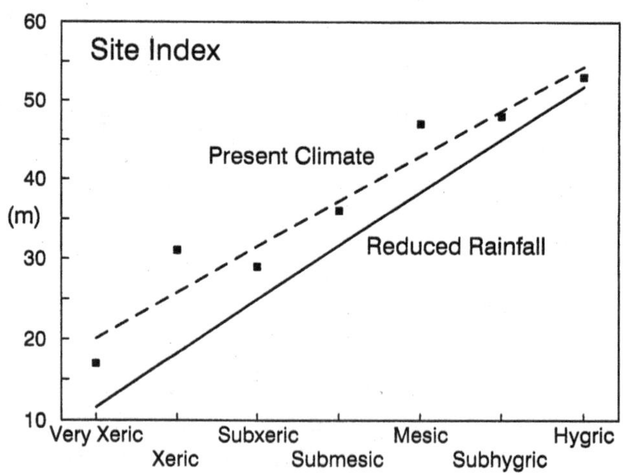

Figure 15.2. Relationship between expected height of the trees at 100 years (site index) and current site moisture class for Douglas fir stands. The dashed line and points show the current situation. The solid line shows the predicted site index for the present site moisture class after a 20% reduction in rainfall.

occurrence of consecutive dry years can increase the susceptibility of the trees to insects and disease. A mortality model (e.g., LeBlanc & Foster, 1992) could be incorporated to determine if tree death is induced at these sites. As was the case in the seedling analysis, the calculation of transpiration has the physiology implicit rather than explicit. As noted earlier, the changes in stomatal conductance and leaf area in response to increased carbon dioxide concentrations are likely to effect transpiration by less than 10%. Increasing atmospheric concentrations of carbon dioxide could act as a "fertilizer," because the photosynthetic rate increases with an increase in the carbon dioxide concentration. This could offset some of the above predicted decrease in growth, though not mortality induced by drought. Increased insect grazing of foliage, changes in partitioning of photosynthate, long-term acclimation, and nutrient deficiencies may limit gains that can be made (Eamus & Jarvis, 1989; Running & Nemani, 1991; Pastor & Post, 1993). Many Douglas-fir sites in British Columbia are nitrogen limited.

Changes in precipitation may be accompanied by changes in temperature. This may increase the length of the growing season and the need for water, and increase respiration. Photoperiod response will not be changed, thus trees may not be able to take full advantage of earlier warming. Temperature increases in the winter and spring may reduce the ability of the trees to meet their chilling requirement. McCreary et al. (1990) found over a range of ecotypes that there is sufficient chilling for Douglas-fir if the mean December to February temperature is below 9°C. The mean December to February temperature in the warmest part of the Douglas-fir zone in British Columbia is 4°C, though over the last 94 years, 14 years exceeded 5°C. However, most of area in British Columbia occupied by Douglas-fir is 1°C or more cooler than the warmest area and it would require a warming of over 4°C in any year to exceed the threshold. This is greater than is predicted by climate models (Boer et al., 1992).

Management Implications and Response

Changes in the rainfall during the summer will affect Douglas-fir growth, and thus other components of the ecosystem. An approximation of potential response is that each site moves down one site class. Management options include longer rotations, reduced harvest rates, or smaller volumes or piece sizes of wood. Other species such as lodgepole pine may be preferable on the drier sites.

Monitoring

As before, assessing how the trees and understory are responding over time is required. Inventories are part of forestry activities, but there is an enhanced need to regularly conduct them and to monitor other variables as well.

Conclusions

The analyses presented in this paper show that it is possible to develop responses to future unknown climates. They also provide information to aid current forest management. It is unlikely that all potential problems will be addressed. Surprises lie ahead, but partial preparation is better than no preparation and a good monitoring program would help detect when action is required. Although development of responses is possible, this is no reason for society to ignore options to reduce the risk associated with rapid climate changes. There will be significant costs to forest management in responding to climate change. A mix of actions will be required now and in the future. Immediate responses include deciding what degree of change in the resource constitutes a problem, determining possible solutions, and initiating monitoring programs to determine when intervention is required. The examples presented here are only a fragment of the assessments that can be done. I have concentrated on the dominant tree species of a managed forest. Analyses are required to assess other tree species, the nontimber resources of managed areas (e.g., understory, wildlife and water) and unmanaged areas. Current forest management initiatives to maintain biodiversity, reduce fragmentation and preserve habitat, and provide wildlife corridors will also aid managing for future climate change. Provenance trial where trees are grown hundreds of kilometers from their source provide information on a species ability to grow under a wide range of conditions.

Reductions in rainfall are a greater threat to the Douglas-fir forests in British Columbia than project changes in air temperature. Different ecotypes, planting stock, or even species may be required for the driest sites. The increased stress at these sites is likely to increase vulnerability to insects and disease. Management techniques developed for dry sites in the southern interior of British Columbia may provide guidance as to techniques to be used at these sites. Douglas-fir provenance trials should be assessed for provenances that have the ability to perform well in a wide range of climatic regimes. The moist, high productivity coastal Douglas-fir sites are less susceptible to the changes in rainfall and temperature predicted by global climate models, though reductions in growth can be expected. Increasing rainfall is likely to have the greatest benefit to forestry on the drier sites, though benefits may be offset by the increased cost of road and culvert maintenance in the wetter areas.

Acknowledgements. I thank a number of colleagues for comments of this paper. The opinions expressed in this paper are those of the author and do not necessarily reflect the views of the British Columbia Ministry of Forests.

References

Ausubel, J.H. 1991. A second look at the impacts of climate change. *Am Scientist*, 79, 210–221.

Baker, D.G., Ruschy, D.L., and Skaggs, R.H. 1993. Agriculture and the recent "benign climate" of Minnesota. *Bull Am Meteor Soc*, 74, 1035–1040.

Boer, G.J., McFarlane, N.A., and Lazare, M. 1992. Greenhouse gas-induced climate change simulated with the CCC second-generation general circulation model. *J Climate*, 5, 1045–1077.

Boyle, T.J.B. 1995. Biodiversity of Canadian forests, with particular reference to the west coast forests. In R.G. Lawford, P. Alaback, and E.R. Fuentes (eds.), *High-Latitude Rain Forests* (pp. 353–378). New York: Springer-verlag, 1995.

Brubaker, L.B. 1992. Climate change and the origin of old-growth Douglas-fir forests in the Puget Sound Lowland. In G. Wall (ed.), *Implications of Climate Change for Pacific Northwest Forest Management* (pp. 5–18). Waterloo, ON: University of Waterloo, Department of Geography, Occassional Paper No. 15.

Clark, J.S. 1991. Ecosystem sensitivity to climate change and complex responses. In R.L. Wyman (ed.), *Global Climate Change and Life on Earth* (pp. 65–98). New York, Chapman and Hall.

Cline, W.R. 1992. *Global Warming: The Economic Stakes*. Washington DC: Institute for International Economics.

de Bruin, H.A.R., and Jacobs, C.M.J. 1993. Impact of CO_2 enrichment on the regional evapotranspiration of agro-ecosystems, a theoretical and numerical modelling study. *Vegitatio*, 104/105, 307–318.

Dixon, R.K., Brown, S. Houghton, R.A., Solomon, A.M., Trexler, M.C., and Wisniewski, J. 1994. Carbon pools and flux of global forest ecosystems. *Science*, 263, 185–190.

Eamus, D., and Jarvis, P.G. 1989. The direct effects of increase in the global atmospheric CO_2 concentration on natural and commercial temperate trees and forests. *Adv Ecol Res*, 19, 1–55.

Farnum, P. 1992. Forest adaptation to global climate change through silvicultural treatments and genetic improvement. In G. Wall (ed.), *Implications of Climate Change for Pacific Northwest Forest Management* (pp. 81–84). Waterloo, ON: University of Waterloo, Department Geography, Occasional Paper No. 15.

Franklin, J.F. et al. 1992. Effects of global climatic change in Northwestern North America. In R.L. Peters and T.E. Lovejoy (eds.), *Global Warming and Biological Diversity* (pp. 244–257). New Haven: Yale University Press.

Giles, D.G., Black, T.A., and Spittlehouse, D.L. 1985. Determination of growing season soil water deficits on a forested slope using water balance analysis. *Can J For Res*, 15, 107–114.

Graham, R.L., Turner, M.G., and Dale, V.H. 1990. How increasing CO_2 and climate change effect forests. *BioScience*, 40, 575–587.

Houghton, J.T., Callander, B.A., and Varney, S.K. (eds.). 1992. *Climate Change 1992: The Supplementary Report to the IPCC Scientific Assessment.* Cambridge, England: Cambridge University Press.

LeBlanc, D.C., and Foster, J.R. 1992. Predicting effects of global warming on growth and mortality of upland oak species in the midwestern United States: a physiologically based dendroecological approach. *Can J For Res*, 22, 1739–1752.

Ledig, F.T., and Kitzmiller, J.H. 1992. Genetic strategies for reforestation in the face of global climate change. *For Ecol Manag*, 50, 153–169.

Leverenz, J.W., and Lev, D.J. 1987. Effects of carbon dioxide-induced climate changes on the natural range of six major commercial tree species in the western United States. In W.E. Shands and J.S. Hoffman (eds.), *The Greenhouse Effect, Climate Change, and U.S. Forests* (pp. 123–155). Washington, DC: The Conservation Foundation.

Malanson, G.P. 1993. Comment on modeling ecological response to climatic change. *Clim Change*, 23, 95–109.

McCreary, D.D., Lavender, D.P., and Hermann, R.K. 1990. Predicted global warming and Douglas-fir chilling requirements. *Ann Sci For*, 47, 325–330.

Meidinger, D.V., and Pojar, J. (eds.) 1991. *Ecosystems of British Columbia.* Victoria: Ministry of Forests, Special Report Series Publication No. 6.

Michaels, P.J., and Stooksbury, D.E. 1992. Global warming: a reduced threat?. *Bull Am Meteor Soc*, 73, 1563–1577.

Pastor, J., and Post, W.M. 1993. Linear regressions do not predict the transient responses of eastern North American forests to CO_2-induced climate change. *Clim Change*, 23, 111–119.

Peters, R.L. 1990. Effects of global warming on forests. *For Ecol Manag*, 35, 13–33.

Pollard, D.F.W. 1991. Forestry in British Columbia: planning for the future climate today. *For Chron*, 67, 336–341.

Reifsnyder, W.E. 1989. A tale of ten fallacies: The skeptical enquirer's view of the carbon dioxide/climate controversy. *Agric For Meteorol*, 47, 349–371.

Robertson, E.O., Jozsa, L.A., and Spittlehouse, D.L. 1990. Estimating Douglas-fir wood production from soil and climate data. *Can J For Res*, 20, 357–364.

Rosenberg, N.J., McKenney, M.S., and Martin, P. 1989. Evapotranspiration in a greenhouse-warmed world: A review and simulation. *Agric For Meteorol*, 47, 303–325.

Running, S.W., and Nemani, R.R. 1991. Regional hydrologic and carbon balance responses of forests resulting from potential climate change. *Clim Change*, 19, 349–368.

Sandenburgh, R., Taylor, C., and Hoffman, J.S. 1987. How forest products companies can respond to rising carbon dioxide and climate change. In W.E. Shands and J.S. Hoffman (eds.), *The Greenhouse Effect, Climate Change, and U.S. Forests* (pp. 247–257). Washington, DC: The Conservation Foundation.

Schneider, S.H. 1990. The global warming debate heats up: an analysis and perspective. *Bull Am Meteor Soc*, 71, 1292–1304.

Schneider, S.H. 1994. Detecting climatic change signals: are there any "fingerprints"? *Science*, 263, 341–347.

Smith, T.M., Shugart, H.H., Bonan, G.B., and Smith, J.B. 1992. Modeling the potential response of vegetation to global climate change. *Adv Ecol Res*, 22, 93–116.

Spittlehouse, D.L. 1985. Determination of the year-to-year variation in growing season water use of a Douglas-fir stand. In B.A. Hutchison and B.B. Hicks (eds.), *The Forest-Atmosphere Interaction* (pp. 235–254). Dordrecht, The Netherlands: D. Reidel.

Spittlehouse, D.L., and Childs, S.W. 1990. Evaluating the seedling moisture environment after site preparation. In S.P. Gessel, D.S. Lacate, G.F. Weetman, and R.F. Powers (eds.), *Sustained Productivity of Forest Soils* (pp. 80–94). Proceedings of the Seventh North America Forests Soils Conference. Vancouver: University of British Columbia, Faculty of Forestry Publication.

Veblen, T.T, and Alaback, P.B. 1994. A comparative review of forest dynamics and disturbance in the temperature rainforests of North and South America. In

R.G. Lawford. P. Alaback, and E.R. Fuentes (eds.), *High-Latitude Rain Forests* (pp. 173–213). New York: Springer-verlag, 1995.

Wall, G. (ed.). 1992. *Implications of Climate Change for Pacific Northwest Forest Management*. Waterloo, ON: University of Waterloo, Department of Geography, Occasional Paper No. 15.

Waterstone, M. 1993. Adrift in a sea of platitudes: why we will not resolve the greenhouse issue. *Enviro Manag*, 17, 141–152.

Watson, H.L., Bach, M.C., and Goklany, I.M. 1992. Global vs climate change. In G. Wall (ed.), *Implications of Climate Change for Pacific Northwest Forest Management* (pp. 97–109). Waterloo, ON: University of Waterloo, Department of Geography, Occasional Paper No. 15.

Woodman, J.N. 1987. Potential impact of carbon dioxide-induced climate changes on management of Douglas-fir and western hemlock. In W.E. Shands and J.S. Hoffman (eds.), *The Greenhouse Effect, Climate Change, and U.S. Forests* (pp. 277–283). Washington, DC: The Conservation Foundation.

16. A Comparison of the Ecology and Conservation Management of Cool Temperate Rainforest in Tasmania and the Americas

Michael J. Brown and Jennifer Read

Cool temperate rainforests reach their greatest extent in the Pacific Northwest of America and on the western coasts of South America, New Zealand, and Tasmania. The purpose of this book is to focus on the comparative responses to global change of ecological systems in North and South America. However the floristic affinities of the forests of the east and west Pacific in the Southern Hemisphere, the generality of climate change, ozone depletion, and atmospheric pollution, together with the different patterns of land use, provide a substantial basis for a longitudinal as well as latitudinal comparison of the effects of global change.

Much of the temperate forest of Australia and Tasmania is predominantly sclerophyllous and is dominated by *Eucalyptus*. These sclerophyll forests are found mainly in drier areas, but they extend into high-rainfall areas when anthropogenic, topographic, and climatic factors interact to produce fire return intervals of less than 80 to 100 years. Their floristic composition and ecology, as well as the management requirements for conservation and timber harvesting, are generally different from those of the rainforests in Australia, New Zealand, and the Americas. These sclerophyll forests will not be considered further here, except when their ecology impinges on that of the rainforests.

Rainforests dominated by *Nothofagus* spp. are confined to Australasia and southern South America. There is an estimated one million ha of such forest in Australia (RAC, 1991) and most of this occurs in Tasmania

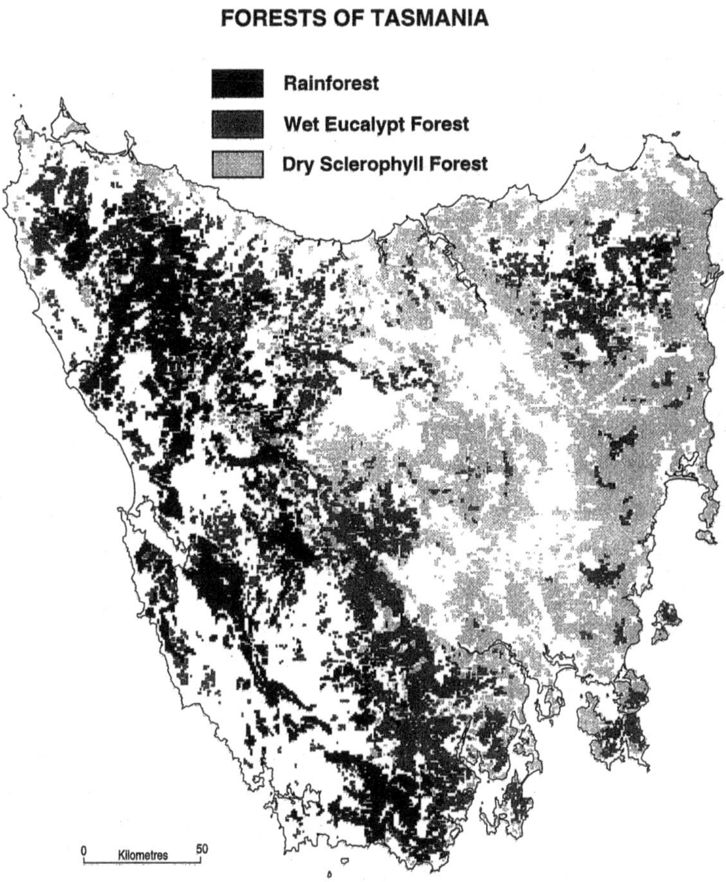

Figure 16.1. Distribution of forests in Tasmania.

(Brown & Hickey, 1990). The forests have high cultural and economic value, as well as great significance for science and conservation. In Tasmania they represent a characteristic and much-cherished facies of the landscape. They are found in the high rainfall areas of the west and northwest, although significant patches are found also in the northeast highlands and smaller outliers are found in gullies and cloud-trapping uplands of eastern Tasmania (Figure 16.1).

This chapter gives an overview of the evolution, ecology, management, and conservation of the Tasmanian rainforests and provides some tentative comparisons with the related forests in the Americas. We also evaluate of the management, conservation, and reservation status of the Tasmanian forests in the light of land use change. Where appropriate, we compare and contrast their management with their analogs in the Americas.

Definition and Extent

Tasmania lies at 39.5 to 43.5° S and rainforest is found within areas where mean annual rainfall ranges from 1300 mm to 3500 mm and mean annual temperatures from 5 to 12°C (Gibson, Brown, & Davies, 1991). Thus the forests fall within the perhumid category as outlined by Veblen and Alaback (see Chapter 9). The climate is distinctly seasonal with cold winters and uniform to winter-dominated rainfall (Read, in press). In drier areas, the rainforests are found primarily in riverine gullies, southeast slopes, and at higher altitudes in cloud-forests where the annual rainfall is augmented by stripping of moisture from cloud and fog. The effective summer monthly rainfall in such areas is at least 40 mm (Jackson, 1983), and the forests must have some topographic protection from fire.

In common with the rainforests of South America, those in Tasmania are dominated by a mixture of broadleaved trees and conifers. Cool temperate rainforests in Tasmania are characterized as forests with trees greater than 8 m in height dominated by species of *Nothofagus*, *Atherosperma*, *Eucryphia*, and/or the conifers *Athrotaxis* (the only extant southern hemisphere representatives of the Taxodiaceae), *Lagarostrobos*, *Phyllocladus* (both Podocarpaceae), and *Diselma* (Cupressaceae), and with the ability to regenerate in the absence of broad-scale disturbance such as wildfire (Jarman & Brown, 1983). In common with Chile, the genus *Nothofagus*, particularly the evergreen *N. cunninghamii*, is almost ubiquitous in the forests. The vegetation intergrades at high altitudes with subalpine and alpine vegetation. At low altitudes on very poorly drained sites the rainforests intergrade with swamp forests which are often dominated by blackwood (*Acacia melanoxylon*). Over most of their range in the lowlands, the rainforests are the end point of a fire-determined succession from wet eucalypt forests or mixed forest (Gilbert, 1959; Ashton, 1981a,b). The wet eucalypt forests contain giant *Eucalyptus* stems, which can be more than 100 m tall (Ashton, 1981b) over an understory of either broad-leaf sclerophyll trees or a closed canopy of rainforest trees. Such forests are unique to Australia (Ashton, 1981b) and can be viewed either as mature eucalypt forest or as developing rainforests (Jarman & Brown, 1983). Following fire in the mixed forest there is abundant regeneration of eucalypts and a broad-leaf wet sclerophyll understorey which is replaced subsequently (after about 80 to 100 years without fire), by rainforest species. The eucalypts are self-thinning and in old mixed forest stands are few and scattered. The continuum in eucalypt stocking from dense to sparse makes practical demarcation between rainforest and mixed forest very difficult. The Forestry Commission in Tasmania currently adopts a 5% eucalypt crown cover as the practical limit to eucalypt forest.

The total area of rainforest in Tasmania has been estimated at 765,000 ha (Brown & Hickey, 1990), but recent estimates indicate that the Nothofagus-dominated component may be about 600,000 ha. The largest continuous patch is in the north west and has an area of 65,000 ha

(J. Hickey, personal communication). The mixed forests are estimated at 113,000 ha and a further 47,000 ha of "pure" rainforest contains eucalypts at less than 5% cover (Hickey & Savva, 1992).

History

The fossil history of cool temperate rainforests in Tasmania is summarized here from Hill (1990a). There appears to have been a progressive floristic simplification since the Tertiary. Following the break up of Gondwana, the climatic changes of the Tertiary included a decline lit, but increased variability of, temperatures, and decreased rainfall and humidity, with increased seasonability. The rainforest fossil record indicates that species have undergone only limited evolutionary change since then. They appear to have suffered from a lack of migratory routes from which to replenish the vegetation. The Early Eocene flora has a high species richness, but *Eucryphia* is the only recognizable modern high-latitude taxon present. Later Eocene sites indicate complex forest containing some taxa which have modern high-latitude cool temperate affinities. Oligocene deposits again exhibit high species diversities, and include mixtures of species with affinities to modern species from temperate high latitudes and equatorial high altitudes (Hill, 1990b). They include the Tasmanian genera *Lagarostrobos*, *Microstrobos*, *Phyllocladus*, *Athrotaxis*, *Nothofagus* subgenus *Menziesospora*, and genera such as *Nothofagus* subgenus *Fuscaspora* and subgenus *Nothofagus*, *Dacrycarpus*, and *Araucaria* sect. *Columbea*, which are now absent from Tasmania, but are found in South America or New Zealand (Hill, 1990a,b; Wells & Hill, 1989). In a review of the evidence for extensive past distributions of Gondwanan floral elements in the Tertiary, Hill and Carpenter (1991) identify at least six Australian fossils with extant relatives only in South America and a further six South American fossils with modern analogs in Australia. There is a gap in the rainforest fossil record until the later Pliocene to early Pleistocene. Fossil rainforest deposits from this period provide evidence of an essentially modern rainforest vegetation (Hill & Macphail, 1985), and it appears that the current floristic composition and richness were reached prior to the Quaternary glaciations (Hill, 1990a). During the Quaternary, these forests appear to have retreated and expanded in response to glacial and interglacial cooling and warming. The exact number and timing of glaciations is uncertain, but at least three and possibly six have been postulated (Kiernan and Hannan, 1991; Colhoun & Fitzsimmons, 1990; Keirnan, 1989). Macphail (1979) has documented the palynological evidence for advances and contractions of forest during the Holocene, however interpretation of climatic influences on the vegetation is confounded by possible anthropogenic fire-caused changes. The Tasmanian Aborigines have been present for at least 20,000 to 30,000 years (Kiernan, Jones, & Ranson, 1984; Cosgrove, 1989) and evidence of their burning of the vegetation is documented in the vegetation itself (e.g., Bowman & Brown, 1986), as charcoal deposits in the soil (e.g., Macphail, 1979) and in early historical accounts of European contacts (e.g., Plomley, 1966). However, there is little evidence that the Aborigines made direct use of the rainforests for food or for shelter. The impacts of fire on the distribution of the major vegetation types and on rainforest in particular are considered below.

Botanical Characteristics

Floristic Composition and Classification

Apart from the dominant species that characterize the rainforests, the abundant or most common vascular plant species encountered are from the Epacridaceae (*Richea*, *Trochocarpa*, *Cyathodes*, *Archeria*, *Prionotes*), Asteraceae (*Olearia*), Proteaceae (*Agastachys*, *Cenarrhenes*, *Orites* and *Telopea*), Escalloniaceae (*Anopterus*), Cunoniaceae (*Anodopetalum*), and Winteraceae (*Tasmannia/Drimys*) (Jarman, Brown, & Kantvilas, 1984; Jarman, Kantvilas & Brown, 1991). Prominent or widespread fern genera include *Blechnum*, *Dicksonia*, *Hymenophyllum*, and *Polystichum*. Some characteristic lichen genera include *Pseudocyphellaria*, *Psoroma*, and *Menegazzia*. In the bryophytes prominent or widespread species are from such families as Lepidoziaceae (five genera) and Geocalycaceae in the liverworts, and Dicranaceae and Sematophyllaceae in the mosses (Jarman et al., 1991).

Trees, shrubs, ferns, and other cryptogams are the most conspicuous life forms in Tasmanian rainforests (Jarman et al., 1984). The trees and shrubs are mostly single stemmed, but some species, such as *Anodopetalum*, produce dense tangles (Barker, 1992). Buttressed trees and lianes are rare or absent, but scrambling plants (*Trochocarpa*, *Aristotelia*), rosette shrubs (*Richea*, *Dracophyllum*), and tree ferns (*Dicksonia*, *Cyathea*) are distinctive in some forest types. *Prionotes cerinthoides* is the only true angiosperm epiphyte present but there are abundant epiphytic cryptogams (Jarman et al., 1984).

Tree heights range from 5–8 to 30–40 m (except for emergent *Eucalyptus* when present) and all trees and shrubs are evergreen except for the winter deciduous *Nothofagus gunnii* which occupies specialized habitats. Leaves are mostly simple, microphylls or nanophylls; entire or serrated; glabrous or with an indumentum of wax or hair on one surface only (Jarman et al., 1984). The woody angiosperms are generally hermaphrodites, but some genera, such as *Atherosperma*, *Nothofagus*, *Tasmannia* and the conifers, are monoecious or dioecious. Many of the species have no obvious aids to seed dispersal, although a few have succulent fruits (e.g., *Cenarrhenes*) or winged fruits or seed (e.g., *Eucryphia*, *Anopterus*).

The rainforests have been formally classified into 38 floristically defined communities in four broad floristic/structural groups which fall within two alliances (Figure 16.2) (see Jarman et al., 1984, 1991, 1994). These communities reflect climatic, altitudinal, and edaphic gradients

Figure 16.2. Forest profiles of the four main structural/floristic types of rainforest in Tasmania.

within Tasmania, and provide a useful basis for the categorization of conservation and reservation of rainforest. The *Athrotaxis* alliance contains one floristic/structural group (termed open-montane forest), which is confined to high altitudes. Where it contains elements from the associated montane vegetation as well as the more typical rainforest elements.

Most Tasmanian rainforest falls within the *Nothofagus cunninghamii* alliance subdivided into three floristic/structural groups, which largely reflect soil fertility. On the most fertile and well-drained soils, derived from dolerite and basalt, the rainforests have tall well-formed trees of *Nothofagus cunninghamii* +/− *Atherosperma* with a park-like open or ferny understory and abundant epiphytic bryophytes and lichens. These forests have been termed "callidendrous" (= beautiful tree) by Jarman et al. (1984). On poorer sites, the trees retain their form, but there is an increasing complement of other tree and shrub species separated into well-defined structural layers. These forests are termed thamnic (= shrubby) forests. On the poorest siliceous quartzites and conglomerates or in areas of very poor drainage, the forest trees are short and of very poor form. There is no clear separation of layers and vascular plant species richness is high, while the numbers of ferns is generally low. These forests are called implicate forests by Jarman et al. (1984).

Biodiversity and Biogeography of Rainforests

There is a very low diversity of vascular plant species in Tasmania lowland rainforests, but they are comparatively rich in nonvascular plants. There are 70 angiosperm species from 25 families of dicotyledons and 5 families of monocotyledons (Jarman et al., 1991). This represents about 5% of the Tasmanian flora. Another 28 species of uncertain status also occur. There are six species of gymnosperms in four families (58% of the flora) and 49 pteridophytes (50% of the flora) (Jarman et al., 1986). In the lower plant groups, 217 lichens and 168 bryophytes have been recorded, each representing about one third of the Tasmanian native complement (Jarman et al., 1991). The differences in richness are greater at any one site, where for example there may be as few as 3 to 6 angiosperms, but 60 to 100 lichens and 30 to 70 bryophytes. The rainforests support a species-depauperate vertebrate fauna which is mostly adventive from surrounding vegetation especially after disturbances such as fire. However they are rich in invertebrates, such as Coleoptera (Coy, 1990).

The vascular plant species show pronounced regional distributions within Tasmania (Brown, Kirkpatrick, & Moscal, 1983; Kirkpatrick & Brown, 1984). For example, the gallery rainforests of the western rivers have a particularly distinctive flora, which includes *Lagarostrobos franklinii*, *Acradenia frankliniae*, *Epacris mucronulata*, *Leptospermum riparium*, and the lily *Milligania longifolia*. Evidence is accumulating also that there are strong regional differences in the nonvascular flora and invertebrate species compositions of disjunct rainforest patches of the same floristic community occurring on geologically and physiographically matched sites (G. Kantvilas, R. Mesibov, personal communication).

The modern affinities of the flora generally reflect the proximity of Tasmania to other land masses, but reflect also the strong Gondwanan relationships. Thus only 6 of the 32 woody plant genera found in Tasmanian rainforests are endemic. Of these, 13 are found in New Zealand, seven are also found in South America, and 23 (72%) are shared with the Australian mainland (Table 1; Read, in press). Similarly,

92% of the ferns are shared with Australia, 69% with new Zealand, and only 8% (4 species) are shared with South America (Jarman et al., 1986). About two thirds of the lichen flora have an austral cool temperate distribution (Galloway, 1988; Kantvilas, 1990). Only 5% of the rainforest lichens and about 4% of the bryophytes are endemic, in contrast to the spermatophytes where, approximately 75% are endemic.

Ecology

Ecological Succession

The distribution and proportions of rainforests, eucalypt forest, and non-forest communities in the lowland high rainfall areas of Tasmania are determined largely by fire (Gilbert, 1959; Jackson, 1968; Ashton, 1981a,b; Brown & Podger, 1982a; Ellis, 1985; Podger, Bird, & Brown, 1988), with large areas of pure rainforest being found only where average fire frequencies are greater than 400 to 600 years.

Once a closed canopy is established in a rainforest, it preserves a humid interior with rapid decomposition of fine fuels, especially on high-base status soils. It is thus generally resistant to fire. However in older forests, when individual *Nothofagus* stems die, they remain as dead snags for some time and support a column of lichens and bryophytes. In severe drying weather and strong winds, these "ladder fuels" act as traps for spot fires and provide ignition and entry points for rainforest fires (A.B. Mount, personal communication). Such fires can burn fiercely and rapidly traverse long distances. For example the 1982 Savage River fire burned through a strip of rainforest 25 km by 5 km wide in one night (Barker, 1991).

A single wildfire in a rainforest is followed by a succession of mosses and liverworts, such as *Funaria* and *Marchantia* (Cremer & Mount, 1965; Duncan & Dalton, 1982), and a flush of seral vascular species such as *Senecio* spp., *Pteridium esculentum*, *Gahnia grandis*, and *Leptospermum* spp (Hill, 1982; Hill & Read, 1984; Barker, 1991), together with the slowly establishing seedling and vegetative regeneration of the original rainforest elements. The fire-sere species generally have attributes conducive to further fire; they are flammable, have high essential oil contents, and produce rapid fine-fuel accumulation because of their generally short life spans or early twig death. In addition, their litter may have high levels of phenols, which inhibit rapid decomposition. The seral species recover readily from fire either by vegetative growth or from soil-stored or widely disseminated seed. Thus repeated fires further promote these species and kill any remaining rainforest elements. The complete conversion of areas of implicate rainforest to moorland and scrub by repeated firing over a 60 year period has been documented by Podger et al. (1988).

If there are no further fires beyond the initial one, rainforest eventually will reestablish on the site. Thus there are many "pole" stands of even-aged *Nothofagus* forest to be found in areas where rainforest fires have occurred in the past 100 years or more. At the age of 200–400 to 600 years, these stands begin to break up through deaths of *Nothofagus* and other canopy dominants. The mature forest reaches a dynamic equilibrium in which the rainforest elements regenerate through time by gap-phase replacement. A further fire before this stage, but after maturity of the dominants, would follow the same path.

Mixed forests are maintained in a manner similar to rainforests, except that the average fire frequency here is 80 to 400 years (Gilbert, 1959). The eucalypt dominants die after about 400 years and they cannot regenerate under the closed rainforest canopy. The stand is thus converted to rainforest. If fires occur in the period 80 to 400 years, the eucalypts are maintained and can expand into any adjacent patches of burnt rainforest. The encroachment is step wise, with most eucalypt seed falling within a radius of the height of the tree (Gilbert, 1959). However, in the strong winds which prevail in wildfires, seed can be carried much further, and eucalypts have been observed to establish at distances to 200 m from the nearest (dead) trees. There is some documented evidence of rainforest invasion of long unburnt moorland sites in western Tasmania (Podger, 1990). Ellis (1985) documents establishment of rainforest in long unburned areas of grassland, shrublands, and eucalypt forest on fertile soils in north eastern Tasmania, and the old-field succession described by Read and Hill (1983) follows a similar pathway.

Relationships among soil nutrient status, climatic factors, and the fire regime have been described by Jackson (1968) for the vegetation on oligotrophic soils of western Tasmania. The vegetation there ranges from sclerophyllous sedgeland and moorland through sclerophyll scrub and woodland to rainforest. The feedbacks among the vegetation, the physical environment, and fire profoundly influence the successional pathways. Jackson described a stochastic process whereby the different structural forms of vegetation may become fixed through ecological drift, a process analogous to genetic drift. Thus large areas of land in the southwest of Tasmania are climatically and edaphically suited to the growth of rainforest, but support sclerophyll communities, either pure or in mixture with rainforest. About 47% of the lowland regions receiving annual precipitation in excess of 2000 mm and with a precipitation to evaporation ratio of >2.5 are occupied by sclerophylls (Jackson, 1983). These disclimax communities are widespread in Tasmania compared with the vegetation of New Zealand and southern South America. The difference has been attributed to the characteristics of the plants, the low inherent soil fertility, the history of human influences and the postre climate (Jackson, 1968). Jackson's theory has been analyzed using Markov chains; the predicted distributions of the rainforest and other vegetation was generally in accord with the actual distributions (Henderson & Wilkins, 1975; Noble & Slatyer, 1981; Brown & Podger, 1982b). In situ evidence for the process has been reported by Brown and Podger (1982b) and Podger et al. (1988).

Dynamics of the Rainforest Dominants

All three of the dominant angiosperms in lowland rainforest are capable of continuous regeneration with self-replacement (Read & Hill, 1988). The order of dominance of these species is not simply a function of their relative shade tolerance as found in the temperate *Nothofagus* forests of South America (Veblen et al., 1981, 1983) and New Zealand (Wardle, 1984). *Nothofagus cunninghamii* and *Eucryphia lucida* are less shade tolerant than *Atherosperma* (Read 1985; Read & Hill, 1985, 1988), yet *Atherosperma* is subdominant in the forests in which it cooccurs with *Nothofagus*. Both *Nothofagus* and *Eucryphia* regenerate continuously in communities containing *Atherosperma* both by root and basal stem suckers and from seed in canopy gaps. *Atherosperma* regenerates mainly from basal stem suckers and has a very limited germinative capacity (Hickey, Blakesley & Turnesr, 1983), because there is a high proportion of malformed seed. The rate of seedling establishment is low due to high browsing pressure from native herbivores. Individual stems of *Atherosperma* do not exhibit the height-growth potential of *Nothofagus* nor are they as long-lived. Thus the species is usually codominant or subdominant to *Nothofagus*. In rainforest gullies surrounded by eucalypt forest in the drier outlying patches in eastern Tasmania, *Atherosperma* is the sole dominant. In these situations dominance is achieved because *Atherosperma* can disperse more readily than *Nothofagus* and can occupy sites where fires are more frequent. *Nothofagus cunninghamii* is able to regenerate in the absence of large-scale disturbance on optimal sites because there are no other more shade-tolerant species able to compete effectively in canopy gaps (Read & Hill, 1985, 1988). Such gap replacement is a feature also of *Nothofagus* in high-altitude forests in New Zealand and Chile, and at high latitudes in Child (e.g., June & Ogden, 1978; Veblen, Donoso, Schlegel, & Escobar, 1981). However, continuous regeneration of *Nothofagus* is less frequent in the forests with higher species richness that occur on optimal sites in Chile and New Zealand (e.g., Veblen et al., 1981; Wardle, 1983, 1984). In such situations, *Nothofagus* may be maintained by infrequent catastrophic disturbances, such as vulcanism, landslides, and wind throw.

The Tasmanian endemic *Nothofagus gunnii* is the only winter-(cold) deciduous species native to Australia. It is found mostly above 800 m, but extends down to 550 m above sea level (Robertson & Duncan, 1991). It occurs as codominant with *Athrotaxis selaginoides*, *Nothofagus cunninghamii*, and *Phyllocladus aspleniifolius* in closed implicate rainforests (Jarman et al., 1984; Read & Hill, 1988). The populations in closed forests suggest episodic regeneration, consistent with the high-light compensation point and high dark rate of respiration of the species (Read & Hill, 1985, 1988). In upland exposed situations, the species has a competitive advantage conferred by its high cold tolerance and its deciduous habit, but even in relatively open implicate forests, the species sometimes shows skewed-normal rather than the usual reverse—J size class distributions.

The coniferous dominants—*Athrotaxis*, *Lagarostrobos*, and *Phyllocladus*—each exhibit continuous regeneration when they occur in implicate forests or open situations in montane forests where there is no competition from grasses or from grazing (Cullen, 1987; Read & Hill, 1988; Gibson & Brown, 1991; Barker, 1992). In mature, closed forests these species have unimodal or multimodal distributions (e.g., Ogden, 1978; Kirkpatrick & Harwood, 1980), suggestive of initial establishment and extended recruitment after catastrophic disturbance such as landslides (Cullen, 1991), floods (Gibson & Brown, 1991), or wildfire (Gibson, 1988; Read & Hill, 1988). Forests containing mixtures of these species are not necessarily self-replacing. However, canopy composition may remain constant over long periods of time in these forests because of the long life-spans of the trees (several hundred to over 1000 years).

Conservation and Management

In common with much of the world's remaining natural vegetation, there is decreasing opportunity for management of rainforest by benign neglect, even within stringently policed protected areas. For example, rainforest is very vulnerable to damage on a broad scale by wildfire and to fragmentation by such infrastructure developments as roads and power lines. The human pressures both within and outside reserves are such that active monitoring and management of the remaining rainforest are needed to ensure its perpetuity. Much of the Tasmanian rainforest and wet eucalypt forest is old-growth i.e., the forests are ecologically mature and essentially unmodified by humans (RAC, 1991). As well as their value for the maintenance of biodiversity and ecosystem processes, they contain valuable timbers and other economic resources. Management of the forests for their conservation therefore needs to address both sets of values.

Clearing

It is thought that about 15% of the rainforest (Hickey, 1990) and 35% of the wet eucalypt forests (Wells, 1989) have been lost to farming or other use. About two thirds of the blackwood swamp forests of northwestern Tasmania have been cleared for agriculture. About 13,000 ha of this marginal rainforest type remains and is managed primarily for the blackwood (*Acacia melanoxylon*) production. A conservation evaluation of these forests has been completed (Pannell, 1992). Small amounts of rainforest on public lands have been clear felled and burned for sowing to eucalypt forest, and some areas of logged mixed forest and damaged rainforest on basalt soils have been converted to eucalypt on industrial private lands. Only about 4% of the rainforest and mixed forest is privately owned, so that management responsibility and conservation needs can largely be addressed on public land tenures.

While most of the remaining old growth forests are in relatively pristine condition, there have been some extensive and some locally severe impacts associated with access for logging, hydroelectric schemes, mineral exploration and mining, and tourism and recreation.

Logging

Currently there is (and has been) little clear falling of *Nothofagus*-dominated rainforests, although there has been extensive selective logging in the premechanized era of riparian stands of Huon pine (*Lagarostrobos franklinii*) and of King Billy pine (*Athrotaxis selaginoides*) (Horne & Hickey, 1991). Some selective logging of previously logged Huon pine continues, but there is now only salvage logging of King Billy pine from areas to be inundated for hydroelectric schemes, or from road lines (WGFC, 1990). Current policy prescribes a small annual sawlog cut of 500 m^3 pa for Huon pine cut from about 20% of the total Huon pine habitat of 10,000 ha (Peterson, 1990). The supply is expected to last 70 years, including 15 years of currently stockpiled logs (Hickey & Felton, 1991).

The mixed forests are an important source of special timbers, a term applied to noneucalypt species in Tasmania. They are logged on rotation lengths of 80 to 100 years, which is insufficient time for the redevelopment of mature mixed forest. Hickey and Savva (1992) consider that a 200-year timeframe and silvicultural systems alternative to the standard clear fall, burn-and-sow treatments could be developed to promote the reestablishment of commercial rainforest tree species to a millable size. Such a procedure would be appropriate in special timber supply zones.

Logging of sclerophyll forest on the periphery of rainforest also poses hazards to the rainforest from edge effects, which are most pronounced in the small outlying patches of riverine rainforests in drier parts of Tasmania. The effects of such disturbances can effectively be ameliorated in eastern Tasmania by the retention of a 40 m, buffer strip of sclerophyll forest adjacent to the rainforest gullies (Neyland & Brown, 1991).

Hydroelectric Developments

There have been some large scale hydroelectric power developments in Tasmania, and the dams associated with these have flooded unknown areas of forest, including rainforests. However in a study of riparian Huon pine communities, Millington et al. (1979) estimated the loss of forested river flats at 13,500 ha or about 16% of the total area of this habitat. Some minor further losses will have accrued since that time with the building of the King River and Henty–Anthony River schemes.

Mining and Mineral Exploration

Activities associated with mining and mineral exploration have had a devastating effect on the rainforest around Queenstown in western Tasmania, through the use of fire as an "exploration tool," but also through toxic waste production, including SO_2 emissions from smelters and downstream deposits of toxic sediments on riparian vegetation (e.g., Kirkpatrick, 1977). Much of the rainforests of western Tasmania are situated on highly prospective rocks, which can contain for example copper, gold, silver, lead, zinc, and iron, and the pressure for exploration and mining is continuing, albeit with modern techniques which are much more environmentally benign.

Fire

Anthropogenic fire is a major threat to Tasmanian rainforest, as it is also in South America (Veblen and Alaback, Chapter 9). Kirkpatrick and Dickinson (1984) report that 8% of the total area of lowland rainforest in the state has been burned since 1950. If attrition at this rate was uniformly spread, then an area of rainforest could burn on average every 500 years. There are some areas of Nothofagus forest in northwestern Tasmania where this frequency may be realistic, but there have been devastating fires in the conifer communities at rates which clearly do not conform to a "natural" cycle. For example Brown (1988) found that about 30% of King Billy pine (*Athrotaxis selaginoides*) stands had been fire killed in the last 100 years. Given that most of these stands contained individuals 800 to 1000 years old, such an attrition rate is clearly not part of a natural fire cycle. Similarly Cullen and Kirkpatrick (1990b) report that one third of the area which supported pencil pines (*Athrotaxis cupressoides*) has been burnt since 1960. The pencil pines reach ages of 900 to 1300 years (Ogden, 1978), so that such rates again are clearly unsustainable. The rainforest conifers *Athrotaxis* spp. and *Lagarostrobos* and *Nothofagus gunnii* are all extremely fire sensitive and lack mechanisms for recovery from even mild fires. This sensitivity appears to contrast with the South American conifer, *Fitzroya cupressoides*, which does regenerate following light (but not intense) fires (Veblen & Ashton, 1982). *Athrotaxis* spp. and *Lagarostrobos* are mast-year seeders and rely almost totally on seed regeneration from marginal unburnt stems for recovery after fire even though they exhibit vegetative spread in undisturbed forests (Gibson & Brown, 1991; Shapcott, 1991; Cullen & Kirkpatrick, 1990a,b). Other dominant and subdominant rainforest elements can recover after fire, both from seed and from vegetative sprouts (Barker, 1991). If there are no further fires, the rainforest canopy is eventually restored, but there is a relatively high probability of reburn in areas of easy access, especially in the first 20 years after fire (Jackson, 1968; Kirkpatrick, 1977; Hill & Read, 1984; Barker, 1991). Repeated fires eventually removed even the resistant rainforest species (Ellis, 1985; Podger et al., 1988).

Diseases, Weeds, and Pests

There are several fungal pathogens which cause disease in Tasmanian wet forests. *Eucalyptus obliqua* and *E. regnans* dominate mixed forests in a number of humid and fog-prone valleys in northwestern Tasmania. These situations favour the development of pathogenic activity by a leaf spot fungus *Aulographina eucalypti* and can lead to epidemic levels of premature defoliation and shoot death (Palzer, 1978). Eucalypts killed by the disease cannot regenerate without disturbance to the understory, and so the disease may contribute to an accelerated succession to rainforest (Wardlaw, 1990).

Phytophthora cinnamomi is an introduced root-rotting fungus capable of killing a wide range of host plant species in Tasmanian rainforests (Podger & Brown, 1989). The fungus has been isolated from 39 species of host plants, particularly in the Epacridaceae, Eucryphiaceae, and Proteaceae. The fungus is active in rainforest only in disturbed areas such as roadsides and recently burned areas where soil temperatures are elevated above 15°C. It has not been found in undisturbed forest, and Podger and Brown (1989) consider that while no single rainforest species is under threat of extinction, susceptible species which are either shade intolerant or which lack efficient dispersal mechanisms may be at risk of local elimination from fire-disturbed sites or regeneration.

Myrtle wilt (Elliott, Kile, Candy, & Ratkowsky, 1987) is a disease of the rainforest dominant *Nothofagus cunninghamii*, caused by a native fungal pathogen *Chalara australis* (Kile & Walker, 1987). The fungus is found throughout rainforest areas and appears to be air and/or water borne. It infects tree wounds although there is some evidence of underground spread of infection by root graft (Kile, Packham, & Elliott, 1989). The disease is associated with a pinhole borer *Platypus subgranosus* (Howard, 1973), which apparently is attracted to infected trees, and the frass left by the insect provides an early indication of disease. In undisturbed rainforest, *Nothofagus cunninghamii* death and gap formation caused by wilt appear to be in equilibrium with regeneration processes (Packham, 1990; Packham et al., 1990), but disease expression is typically exacerbated by disturbances such as roads or logging (Elliott et al., 1987). *Chalara* species are known also from the *Nothofagus* forests of New Zealand, but do not appear to be a conservation problem there or in South America. However, it may become relevant for forest conservation there given the increasing levels of disturbance which are occurring.

Weedy plant species are not a serious problem in undisturbed rainforests, but they may invade into disturbed vegetation, especially near urban areas along roadsides, in cut-over areas and after fires. Some of these species have the capacity to alter the structure, composition and ecology of the native forest (Duckett, 1990). Potential problem species in this regard include species of *Fuchsia*, *Rubus*, *Digitalis*, *Clematis*, *Leycesteria*, *Genista*, *Sarothamnus*, and *Cortaderia* spp. There have been no concerted approaches to these problem species, except for *Cortaderia*. Following the recent importation of a new species and an effective pollen source, for the existing vegetatively propagated females the species has become a potentially invasive threat. A high profile publicity campaign and active work by land managers have controlled the spread to date (Duckett, 1990).

The rainforests are not generally subject to attack by insect pests. Some concern has been expressed about the effects of honey bees on native insect pollinators and feeders on leatherwood (*Eucryphia lucida*) but studies to date have not shown any direct competitive effect (Ettershank & Ettershank, 1990). The introduced European wasp (*Vespula germanica*) has naturalised in much of the southwest and west of Tasmania, including the World Heritage Area. No detailed studies of its ecology have been made but it is a voracious feeder and preys on many native insect pollinators as well as feeding on nectar.

Vertebrate pests have severely affected the regeneration of pencil pine (Cullen and Kirkpatrick, 1990a,b). These authors attribute seedling loss to sheep and cattle grazing, the introduction of rabbits, and the high numbers of wallabies (*Macropus rufogriseus*) present. High numbers of wallabies and possums (*Trichosurus vulpecula*) are maintained also in areas of forest adjacent to pasture, and contribute to a depletion of rainforest regeneration, both within the forest and in oldfields (e.g., Read & Hill, 1983).

Global Climate Change

The current regional models of global climate change are not sufficiently accurate to predict likely outcomes on Tasmania's cool temperate rainforest. However it is likely that the most vulnerable species will be those that have the combination of a restricted distribution and poor seed dispersal characteristics. Busby (1988) has examined one climate-change scenario and found that most of the range of *Nothofagus cunninghamii* in Victoria and Tasmania would become marginal. However it is likely that even under conservative scenarios, individual effects from altered fire regimes and disease may be substantial (Busby & Brown, 1994). For example, elevated temperatures are conducive to increased activity levels of the pathogenic *Phytophthora cinnamomi* as outlined above, and the changes in predicted rainfall patterns and associated precipitation/evaporation ratios are likely to increase the frequency of fire prone weather conditions in rainforest areas.

Evaluation of Reservation Status

In the early 1980s a moratorium was placed on logging in rainforest until an evaluation of conservation needs was made. The moratorium was put in place because of the ecological fragility of rainforest systems, and because there had been no systematic surverys of the nature of the variability of the forests either for conservation or for production purposes. Subsequent statewide surveys resulted in the formal botanical classification and description of the forests (Jarman et al., 1984, 1991). Inventories have included vascular plants, pteridophytes, bryophytes, and lichens, and further work is being undertaken on the nonvascular plants. In addition to these broad studies, more detailed studies have been conducted of the communities associated with Huon Pine (Gibson et al., 1991; Shapcott, 1991), *Athrotaxis* spp. (Cullen & Kirkpatrick, 1990a,b), relict rainforest patches in eastern Tasmania (Neyland & Brown, 1991), swamp forests (Pannell, 1992) and of epiphytic lichen communities (Kantvilas, 1988; Kantvilas & Minchin, 1989).

These community studies are of great utility in determining the extent to which known occurrences fall within reserves. However they are of limited use in inaccessible or poorly known vegetation, especially in determining the areal extent of any one community. A complementary approach to the problem has been to use environmental domain analysis. The method is described in detail in Hickey and Brown (1989); in essence it consists of locating potential regional rainforest reserve areas occurring on the range of geologies over five altitude classes found within each forested region of the state. Tasmania has been divided into 11 nature conservation regions (Orchard, 1988) and in each of these that contain rainforest, Recommended Areas for Protection (RAPs) were targeted to include 30% of rainforest in each altitude/geological class. This approach is similar to the biogeoclimatic classification adopted for forest management in Canada (Boyle, Chapter 18), except that is has been used in Tasmania specifically to target gaps in the reserve system.

At the formation level, 45% of rainforest is in secure reserves and another 12% is in currently proposed RAPs. Another 20% of the rainforest is in zones recognized to be of high conservation value and in which logging has been deferred for at least ten years until conservation evaluations have been made (Rolley & Brown, 1992). Thus over three quarters of the rain forest is in reserves, RAPs, or deferred forest categories (Jarman et al., 1991).

Most of the rainforest dominants are well reserved; *Nothofagus cunninghamii* has 57% of its areal extent in reserves. The corresponding figures for the other dominants which have been mapped are *N. gunnii*, 95%, *Lagarostrobos*, 76%, *Athrotaxis cupressoides*, 100%, *A. selaginoides*, 84% (Jarman et al., 1991). At the community level, 6 of the 38 communities are not known from reserves, but all of these are expected to be found within RAPs when surveys are completed (Jarman et al., 1991). At the individual species level, all of the vascular plants have populations within reserves, but Shapcott (1991), for example, has demonstrated that current reserves of *Lagarostrobos* do not sample its full genetic variation, even though 76% of the species is reserved. There is insufficient information to make any more than a preliminary statement about the bryophytes and lichens. This is partly because most studies have been conducted within reserves and also because distribution information for the species is very poor (G. Kantvilas, S.J. Jarman, personal communication).

Most of the rainforest reserves in Tasmania probably are sufficiently large to ensure that ecosystem processes and the species within them are maintained, given current management and climates. However the disposition of the reserves in the landscape and the utility of current links and corridors between reserves in the light of land use changes are poorly studied and are subjects of current research. The need for balanced management on and off reserves is recognized as a necessary condition for the continued viability of the rainforests, but debate continues about the optimum management strategies outside reserves, ranging from total preservation to some selective utilization of high-value timbers from sites that are sufficiently resilient to sustain some harvesting.

Conclusion

The rainforests of Tasmania (and those of New Zealand) offer a potential set of "controls" for comparisons with those of the Americas. The components of global change (land use, climate, and atmospheric pollution) are ubiquitous, but their relative intensities probably differ from place to place. The commonalities of the forest ecosystems in the different places (e.g., the abundance of lichens (pollution indicators), bryophytes, and long-lived conifers (climate change indicators) and the susceptibility to economic pressure) argue for their inclusion in a net of sites to monitor global change. At the local level, Tasmanian cool temperate rainforests are subject to the same human pressures for development and utilization that occur in native vegetation elsewhere. However while they are relicts of their former botanical diversity, that which remains is sufficiently pristine and well buffered to ensure that with careful management and monitoring, their multiplicity of values can be retained into the long term, provided that the effects of fire and climate change can be managed!

Acknowledgments. We thank Jean Jarman and Fred Duncan for use of diagrams and their assistance.

References

Ashton, D.H. 1981a. Tall open forests. In R.H. Groves (ed.), *Australian Vegetation* (pp. 121–151). Cambridge, England: Cambridge University Press.

Ashton, D.H. 1981b. Fire in tall open-forests (wet sclerophyll forests). In R.H. Groves (ed.), *Australian Vegetation* (pp. 339–363). Cambridge, England: Cambridge University Press.

Barker, M.J. 1991. The Effect of Fire on West Coast Lowland Rainforest. *Tasmanian National Rainforest Conservation Program, Report No. 7.* Tasmania: Forestry Commission.

Barker, P.C.J. 1992. Autecology of *Phyllocladus* and *Anodopetalum* in Tasmania. *Tasmanian National Rainforest Conservation Program, Report No. 10.* Tasmania: Forestry Commission.

Bowman, D.M.J.S., and Brown, M.J. 1986. Bushfires in Tasmania: a botanical approach to anthropological questions. *Arch Oceania*, 21, 168–171.

Brown, M.J. 1988. *Distribution and Conservation of King Billy Pine*. Tasmania: Forestry Commission.

Brown, M.J., and Hickey, J.E. 1990. Tasmanian forest—genes or wilderness? *Search*, 21, 86–87.

Brown, M.J., Kirkpatrick, J.B., and Moscal, A. 1983. *An Atlas of Tasmania's Endemic Flora*. Hobart: Tasmanian Conservation Trust.

Brown, M.J., and Podger, F.D. 1982a. Floristics and fire regimes of a vegetation sequence from sedgeland-heath to rainforest at Bathurst Harbour, Tasmania. *Aust J Bot*, 30, 659–676.

Brown, M.J., and Podger, F.D. 1982b. On the apparent anomaly between observed and predicted percentages of vegetation types in south west Tasmania. *Aust J Ecol*, 7, 203–205.

Busby, J. 1988. Potential impacts of climate change on Australia's flora and fauna. In G.I. Pearman (ed.), *Greenhouse: Planning for Climate Change*. Melbourne: CSIRO.

Busby, J., and Brown, M.J. 1994. Southern rainforests. In R.H. Groves (ed.), *Australian Vegetation*, 2d ed. Cambridge, England: Cambridge University Press, 131–155.

Colhoun, E.A., and Fitzsimons, S.J. 1990. Late Cainozoic glaciation in western Tasmania, Australia. *Quat Sci Reviews*, 9, 199–216.

Cosgrove, R. 1989. Thirty thousand years of human colonisation in Tasmania—new Pleistocene dates. *Science*, 243, 1706–1708.

Coy, R. 1990. A survey of invertebrate animals in Tasmanian rainforest. In *Tasmanian National Rainforest Conservation Program, Report No. 1*. Tasmania: Forestry Commission.

Cremer, K.W., and Mount, A.B. 1965. Early stages of plant succession following the complete clear felling and burning of Eucalyptus regnans forest in the Florentine Valley, Tasmania. *Aust J Bot*, 13, 303–322.

Cullen, P. 1987. Regeneration patterns in populations of *Athrotaxis selaginoides* D. Don. from Tasmania. *J Biogeog*, 14, 39–51.

Cullen, P. 1991. Regeneration of *Athrotaxis selaginoides* and other tree species in landslide faces within rainforest in Tasmania. In M.R. Banks, S.J. Smith, A.E. Orchard, and G. Kantvilas (eds.), *Aspects of Tasmanian Botany* (pp. 191–200). Hobart: Royal Society of Tasmania.

Cullen, P., and Kirkpatrick, J.B. 1990a. The ecology of *Athroatxis* D. Don. (Taxodiaceae) 1. Stand structure and regeneration of *A. cupressoides*. *Aust J Bot*, 36, 547–560.

Cullen, P., and Kirkpatrick, J.B. 1990b. The ecology of *Athrotaxis* D. Don. (Taxodiaceae) 2. The distributions and ecological differentiation of *A. cupressoides* and *A. selaginoides*. *Aust J Bot*, 36, 561–573.

Duckett, T. 1990. Invading alien plants. *Tasforests*, 2, 121–123.

Duncan, D., and Dalton, P.J. 1982. Recolonisation by bryophytes following fire. *J Bryol*, 12, 53–63.

Elliott, H.J., Kile, G.A., Gandy, S.G., and Ratkowsky, D.A. 1987. The incidence and spatial pattern of *Nothofagus cunninghamii* (Hook.) Oerst. attacked by *Platypus subgranosus* Schedl. in Tasmania's cool temperate rainforest. *Aust J Ecol*, 12, 125–138.

Ellis, R.C. 1985. The relationships among eucalypt forest, grassland and rainforest in a highland area in north-eastern Tasmania. *Aust J Ecol*, 10, 91–109.

Ettershank, G., and Ettershank, J.A. 1990. Insects associated with flowers of Tasmanian leatherwoods (*Eucryphia* spp.) In *Tasmanian National Rainforest Conservation Program Report, No. 1*. Tasmania: Forestry Commission.

Galloway, D.J. 1988. Plate tectonics and the distribution of *cool temperate* Southern Hemisphere macrolichens. *Bot J Linn Soc*, 96, 45–55.

Gibson, N. 1988. A description of Huon pine *Lagarostrobos franklinii* (Hook. f.) Quinn (Podocarpaceae) forests of the Prince of Wales and King Billy ranges. *Pap Proc R Soc Tasm*, 122, 127–133.

Gibson, N., and Brown, M.J. 1991. The ecology of *Lagarostrobos franklinii* (Hook. f.) Quinn (Podocarpaceae) in Tasmania. 2. Population structure and spatial pattern. *Aust J Ecol*, 16, 223–229.

Gibson, N., Brown, M.J., and Davies, J.L. 1991. The ecology of *Lagarostrobos franklinii* (Hook. f.) Quinn (Podocarpaceae) in Tasmania 1. Distribution, floristics and environmental correlates. *Aust J Ecol*, 16, 215–222.

Gilbert, J.M. 1959. Forest succession in the Florentine Valley, Tasmania. *Pap Proc R Soc Tasm*, 93, 129–151.

Henderson, W., and Wilkins, C.W. 1975. The interaction of bushfires and vegetation. *Search*, 6, 130–133.

Hickey, J.E. 1990. Change in rainforest vegetation in Tasmania. *Tasforests*, 2, 143–149.

Hickey, J.E., Blakesley, A.J., and Turner, B. 1983. Seedfall and germination of *Nothofagus cunninghamii* (Hook.) Oerst., *Eucryphia lucida* (Labill.) Baill. and *Atherosperma moschatum* Labill. Implications for regeneration practice. *Aust For Res*, 13, 21–28.

Hickey, J.E., and Brown, M.J. 1989. Planning for regional biological conservation of Tasmania's forest vegetation types. *Proc Bienn Conf 1989 I.F.A.*, Leura, N.S.W., Australia.

Hickey, J.E., and Felton, K.C. 1991. Management of Tasmania's cool temperate rainforest. In F.H. McKinnell, E.R. Hopkin, and J.E.D. Fox (eds.), *Forest Management in Australia*. Sydney: Survey Beatty.

Hickey, J.E., and Savva, M. 1992. *The Extent, Regeneration and Growth of Tasmanian Lowland Mixed Forest*. Tasmania: Forestry Commission.

Hill, R.S. 1982. Rainforest fire in western Tasmania. *Aust J Bot*, 30, 583–589.

Hill, R.S. 1990a. Sixty million years of change in Tasmania's climate and vegetation. *Tasforests*, 2, 89–98.

Hill, R.S. 1990b. Evolution of the modern high latitude Southern Hemisphere flora: evidence from the Australian macrofossil record. In J.G. Douglas and D.C. Christophel (eds.), *Proceedings of the Third International Organisation of Palaeobotany Conference* (pp. 31–42). Melboure, 1988.

Hill, R.S., and Carpenter, R.J. 1991. Extensive past distributions for major Gondwanic floral elements: macrofossil evidence. In *Aspects of Tasmanian Botany: A Tribute to Winifred Curtis* (pp. 239–247). Hobart: Royal Society of Tasmania.

Hill, R.S., and Macphail, M.K. 1985. A fossil flora from rafted Plio-Pleistocene mudstones at Regatta Point, Tasmania. *Aust J Bot*, 33, 497–517.

Hill, R.S., and Read, J. 1984. Post-fire regeneration of rainforest and mixed forest in western Tasmania. *Aust J Bot*, 32, 481–493.

Horne, R., and Hickey, J.E. 1991. Ecological sensitivity of Australian rainforests to selective logging. *Aust J Ecol*, 16, 119–129.

Howard, T.M. 1973. Accelerated tree death in mature *Nothofagus cunninghamii* Oerst. forests in Tasmania. *Vic Nat*, 90, 343–345.

Jackson, W.D. 1968. Fire, air, water and earth: An elemental ecology of Tasmania. *Proc Ecol Soc Aust*, 3, 9–16.

Jackson, W.D. 1983. Tasmanian rainforest. In R. Blakers and P. Robertson (eds.), *Tasmania's Rainforest, What Future?* (pp. 9–39). Hobart: Australian Conservention Foundation.

Jarman, S.J., and Brown, M.J. 1983. A definition of cool temperate rainforest in Tasmania. *Search*, 14, 81–87.
Jarman, S.J., Brown, M.J., and Kantvilas, G. 1984. *Rainforest in Tasmania.* Tasmania: National Parks and Wildlife Service.
Jarman, S.J., Kantvilas, G., and Brown, M.J. 1986. The ecology of pteridophytes in Tasmanian cool temperate rainforest. *Fern Gazette*, 13, 77–86.
Jarman, S.J., Kantvilas, G., and Brown, M.J. 1991. Floristic and ecological studies in Tasmanian rainforest. *Tasmanian National Rainforest Conservation Program, Report No. 3*. Tasmania: Forestry Commission.
Jarman, S.J., Kantvilas, G., and Brown, M.J. 1994. Phytosociological studies in Tasmanian cool temperate rainforest. *Phytocoenologia*, 22, 355–390.
June, S.R., and Ogden, J. 1978. Studies on the vegetation of Mt Colenso, New Zealand. 4. An assessment of the processes of canopy maintenance and regeneration strategy in a red beech (*Nothofagus fusca*) forest. *NZJ Ecol*, 1, 7–15.
Kantvilas, G. 1988. Tasmanian rainforest lichen communities: a preliminary classification. *Phytocoenologia*, 16, 391–428.
Kantvilas, G., and Minchin, P.R. 1989. An analysis of epiphytic lichen communities in Tasmanian cool temperate rainforest. *Vegetatio*, 84, 99–112.
Kantvilas, G. 1990. Succession in rainforest lichens. *Tasforests*, 2, 167–170.
Kiernan, K. 1989. Multiple glaciation of the upper Franklin River Valley, Western Tasmania Wilderness World Heritage Area. *Aust Geog Stud*, 27, 208–233.
Kiernan, K., and Hannan, D. 1991. Glaciation of the upper Forth River catchment, Tasmania. *Aust Geog Stud*, 29, 155–173.
Kiernan, K., Jones, R., and Ranson, D. 1984. New evidence from Fraser Cave for glacial age man in south west Tasmania. *Nature*, 301, 28–32.
Kile, G.A., Packham, J.M., and Elliott, H.J. 1989. Myrtle wilt and its possible management in association with human disturbance of rainforest in Tasmania. *N Z For Sci*, 18, 166–186.
Kile, G.A., and Walker, J. 1987. *Chalara australis* sp. nov. (Hyphomycetes), a vascular pathogen of *Nothofagus cunninghamii* (Fagaceae) in Australia and its relationship to other *Chalara* species. *Aust J Bot*, 35, 1–32.
Kirkpatrick, J.B. 1977. The impact of man on the vegetation of the west coast region. In M.R. Banks and J.B. Kirkpatrick (eds.), *Landscape and Man.* Hobart: Royal Society of Tasmania, 151–156.
Kirkpatrick, J.B., and Brown, M.J. 1984. A numerical analysis of Tasmanian higher plant endemism. *Bot J Linn Soc*, 88, 165–183.
Kirkpatrick, J.B., and Dickinson, K. 1984. *Vegetation Map of Tasmania 1: 500,000*. Tasmania: Forestry Commission.
Kirkpatrick, J.B., and Harwood, C.E. 1980. Vegetation of an infrequently burned Tasmanian mountain region. *Proc R Soc Vic*, 91, 79–107.
Macphail, M.K. 1979. Vegetation and climates in Tasmania since the last glaciation. *Quaternary Res*, 11, 306–341.
Millington, R.J., Jones, R., Brown, D., and Vernon, B. 1979. Huon pine—endangered? *Environmental Studies Occupational Paper, No. 9*. Hobart: University of Tasmania.
Neyland, M.G., and Brown, M.J. 1991. Disturbance of cool temperate rainforest patches in eastern Tasmania. *Aust For*, 57, 1–10.
Noble, I.R., and Slatyer, R. 1981. Concepts and models of succession in vascular plant species subject to recurrent fire. In A.M. Gill, R.H. Groves, and I.R. Noble (eds.), *Fire and the Australian Biota* (pp. 311–335). Canberra: Australian Academy of Science.

Ogden, J. 1978. On the dendrochronological potential of Australian trees. *Aust J Ecol*, 10, 339–356.
Orchard, A.E. 1988. A natural regions map for Tasmaniai. *Pap Proc R Soc Tasm*, 122, 47–51.
Packham, J.M. 1990. Myrtle wilt. *Tasmanian Rainforest Conservation Program, Technological Report No. 2*. Tasmania: Forestry Commission.
Packham, J., Elliott, H.J., and Brown, M.J. 1990. Myrtle Wilt. In *Tasmanian National Rainforest Conservation Program, Report No. 1*. Tasmania: Forestry Commission.
Palzer, C. 1978. Defoliation and death in a *Eucalyptus obliqua* forest. *Aust For Res Newsl*, 5, 171.
Pannell, J. 1992. *Swamp Forests of Tasmania*. Tasmania: Forestry Commission.
Peterson, M.J. 1990. *The Distribution and Conservation of Huon Pine*. Tasmania: Forestry Commission.
Plomley, N.J.B. 1966. *Friendly Mission: The Tasmanian Journals and Papers of George Augustus Robinson 1829–1834*. Hobart: Tasmanian Historical Research Association.
Podger, F.D. 1990. Phytophthora, fire and change in regetation. *Tasforests*, 2, 125–128.
Podger, F.D., and Brown, M.J. 1989. Vegetation damage caused by *Phytophthora cinnamomi* on disturbed sites in temperate rainforest in western Tasmania. *Aust J Bot*, 37, 443–480.
Podger, F.D., Bird, T., and Brown, M.J. 1988. Human activity fire and change in the forest at Hogsback Plain, southern Tasmania. In K.J. Frawley and N.M. Semple (eds.), *Australia's Everchanging Forests*. Canberra: Australian Defence Force Academy.
RAC. 1991. *Report of the Resource Assessment Commission Forest and Timber Inquiry*. Canberra: Resource Assessment Commission.
Read, J. 1985. Photosynthetic and growth responses to different light regimes of the major canopy species of Tasmanian cool temperate rainforest. *Aust J Ecol*, 10, 327–334.
Read, J. In press. Pattern and process in Tasmanian rainforest. In J.B. Reid, R.S. Hill, and M.J. Brown (eds.), *Vegetation of Tasmania*. Hobart: University of Tasmania.
Read, J., and Hill, R.S. 1983. Rainforest invasion onto Tasmanian old-fields. *Aust J Ecol*, 8, 149–161.
Read, J., and Hill, R.S. 1985. Dynamics of *Nothofagus* dominated rainforest on mainland Australia and lowland Tasmania. *Vegetatio*, 63, 67–78.
Read, J., and Hill, R.S. 1988. The dynamics of some rainforest associations in Tasmania. *J Ecol*, 76, 558–584.
Robertson, I.D., and Duncan, F. 1991. *Distribution and Conservation of Deciduous Beech*. Tasmania: Forestry Commission.
Rolley, E.R., and Brown, M.J. 1992. Forest allocation and conservation in Tasmania, Can both win? In: *Papers on Conflict Resolution: Fourth World Congress on National Parks and Protected Areas; Venezuela*. Switzerland.
Shapcott, A. 1991. Studies in the population biology and genetic variation of Huon pine (*Lagarostrobos franklinii*). *Tasmanian National Rainforest Conservation Program, Report No. 4*. Tasmania: Department of Parks, Wildlife, and Heritage.
Veblen, T., and Ashton, D.H. 1982. The regeneration status of *Fitzroya cupressoides* in the Cordillera Pelada, Chile. *Biol Cons*, 23, 141–161.
Veblen, T.T., Donoso, Z.C., Schlegel, F.M., and Escobar, R.B. 1981. Forest dynamics in south-central Chile. *J Biogeog*, 8, 211–247.

Veblen, T.T., Schlegel, F.M., and Oltermari, J.V. 1983. Temperate broad-leaved evergreen forests of South America. In J.D. Ovington (ed.), *Temperate Broad-Leaved Evergreen Forests* (pp. 5–31). Amsterdam: Elsevier.

Wardle, P. 1983. Temperate broad-leaved evergreen forests of New Zealand. In J.D. Ovington (ed.), *Temperate Broad-Leaved Evergreen Forests* (pp. 33–72). Amsterdam: Elsevier.

Wardle, J.A. 1984. The New Zealand Beeches, Ecology, Utilisation and Management. Wellington: New Zealand Forest Service.

Wardlaw, T. 1990. Changes in forest health associated with short-term climatic fluctuation. *Tasforests*, 2, 107–110.

Wells, P. 1989. *Conservation Status of Wet Eucalypt Forest in Tasmania*. Tasmania: Working Group for Forest Conservation.

Wells, P., and Hill, R.S. 1989. Fossil imbricate-leaved Podocarpaceae *from Tertiary* sediments in Tasmania. *Aust Sys Bot*, 2, 387–423.

WGFC. 1990. *Recommended Areas for the Protection of Rainforest, Wet Eucalypt and Dry Sclerophyll Forest*. Tasmania: Working Group for Forest Conservation.

17. Logging Effects on the Aquatic Ecosystem: A Case Study in the Carnation Creek Experimental Watershed on Canada's West Coast

Eugene D. Hetherington

The Carnation Creek Experimental Watershed Project was initiated in 1970 to investigate the impacts of forest harvesting on vegetation, soil, water, and fish in a typical coastal rainforest salmonid stream ecosystem. A major emphasis was placed on the stream system and the production of salmon. Monitoring of hydrometeorological parameters, vegetation changes, and forestry activities provided necessary information for interpretation of fisheries and stream-related data. The project was originally scheduled for completion in 1986, but was continued because of recognition that important changes in the stream channel and in fish populations are still occurring. In the context of climate change and the extremely limited amount of information on streamflow, precipitation, and fish life cycles in mountainous coastal watersheds, the relatively long length of the Carnation Creek data record is becoming increasingly valuable.

Study Area

Carnation Creek drains a 1000 ha mountainous watershed located on the west coast of Vancouver Island, British Columbia, Canada, 49° N and 125° W (inset, Figure 17.1). The watershed lies in the transition area between perhumid and seasonal temperate rainforests, having some tree species typical of the seasonal zone but temperatures consistent with the perhumid zone (Veblen & Alaback, Chapter 9). The preharvest forest cover included Sitka spruce (*Picea sitchensis*), red alder (*Alnus rubra*), western hemlock (*Tsuga heterophylla*), amabilis fir (*Abies amabilis*), Douglas fir (*Pseudotsuga menziesii*), and western red cedar (*Thuja plicata*). After harvesting, Sitka spruce and Douglas fir were planted, while western hemlock and red alder regenerated naturally.

The basin has been glaciated and has steep slopes with elevations ranging from sea level to 900 m. Volcanic bedrock is overlain with Orthic Ferro-Humic podzol soils which are shallow (about 0.7 m mean depth), highly permeable, coarse colluvial materials of gravely loam and loamy sand texture with an organic surface layer. The lower 3 km of Carnation Creek flows through a 50 to 200 m-wide floodplain.

The climate features mild, wet winters with frequent rainstorms, and cool, dry summers. Mean mothly air temperatures vary from 3°C in January to 15°C in August. Annual precipitation (mostly rain) has ranged from 2100 to over 5000 mm, and averages 3400 mm. Snow occurs briefly most years at higher elevations, but seldom at sea level. Streamflow is flashy due to intense rainfall and rapid runoff, and discharges at the main weir have ranged from 0.02 to 64 m^3/s.

The principal species of fish are chum salmon (*Oncorhynchus keta*), which spawn in the lower 500 m of stream, and coho salmon (*O. kisutch*) which utilize the first 3.1 km of stream accessible to the anadromous fish.

Carnation Creek Project

The original objectives of the Carnation Creek project (Hartman & Scrivener, 1990) were (1) to develop a better understanding of how undisturbed coastal rainforest-salmonid stream ecosystems work; (2) to explain and quantify the impacts of timber production activities on stream environments and their capacity to produce salmonid fishes; and (3) to provide continuous input to the further development of integrated resource management guidelines. An emphasis has been placed on understanding processes to maximize the transferability and application of results, and evaluation of recovery of the stream system and fish production from logging impacts. New objectives for the project are being developed

The initiating and lead agency until 1986 was the federal Department of Fisheries and Oceans, which provided continuity of leadership and most of the funding for field operations. Participating organizations included both federal and provincial fisheries, forestry and water resource agencies, universities, and MacMillan Bloedel Ltd., which carried out forestry operations. Steering and working groups plus a project coordinator provided direction, support, and coordination for research and monitoring activities. Since 1986, other agencies have contributed more directly to funding field operations.

Study Design

The original project had three phases: prelogging (1971–1975), active logging (1976–1981), and postlogging (1982–1986). A total of 40% of the watershed area above the main weir was clearcut in patches ranging from 7 to 64 ha (which merged to form a continuous clearcut) using high-lead and grapple cable yarding techniques. Monitoring has continued in the watershed since 1986, and an additional 11% was logged up to 1993. Silvicultural treatments included prescribed burning of slash, reforestation, some scarification, and use of herbicides in the floodplain. The experiment was designed to evaluate the impacts of three streamside forest management treatments on Carnation Creek (1) a leave strip of variable width (between the main weir B and the western edge of the 76–77 opening in (Figure 17.1)), (2) an intensive treatment where both sides of the stream were logged and the stream channel was disturbed by debris and yarding (76–77 opening), and (3) a careful treatment leaving riparian brush and avoiding channel bank and bed disturbance (78–79 opening). There were no separate controls for these treatments, but four other streams were monitored as external controls for stream biota. Within the constraints of experimental design, normal planning procedures and logging operations were followed. However, treatment 2 is no longer permitted for salmon streams. An additional herbicide treatment experiment took place in 1984–1986 (Reynolds, 1989).

Streamflow was measured at 5 weirs and precipitation at 10 sites (Figure 17.1), groundwater levels were measured at a number of locations (Hetherington, 1982, 1989) and two unlogged tributaries (C and E) served as hydrological controls (Figure 17.1). Additional facilities included a major weather station at site A, thermographs for air and water temperatures at 15 sites, an automatic sediment sampler at B weir, and a permanent fish counting fence (Figure 17.1). A permanent field camp was staffed year round until June 1990.

The impacts of forestry operations were evaluated by studies of both terrestrial and stream processes (Hartman & Scrivener, 1990). Instream studies covered fish life stages, population dynamics, and behavior; basic stream productivity, including aquatic insects, detritus, and algae as well as fish; fish habitat, including the role of side channels and large woody debris; changes in gravel quality, channel morphology, and woody debris; suspended sediment; water temperature and chemistry; stream discharge and possible changes after logging; and bedload movement. Terrestrial studies included soil and vegetation classification; soil disturbance; revegetation patterns; landslides; hillslope, and floodplain groundwater

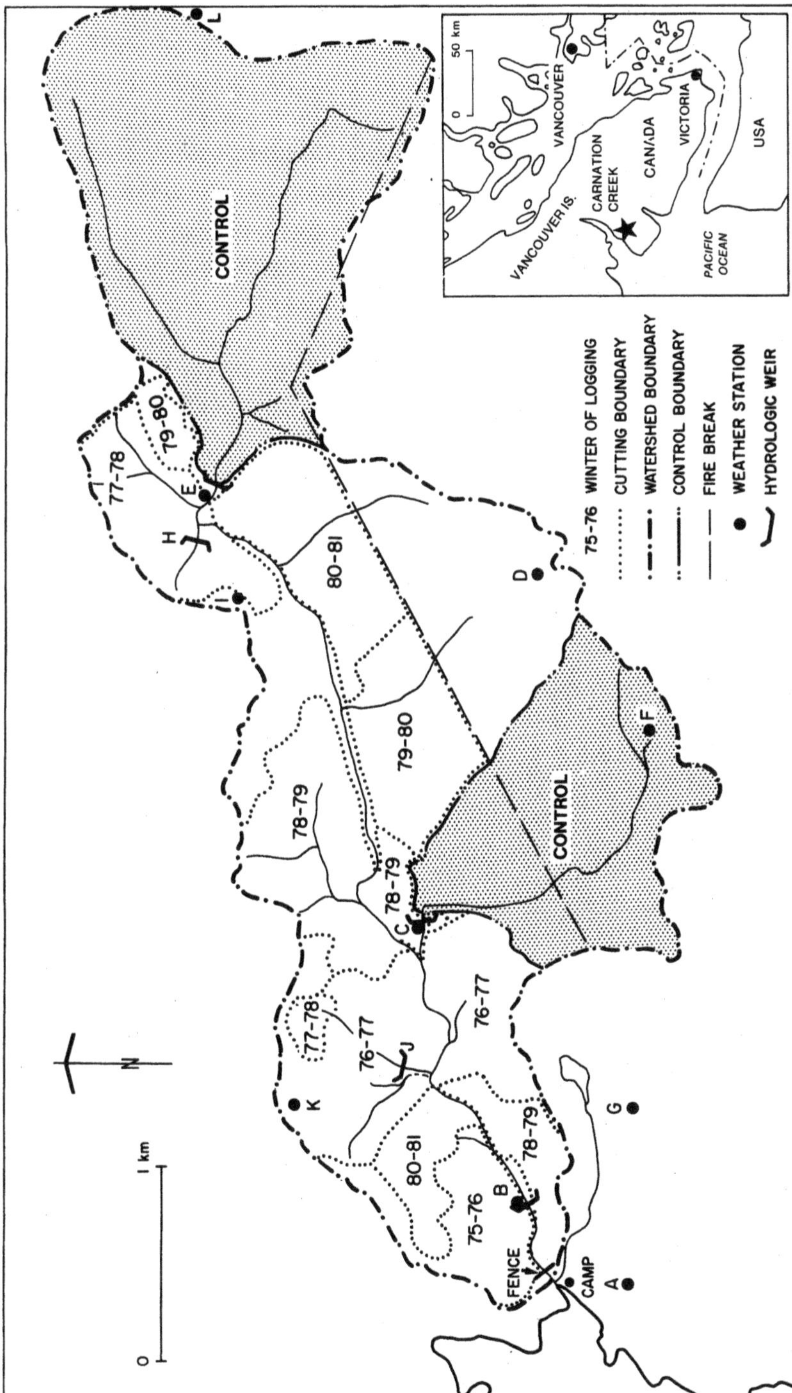

Figure 17.1. Carnation Crreek Experimental Watershed showing location of hydrometeorological stations and annual clearcut logging boundaries.

levels; soil moisture; subsurface flow processes; precipitation chemistry; and precipitation interception.

Summary of Research Results

More than 160 publications have documented the results and progress of research in the Carnation Creek watershed over the past 20 years (Poulin & Scrivener, 1988). The findings have been summarized and integrated to show the interrelationships among the various hydrological, fluvial-geomorphological, thermal, chemical, and biological components, and the related changes that have taken place as a result of logging (Hartman & Scrivener, 1990). Some of the main findings are reviewed here to illustrate the nature of the logging impacts.

Hydrology

Forest harvesting has had variable effects on the hydrological processes within the watershed. Slope water behaviour was studied in the tributary watersheds H and J before and after logging, as were summer groundwater levels in the lower floodplain. Postlogging changes in streamflow have been assessed in preliminary analyses for the main stream at B weir (40% clearcut) and for tributaries H (90% clearcut) and J (100% clearcut).

Water Yield, Summer Low Flows, and Floodplain Groundwater Levels

Precipitation data collected beneath the forest canopy and in an adjacent clearcut indicate a net loss to evaporation of the intercepted rainfall during both winter and summer. Canopy collection of "fog-drip" precipitation occasionally exceeds interception losses, although fog-drip amounts at Carnation Creek are less than those reported by Harr (1982) in one small watershed in Oregon.

Reduced evapotranspiration losses appear to be the main cause of increases in annual water yield after logging in both clearcut watersheds H and J of at least 14% initially (Hetherington, 1982; Hartman & Scrivener, 1990). The increases had only diminished by about 30 to 50% after 10 years. Increases in yield after harvesting have been documented both for other coastal watersheds in the United States (Harr, Fredricksen, & Rothacher 1979; Keppeler & Ziemer, 1990) and world wide (Bosch & Hewlett, 1982). Summer low flows also increased initially in H watershed (Hetherington, 1982). Such initial low flow increases have also been found in Oregon (Harr et al., 1979) and California (Keppeler & Ziemer,

1990). Inaccurate low-flow measurements preclude reliable evaluation of water yield changes in the main stream at B weir.

In the lower Carnation Creek floodplain, summer groundwater levels were up to 50 cm higher after harvesting of both the floodplain and the adjacent side slopes (Hetherington, 1982). Despite vigorous regrowth of deciduous vegetation on the floodplain, the increases in summer groundwater levels had barely diminished after 10 years. This result is believed to be due to the maintenance of higher seepage from the side slopes where vegetation growth has been slower (Hetherington, 1988).

Slope Groundwater Levels

At one piezometer site on a clearcut slope in H watershed above any road influence, storm groundwater peaks were up to 20 cm higher after logging than before (Hetherington, 1982). The suspected cause of this change was surface soil disturbance that possibly caused local modifications in the pathways of water movement into and through the soil; that is, less water was moving through the macrochannels and more through the soil matrix resulting in a short-lived but significant localized rise in the water table. At another piezometer site below a road, groundwater peaks were up to 20 cm lower due to interception and lateral diversion of subsurface seepage by the road (Hetherington, 1982). Changes in slope groundwater regimes due to road diversion of surface runoff and excessive soil disturbance were presumed to have been major causes of two small landslides near H watershed in 1982 (Hetherington, 1988).

Peak Flows

Preliminary results regarding the question of harvesting effects on peak streamflows have been variable. The analyses were based on rain-only conditions because the vast majority of storms fall into this category. In tributary H, peak flows appear to have increased after road construction (Hetherington, 1982). This result suggests that the roads may have accelerated flow to the stream channel by converting some subsurface seepage into surface runoff. These changes were sustained for at least three years after harvesting. Road-related peak flow increases in small watersheds have been observed in Oregon (Harr et al., 1979) and for small storms in California (Wright, Sendek, Rice, & Thomas, 1990).

At B weir on Carnation Creek, rain-only peak flows showed no significant changes (Hetherington, 1982). The highly permeable soils retain high infiltration capacities after disturbance so surface runoff in the clearcuts apart from the roads remains negligible or nonexistent Thus, in the larger watershed where roads have a smaller relative effect than in the very small tributary drainages and where only 40% (versus 90%) of the area was clearcut, it is not surprising that changes in peak flows were not detected. Mid-winter rain-only peak flows in Oregon (Harr et al., 1979) and those during larger storms in California (Wright et al., 1990) were also not significantly changed after harvesting. However, peak flows during a major rain-on-snow event at Carnation Creek in 1982 (after logging) were higher than might be expected from rain only. Data from Oregon also indicate that harvesting probably does increase rain-on-snow peaks in the transient snow zone where the snowpack is relatively shallow and intermittent (Harr, 1986).

Erosion and Sediment Production

Despite steep slopes and high rainfall, erosion on the logged slopes and input of sediment to Carnation Creek has been modest (Hetherington, 1988; Scrivener, 1988). Sediment input to small tributary streams was monitored during road construction (Ottens & Rudd, 1977), and minor ditch erosion has occurred on steeper sections of the road network. A few landslides, some very small, and debris torrents in gullies have occurred during major storms after logging in both clearcut and uncut areas (Hartman & Scrivener, 1990; Hetherington, 1988). Only a few of these events actually put soil and debris into the stream. Much of the extra sediment observed in the stream after logging has come from within the stream channel itself, as noted below.

Stream Channel

The morphology and structure of the main steam channel changed dramatically following streamside logging primarily due to deterioration of stream banks, destabilization and removal of instream logs, and introduction of small organic debris. In the intensively disturbed section of the stream, channel width increased up to 8 m after logging (Powell, 1988) with a concurrent 2.7-fold increase in volume of eroded stream bank sediments (Scrivener, 1988). Sand content of spawning gravels in lower Carnation Creek increased by 5.8% and was still accumulating 10 years after logging (Scrivener, 1988). Bedload movement at B weir increased from 251 t/yr before logging, to 289 t/yr during logging and 270 t/yr after logging (Hartman & Scrivener, 1990). This increase came from mobilization of stored gravel within the channel plus extra input by landslides and debris torrents. The channel has straightened in several sections and has fewer pool-riffle stages.

Water Quality

Carnation Creek stream water is very low in dissolved nutrients and minerals and is weakly acidic (pH 6.0–7.1) (Hedin & Hetherington, Chapter 4; Scrivener, 1975). Minor changes in stream water quality occurred following harvesting and prescribed burning in the watershed (Hartman & Scrivener, 1990). Average stream-water temperatures increased in the summer (by 3.2°C in August) and also in the winter (by 0.8°C in January), primarily as a result of removal of the streamside forest canopy. Electrical conductivity (representing major cations and anions) increased from 20 to 35 and 40 μmho/cm^2 during higher flows at B and J weirs, respectively, with smaller changes at lower flows. The increases persisted for 2 to 5 years. Nitrate-N concentrations increased 2-fold at B weir (40% clearcut) and 7-fold in a floodplain tributary (100% clearcut and burned) from mean values of 0.18 mg/l. These changes persisted up to 3 years. Phosphate-P concentrations remained unchanged at very low levels (mean value 0.004 mg/l), except for increases to a maximum of 0.027 mg/l in the floodplain tributary immediately after burning. Forest soils tend to retain phosphorus and at Carnation Creek the phosphorus concentrations in the stream water were only half those in the incoming rain water (Scrivener, 1975). Phosphorus is the limiting nutrient for algae growth at Carnation Creek and other coastal streams (Stockner & Shortreed, 1978). Suspended sediment concentrations up to 1000 mg/l were measured in small tributaries during road construction in wet weather (Ottens & Rudd, 1977). However, suspended sediment levels and yields in Carnation Creek remained low and changed little during and after road construction and logging. At B weir, annual sediment yields ranged from 17 to 44 t/km^2 prior to any disturbance, 13 to 33 t/km^2 during logging, and 11 to 34 t/km^2 after logging (Hartman & Scrivener, 1990), whereas peak suspended sediment concentrations during storm runoff were mostly less than 75 mg/l with a maximum observed value of 560 mg/l (Tassone, 1988).

Fish production

The research has revealed a complex variety of both positive and negative impacts on fish in relation to different life stages of the fish, different time scales and differences among species (Hartman & Scrivener, 1990). For example, salmon egg-to-fry survival rates were reduced by the deterioration in spawning gravel quality. Lower densities of the very young coho (fry) permit better survival rates. Warmer stream water causes earlier emergence of the fry in the spring, which exposes the fry to spring storms which can wash them out to sea but also lengthens the growing season and permits the surviving fish to grow larger during their first summer. Increased nutrient levels and warmer water also promote higher growth rates and higher overwinter survival. The importance of side-channels in the floodplain as overwinter habitat for young salmon was a major finding (Bustard & Narver, 1975). Logging-related changes in large organic debris, channel morphology, and streambed stability depressed young fish survival and production.

What has been the net effect of logging-related effects on young salmon life stages on the numbers of adult salmon returning to spawn in the creek? This question is more difficult to answer because of the impacts of ocean conditions and fishery exploitation on salmon survival outside the confines of the creek. Through the use of models that can partition the various factors, it appears that logging has had only a small negative effect (−5%) on returning numbers of adult coho but a larger adverse impact (−26%) on the chum salmon (Scrivener, 1990). In contrast to the coho which rear in the stream, the chum leave the creek after emerging in the spring and appear to have been mainly affected by the deterioration of spawning gravel quality. Annual variability in returning populations has increased since logging. The results suggest that logging impacts on the salmon are strongly conditioned by environmental and climatic factors.

Summary

Long-term studies are necessary for a proper understanding and quantification of the effects of land use practices such as logging on the complex physical and biological processes of stream ecosystems. Variations in climate and time required for recovery from disturbance are but two of the many factors that require a lengthy record to be properly evaluated. These factors alone are sufficient to justify continued monitoring at Carnation Creek. Other supportive arguments include the absence of other similar long-term studies in coastal British Columbia, the recognition that long-term records are required to document climate change and its implications for natural ecosystems, and the value of long-term records for extrapolating short-term hydrometeorological measurements. Long-term data records are scarce, particularly in mountainous areas, and are becoming more and more valuable as we attempt to understand and define the impacts of both man's activities and broader climatic changes on forested ecosystems.

References

Bosch, J.M., and Hewlett, J.D. 1982. A review of catchment experiments to determine the effect of vegetation changes on water yield and evapotranspiration. *J Hydrol*, 55, 3–23.

Bustard, D.R., and Narver, D.W. Preferences of juvenile coho salmon (*Oncorhynchus kisutch*) and cutthroat trout (*Salmo clarki*) relative to simulated alteration of winter habitat. *J Fish Res Board Can*, 32, 681–687.

Harr, R.D. 1982. Fog drip in the Bull Run municipal watershed, Oregon. *Wat Resour Bull*, 18(5), 785–789.

Harr, R.D. 1986. Effects of clearcutting on rain-on-snow runoff in western Oregon: a new look at old studies. *Wat Resour Res* 22(7), 1095–1100.

Harr, R.D., Fredricksen, R.L., and Rothacher, J. 1979. *Changes in Streamflow Following Timber Harvesting in Southwest Oregon*. Portland, Or: U.S.D.A.,

Forest Service, Pacific Northwest Forest and Range Experiment Station, Research Paper PNW-249.

Hartman, G.F., and Scrivener, J.C. 1990. Impacts of forestry practices on a coastal stream ecosystem, Carnation Creek, British Columbia. *Can Bull Aquat Sci*, 223, 1–148.

Hetherington, E.D. 1982. Effects of forest harvesting on the hydrologic regime of Carnation Creek Experimental Watershed: A preliminary assessment. In *Hydrological Processes of Forested areas: Proceedings of the Canadian Hydrology Symposium '82, Fredericton, N.B.* (pp. 247–267). Ottawa: National Research Council of Canada.

Hetherington, E.D. 1988. Hydrology and logging in the Carnation Creek watershed—what have we learned? In T.W. Chamberlin (ed.), *Proceedings of the Workshop: Applying 15 years of Carnation Creek results, 1987* (pp. 11–15) Nanaimo, BC: Pacific Biological Station.

Hetherington, E.D. 1989. Carnation Creek floodplain hydrology: September 1984–September 1985. In P.E. Reynolds (ed.), *Proceedings of the Carnation Creek Herbicide Workshop, 1987* (pp. 27–44). Canada/British Columbia Economic and Regional Forest Resource Development Agreement, Report No. 063.

Keppeler, E.T., and Ziemer, R.R. 1990. Logging effects on streamflow: water yield and summer low flows at Caspar Creek in northwestern California. *Wat Resour Res*, 26(7), 1669–1679.

Ottens, J., and Rudd, J. 1977. *Environmental Protection Costs in Logging Road Design and Construction to Prevent Increased Sedimentation in the Carnation Creek Watershed.* Pacific Forest Research Center, Victoria, BC: Canadian Forest Service Report BC-X-155.

Poulin, V.A., and Scriviener, J.C. 1988. *An Annotated Bibliography of the Carnation Creek Fish–Forestry Project: 1970 to 1988.* Canadian Technical Report of Fisheries and Aquatic Sciences, No. 1640.

Powell, L.H. 1988. Stream channel morphology changes since logging. In T.W. Chamberlin (ed.), *Proceedings of the Workshop: Applying 15 years of Carnation Creek Results, 1987* (pp. 16–25). Nanaimo, BC: Pacific Biological Station.

Reynolds, P.E. (ed.), 1989. *Proceedings of the Carnation Creek Herbicide Workshop, 1987.* Canada/British Columbia Economic and Regional Forest Resource Development Agreement, Report No. 063.

Scrivener, J.C. 1975. *Water, Water Chemistry and Hydrochemical Balance of Dissolved Ions in Carnation Creek Watershed, Vancouver Island, July 1971–May 1974.* Fisheries and Marine Service, Technical Report 564—Environment Canada.

Scrivener, J.C. 1988. Changes in composition of the streambed between 1973 and 1985 and the impacts on salmonids in Carnation Creek. In T.W. Chamberlin (ed.), *Proceedings of the Workshop: Applying 15 years of Carnation Creek Results, 1987* (pp. 59–65). Nanaimo, BC: Pacific Biological Station.

Scrivener, J.C. 1989. Comparative changes in concentration of dissolved ions in the stream following logging, slash burning, and herbicide application (glyphosate) at Carnation Creek, British Columbia. In P.E. Reynolds (ed.), *Proceedings of the Carnation Creek Herbicide Workshop 1987* (pp. 197–211). Canada/British Columbia Economic and Regional Forest Resource Development Agreement, Report No. 063.

Scrivener, J.C. 1990. Complexity of ecosystem processes and the need for long-term Studies—the Carnation Creek experience. In E.M.P. Chadwick (ed.), *Proceedings of the Gulf Habitat Science Workshop, 1989* (pp. 18–37). Moncton, NB: Canadian Industry Report of Fisheries and Aquatic Sciences, No. 206.

Stockner, J.G., and Shortreed, K.R.S. 1978. Enhancement of autotrophic production by nutrient addition in a coastal rain-forest stream on Vancouver Island. *J Fish Res Board Can*. 35, 28–34.

Tassone, B.L. 1988. Sediment loads from 1973 to 1984 08HB048 Carnation Creek at the mouth, British Columbia. In T.W. Chamberlin (ed.), *Proceedings of the Workshop: Applying 15 years of Carnation Creek Results, 1987* (pp. 46–58). Nanaimo, BC: Pacific Biological Station.

Wright, K.A., Sendek, K.H., Rice, R.M., and Thomas, R.B. 1990. Logging effects on streamflow: Storm runoff at Caspar Creek in Northwestern California. *Wat Resour Res*, 26(7), 1657–1667.

18. Biodiversity of Canadian Forests, with Particular Reference to the West Coast Forests

Timothy J.B. Boyle

One of the issues that may have the greatest impact on forestry practices and the forests of the future is the conservation of biodiversity. There is considerable unease in the modern forestry community about the concept of biodiversity, and the consequences of its loss or of the actions required for its conservation. The weight of public opinion is firmly in favor of conserving biodiversity, as is demonstrated by recent events in the Pacific Northwest of the United States concerning the preservation of the northern spotted owl (*Strix occidentalis*), and its coverage in scientific (e.g., Heilman, 1990) and popular publications (e.g., Findley, 1990; Watts, 1991). Recent public opinion polls in Canada have also indicated support for the conservation of biodiversity over strictly commercial uses.

The forests of Canada's west coast are of great importance to its overall economy. British Columbia, which contains 38% of Canada's productive, nonreserved timber volume, contributed $6.1 billion to the Canadian economy in 1989, and in the same year accounted for 44% of the value of forest product exports and almost half of the forest sector balance of trade in forest products (Forestry Canada, 1990). Beyond pure economics, Canada's west coast forests differ in a number of important respects from those in the rest of the country. Productivity of the coastal rain forests is many times greater than boreal and temperate forests in the rest of the country. This is due to the equability of the climate and the stable floristic history of the region, which has allowed close adaptation to the environment (Alaback, Chapter 7). West coast forests are also unique in terms of their successional history, which is dominated by such small-scale disturbances as individual tree deaths and local wind throws. This results in the maintenance of a very high proportion of the forest in "old-growth" conditions, with high-structural diversity. Finally, the physiographic complexity of the region results in very high ecosystem diversity, and consequently beta species diversity. Whilst this paper touches on biodiversity in all of Canada's forests, the forests of the west coast are given special attention.

Status of Biological Diversity in Canada's Forests

Biodiversity covers the entire range of the hierarchy of biological organization, and this results in an inherent problem in its measurement and monitoring. Biodiversity is usually divided into a three-to-five level hierarchy (e.g., Cohen & Potter, 1993); here four levels are recognized: genetic, species, community/ecosystem, and landscape diversity. There are also a multitude of other ways of measuring or defining biodiversity. Rather than considering biodiversity in terms of hierarchical organization, it is also possible to characterize it in terms of ecological processes or functions. Thus it is possible to assess trophic diversity, structural diversity, or functional diversity (Franklin, 1988); and evolutionary, or historical diversity (Arroyo et al., Chapter 8). However, although this latter group of terms may have great value for certain applications or for examining particular aspects of biological systems, the genetic-species-ecosystem-landscape approach is conceptually simpler and easier to apply at multiple spatial scales.

Genetic Diversity

Most west coast conifer species have high genetic diversity, typical of tree species generally (Hamrick, Mitton, & Linhart, 1981), with the notable exception of western red cedar (*Thuja plicata*) (Copes, 1981). For example, estimates of among-population genetic variation for Sitka spruce (*Picea sitchensis*) (Yeh & El-Kassaby, 1980) were typical of those for all conifers (Hamrick et al., 1981). Two species of west coast conifer that have been studied more intensively than any other are Douglas fir (*Pseudotsuga menziesii*) and lodgepole pine (*Pinus contorta*). These species both have large latitudinal ranges as well as coastal and inland varieties. For both species, there is no indication of differences in genetic structure among coastal and inland varieties (El-Kassaby, 1990), apart from one study of inland Douglas fir from the southern end of its range in the United States, where among-population differences are much larger than elsewhere (Li & Adams, 1988). El-Kassaby and Sziklai (1982) reported a relatively high level of population differentiation within a small geographic area of coastal Douglas fir in British Columbia. This they interpreted to be the result of large elevational gradients for Douglas fir habitats in the sampled region.

Typically, vertebrates have lower levels of genetic diversity than invertebrates, which in turn are less heterozygous than plants (Nevo, 1987). Because animals are more mobile than plants, differentiation tends to be detectable only over much larger areas than for plants (Liu & Godt, 1983). Paradoxically, barriers to vertebrate migration tend to be more significant than for plants that rely on wind for pollination and/or seed dispersal. Consequently, genetic differentiation in vertebrates is more commonly recognized at the subspecies levels (see below for further discussion on west coast fauna).

Species Diversity

There is generally a gradient in tree species diversity from north to south, with exceptional areas of diversity in southern Ontario, and southern and western British Columbia (Hosie, 1979). In the enormous boreal forest region in Canada, only 38 species of tree are found, compared with 36 species in the much more restricted Coastal forest region of British Columbia and 79 species in the Deciduous forest region of southern Ontario.

Microorganisms and invertebrates have received little attention from taxonomists and ecologists despite their species diversity and their major roles in ecosystem functioning. There have been few studies on these groups of organisms in Canada's west coast forests. A current study of canopy arthropods in old-growth, second growth, and clear-cut sites in the Carmanah Valley of Vancouver Island represents the first of its kind in temperate rain forests (Winchester, 1993). Analyses of the distribution of phytophagous and predator/parasitoid arthropods indicate significant differences among individual trees for both groups in terms of species densities, as well as among the different sites. The phytophagous arthropods were also affected by seasonal changes. Although there were no vertical distribution differences for these species, there were highly significant differences for mites. In particular, the canopy soil litter associated with moss mats that develop in the old-growth canopy are dominated by Collembola and Acari, with significant differences at the generic level, compared with forest floor communities. Due to the absence of such moss mats in second growth forests, these canopy communities are obviously not present.

Nonvascular plants have also been relatively little studied in comparison with vascular plants. About half of the 1013 species of lichens and associated fungi of British Columbia have been reported from five or fewer sites (Goward, 1991). Although many of these species may indeed be rare, without doubt many also exist at other sites that have not yet been reported. It has been estimated that 60 to 70% of British Columbia's rare and endangered lichens are old-growth dependent (Goward, 1990). Whereas a 100-year-old coastal western hemlock (*Tsuga heterophylla*) forest may contain 8 to 10 species of epiphytic macrolichen species, a 250-year-old forest may have more than three times this number and will still not have reached its maximum complement (Goward, 1991). One reason for this is that lichens require a high degree of environmental stability, such as is only found in old-growth forests. Although air pollution, well known to be a major cause of lichen decline in Europe, may be locally significant in British Columbia in causing lichen decline in southeastern Vancouver Island and the southern mainland, cutting of old-growth forest is likely to pose a far greater threat to lichens, especially those characteristic of the old-growth coastal western hemlock forest (Goward, 1990, 1991).

A proportionately large number of bryophyte species are widely distributed, with only about 23% of the 1221 North American species being confined to the North American continent, including Mexico (Schofield, 1980). However, the abundant rainfall and frequent fog of coastal British Columbia make this region particularly rich in bryophyte species (Schofield, 1988). The favorable environment, combined with the survival of many species in glacial refugia in the region, also results in more endemics and widely disjunct species of bryophytes than elsewhere in Canada.

In the Coastal forest region, the vascular flora has been most intensively studied in the Queen Charlotte Island; a detailed account has been published by Calder and Taylor (1967). Of 593 taxa, 11 are endemic to the islands, an additional 6 are restricted to the islands and the adjacent mainland coast of Alaska and northern British Columbia, and 77 more are found only on the Pacific coast of North America (Calder & Taylor, 1967). The human impact on the flora of Canada and the west coast is illustrated by considering the number of introduced taxa. For Canada as a whole, 22% of the species are introduced (Environment Canada, 1991). The figures are similar for British Columbia (21%, Taylor & MacBryde, 1977) and the Queen Charlotte Islands (20%, Calder & Taylor, 1967). However, these proportions are more startling when different life habits are considered. In British Columbia, 47% of annual taxa and 48% of biennials are introduced, compared with only 14% of perennials (Taylor & MacBryde, 1977). In the Queen Charlotte Islands, 51 of the 69 annual taxa and 8 of 13 biennials are introduced, compared with only 11% of the 511 perennial taxa (Calder & Taylor, 1967).

As for many species groups, the total number of species of breeding birds and mammals in Canada is subject to interpretation of what constitutes a subspecies as opposed to a distinct species. For example, Bunnell and Williams (1980) report 163 species of mammals in Canada, excluding

Cetacea (Whales). In contrast, British Columbia Environment (1991) estimates 163 species excluding both Cetacea and Pinnepedia (seals), while Environment Canada (1991) gives 193 as the total number of species including these two groups. For breeding birds, Bunnell and Williams (1980) report 288 species for British Columbia, compared with 295 reported by British Columbia Environment (1991).

The number of mammalian species in Canada tend to increase from around 30 at the treeline, to about 60 at the Canada–US border, with up to 80 in southern British Columbia (Simpson, 1969). Bunnell and Williams (1980) report 112 species of land mammals for British Columbia, which amounts to 74% of all Canadian species. Range maps given in Banfield (1974) indicate that a little over 50% of all Canadian species are found in the forests of the west coast. Of Canadian breeding bird species, 70% breed in British Columbia (Bunnell & Williams, 1980; British Columbia Environment, 1991). The comparative species diversity of amphibians and reptiles is not quite so impressive. Fifteen species of reptile are found in British Columbia, representing 37% of the Canadian total of 41 species, while there are 18 species of amphibian, 44% of the Canadian total (Orchard, 1980; British Columbia Environment, 1991; Orchard, 1992; Environment Canada, 1991).

Not only is species diversity highest in British Columbia, but the number of species found exclusively in the province are much higher than for other provinces. For breeding birds, 35 species are exclusive to British Columbia, representing 55% of all species exclusive to one of Canada's political jurisdictions, and for mammals 24 species are exclusive, 73% of the Canadian total (Bunnell & Williams, 1980; British Columbia Environment, 1991). The environmental gradients of precipitation and temperature caused by mountainous terrain account for this high species diversity. Banfield (1974) also notes that during the last glaciation, refugia in the Queen Charlotte Islands would have allowed some species to persist, and would have served as centres for post-glacial migration.

Beta species diversity is related to both ecosystem diversity, discussed below, and topographical diversity. Clearly, beta diversity is greater in landscapes having steep environmental gradients, such as British Columbia, than in landscapes having more subtle gradients, such as northern Ontario.

Rare and Endangered Species

The Committee on the Status of Endangered Wildlife in Canada (COSEWIC) lists 195 species of animals and plants as being vulnerable, threatened, endangered, or extirpated in Canada in 1990 (Canadian Wildlife Service, 1990). Endangered species are those that are in imminent danger of becoming extinct or extirpated in Canada. Threatened species are those which are likely to become endangered in the near future if existing stresses are not reduced, and vulnerable species are those which, while not under any immediate threat, are at risk for reasons of small population size or other factors.

The status of wildlife species in the west coast forests is generally better than elsewhere in Canada. Only nine species in the "Pacific/Mountain" life zone are listed as threatened or endangered, and only in the Arctic life zone are there fewer listed species (Environment Canada, 1991). Three plant species from the Pacific/Mountain life zone are listed as threatened or endangered, but only one of these is found in the coastal rain forests: the giant helleborine (Epipactis gigantea). In an inventory of more than 3100 vascular plant taxa of British Columbia, Taylor and MacBryde (1977) classify the taxa subjectively, based on their perceived status. They classify giant helleborine as "rare," but classify 12 other taxa as "endangered," of which 7 are found in the coastal rain forests (Table 18.1). The redesignation of rare and endangered vascular plant taxa in British Columbia is currently being undertaken by the Conservation Data Centre of the Nature Conservancy of Canada (G. Douglas, personal communication). At present, about 630 taxa have received a preliminary designation as being rare, but this number will probably decrease as more records are accumulated. This enormous task of dealing with vascular plants will have to be finished before nonvascular plants can be considered. COSEWIC also does not consider the status of nonvascular plants. However, six species of British Columbian lichens have so far been assigned a "conservation status" by the Rare Lichens Project of the Smithsonian Institution (Goward, 1990). As one of four criteria used in determining conservation status is global range, many additional rare British Columbian species do not qualify for the Smithsonian listing.

Many woodland species of animal, bird, and fish are also on the endangered species list. Some of these, in contrast to the endangered tree species, are not represented by large populations elsewhere. In British Columbia, the Ministry of the Environment has categorized wildlife species based on scores for six criteria: abundance, distribution, habitat integrity, population trend, reproductive potential, and national/interna-

Table 18.1. Endangered vascular plant taxa of the coastal rain forests of British Columbia. From Taylor and MacBryde (1977)

Common name	Botanical name	COSEWIC status
Adder's tongue	*Ophioglossum vulgatum*	Not listed
Alaskan orache	*Atriplex alaskensis*	Not listed
Austin's phantom orchid	*Eburophyton austiniae*	Not listed
Macoun's meadowfoam	*Limnanthes macounii*	Vulnerable
Pacific rhododendron	*Rhododendron macrophyllum*	Not listed
Seaside bird's-foot trefoil	*Lotus formosissimus*	Not listed
Sour clover	*Trifolium fucatum*	Not listed

tional status (Munro & Low, 1980; British Columbia Environment, 1991). Those species included in the "red list" on the basis of scores for these criteria are considered to be endangered or threatened, while species on the "blue list" are sensitive or vulnerable. Seventeen species of mammals appear on the red and blue lists (6 on the red and 11 on the blue), together with 30 species of birds (5 red, 25 blue), 1 reptile (red), and 2 amphibians (both blue), This list encompasses all of the COSEWIC listed species, except for the Queen Charlotte Islands' ermine (*Mustela erminea haidarum*), which is COSEWIC-listed as vulnerable.

Ecosystem and Landscape Diversity

The quantification of habitat or ecosystem diversity has received far less attention than species or genetic diversity (Christensen & Peet, 1984). Also, the frame of reference for ecosystem diversity must be larger than the ecosystem itself, and will therefore usually be completely arbitrary (e.g., provincial). Nevertheless, some studies of ecosystem diversity have been conducted (e.g., Romme, 1982). In order to enable comparison of measurements of ecosystem diversity, however, the ordination and classification procedures must be identical, or at least comparable, and this is rarely the case.

The measurement of ecosystem diversity within a landscape again encounters the problem of scale, along with problems related to pattern and shape. Most methods of estimating spatial statistics, such as autocorrelation, require classification of types through procedures such as ordination (Turner, O'Neill, Conley et al., 1991). However, image textural measures (originally derived for engineering applications), which do not require a priori classification, have recently been used to quantify the spatial arrangement of landscapes (Musick & Grover, 1991).

The mountainous nature of British Columbia, and associated steep environmental gradients, mean that ecosystem diversity in the Coastal forest is the highest in Canada. Several different ecosystem classification systems have been developed in Canada. These reflect the needs of the individual agencies responsible for their development. In British Columbia, two systems have been used for various purposes in recent years: the biogeoclimatic system, and the biophysical classification system.

The Biogeoclimatic Ecosystem Classification System

The biogeoclimatic ecosystem classification (BEC) system of British Columbia, which has been adapted for use in Alberta, is a hierarchical system that integrates ecosystem variation at regional and local levels and over time (Pojar, Klinka, & Meidinger, 1987). At the local level, ecosystems are organized according to similarities in vegetation and site. At the regional level, the ecosystems are categorized according to vegetation-inferred climatic space. Zonal climax ecosystems define a subzone of the biogeoclimatic units. In the coastal forest region there are two major biogeoclimatic units, the "coastal western hemlock" zone (CWH), which is divided into ten subzones, and the "coastal Douglas fir" zone (CDF). For forest management purposes, "site units" based on environmental characteristics, such as moisture and nutrient status, are identified at the local level, and used for developing silvicultural prescriptions, for range management, and for integrating recreation into multiple-use management (MacKinnon, Meidinger, & Klinka, 1992).

The Biophysical Classification System

A biophysical classification system was developed by the wildlife branch of British Columbia's Ministry of the Environment, originally for the purpose of classifying ecosystems for wildlife habitat (Demarchi, 1992). It is a hierarchical system, with a strong emphasis on physical attributes and successional, as well as climax, vegetation. At the highest level of the hierarchy are ecodomains, which are global in context. These are subdivided into ecodivisions, ecoprovinces, of which there are 10 in British Columbia, ecoregions, and ecosections.

A hierarchical biophysical system has been design by the Canadian Committee on Ecological Land Classification, as a consistent classification system for the national level. The system recognizes 15 "ecozones," 45 "ecoprovinces," 177 "ecoregions," and 5428 "ecodistricts." The biophysical characteristics of these hierarchical classes also result in different landscape types. The ecoregions provide a scale of detail that is both meaningful and manageable on a national scale, and are now widely used by Environment Canada for data base and other applications (Turner, Rubec, & Wiken, 1992). Although developed independently from the British Columbia system, the two systems were recently harmonized with the adoption of the British Columbia units by Environment Canada (D. Demarchi, British Columbia Environment, and H. Hirvonen, Environment Canada, personal communication).

The biogeoclimatic and biophysical classification systems were developed independently, and are based on a different set of characteristics. The BEC system uses primarily climax vegetation types to infer units that are homogeneous in terms of climate and soil properties. In contrast, the biophysical system places major emphasis on landforms (Demarchi, 1992). The result is that the ecoregions and ecosections of the biophysical system are geographically contiguous, whereas the site units of the BEC system are not necessarily contiguous and have a strong topographical element (Demarchi, 1991). The two systems have now effectively been integrated, with ecosection being subdivided by biogeoclimatic units. This integration of the two systems is widely used for many applications, including ecological reserve system design, a provincial old-growth strategy, and threatened and endangered species cataloguing, in addition to the silvicultural and habitat monitoring functions for which they were originally designed (D. Demarchi, British Columbia Environment, personal communication).

Conservation of Biodiversity

For the whole of Canada, an estimated 4% of productive forest land is categorized as having a highly protected status: land in National Parks, ecological reserves, and other areas in which forest management operations are prohibited (Lowe, 1991). This percentage corresponds with the estimate of protected areas by the World Commission on the Environment and Development (WCED); their call for a tripling of protected areas (World Commission on the Environment and Development, 1987) led to the federal government's target of protecting 12% of Canada's forest land. The figures for much of British Columbia are generally somewhat higher than for the national average, although the degree of protection varies widely by forest region. For example, 18.4% of the subalpine forests is estimated to be protected, compared with only 1.2% for the Columbia forest region that includes coastal rainforests (State of the Environment reporting/Canadian Forest Service, unpublished data). The figure for the coastal forest region is 7.7%. The degree of protection afforded by different categories of reserve and park is highly variable. The World Conservation Union (IUCN) utilizes five categories of protection in classifying conservation areas (World Conservation Union, 1990), and these categories are now recorded in the National Conservation Area Data Base created by Environment Canada and the Canadian Council on Ecological Areas. Under the federal government's Green Plan, the Canadian Forest Service is promoting the extension of the forested ecological reserve network to ensure adequate representation of all forest ecosystems in protected areas, and this coincides with a goal of the provincial government of British Columbia (see below for further discussion).

A considerable body of literature has been devoted to the topic of the design and management of ecological reserves (for reviews, see Harris, 1984; Hunter, 1990). The effectiveness of ecological reserves will be affected by their size, shape, and distribution, and by the management of surrounding lands. Small reserves will conserve fewer species than larger ones, approximately circular reserves will minimize edge effects, and isolated, scattered reserves will not be as effective as closely spaced or linked reserves. Sensitive management of the area surrounding the reserve can also increase its value. At present these considerations and others have received little attention in the establishment of forest ecological reserves in Canada. For this reason, the target figure of 12% may be less important than the choice and design of reserves, and the enlightenment of forestry practices outside the reserves.

In addition to the World Commission on Development and the Environment proposal that the extent of protected areas throughout the world should be trebled, similar targets have been proposed by IUCN, UNEP and WWF (World Conservation Union et al., 1991), and the Caracas declaration of the Fourth World Congress on National Parks and Protected Areas. Whatever may constitute an appropriate target, there is no question that, in order to protect the welfare of present and future generations at a time of increasing human pressure on natural resources, additional areas of land need to be protected. It is only sensible to try to ensure that the areas protected are those which will provide the greatest future value per unit area. The challenge lies in developing suitable procedures for identifying such areas.

In the absence of reliable information on future value of natural resources, the logical approach is to protect areas of high diversity and areas that contain particularly rare, or unique, resources. Such an approach has often been used, for example by Hopper and Burgman (1983), Moran and Hopper (1983), and Sampson, Hopper, and James (1988) for various species of eucalypts. These authors used morphological and allozyme data from discrete populations of eucalypt species to quantify the distinctness of the sampled populations already within protected areas. Although such an approach is useful for species of known or potential commercial value, it is practicable only for a small fraction of all species. In any case, the conclusion reached for different species would soon result in conflicting recommendations and an unrealistically large area being proposed for protection.

An alternative approach is provided by "gap analysis" (Scott, Csuti, Smith et al., 1991), which is greatly facilitated by modern Geographic Information Systems (GIS) technology. Areas of high conservation value (e.g., old-growth forest) for different ecoregions can be compared with existing protected areas and with areas under conflicting land use demands. Relative diversity information (if available) can also be used to make recommendations for areas that do not meet all criteria. For example, an area of pristine forest in a high diversity ecoregion might be recommended for protection even if the potential mineral value was quite high, whereas a similar area in a low diversity ecoregion would only be recommended if the mineral value were much lower or nil.

Other aspects that must be considred in the identification of protected areas are the size, shape, and distribution of the areas and linkages (corridors) among them. These issues have received much attention (e.g., Shafer, 1990), particularly in relation to island biogeography theory and species-area relationships. Intuitively, large reserves will conserve biodiversity more efficiently than small reserves, and it is better to minimize the length of boundary per unit area (leading to more circular, rather than elongated reserves), and to reduce the distance among reserves. In a world of conflicting interests, where compromises must be made, the problem lies in how much, and in relation to which feature of the reserve to compromise.

Much of the concern with problems of species and genetic conservation stems from the impact of human activities on natural ecosystems, and nonsustainable exploitation of such ecosystems. It is only natural, therefore, that the impact of various forestry operations on forest ecosystems has been the subject of much debate and not a little acrimony. Recent events in the Carmanah valley; on South Moresby Island; and various other areas in British Columbia have received much public attention and, in some cases, resulted in court battles. It is now widely accepted that protected areas alone cannot constitute a fully effective conservation strategy, forest management outside protected areas must also incorporate the objective of maintaining biodiversity.

Threatened Forest Ecosystems

Pojar (1980) discussed threatened forest ecosystems in British Columbia, and drew a distinction between those that are truly threatened, and those that are merely depleted (i.e., capable of regeneration). In the coastal zone, the most threatened ecosystems are associated with garry oak (Quercus garryana) in the Mediterranean climate of sourtheastern Vancouver Island and the islands of the Gulf of Georgia. Other lowland forest ecosystems in the same area, as well as bottomland and bog ecosystems of the lower Fraser valley, have also been almost completely extirpated by a combination of logging and development. Logging is also reducing lowland Douglas fir, western red cedar, and yellow cedar (*Chaemacyparis nootkatensis*) forests, and alluvial Sitka spruce forests to a threatened state. Finally, the introduction of Sitka black-tailed deer (*Odocoileus hemionus sitkensis*) to the Queen Charlotte Islands in the early years of this century has resulted in major changes to, and probably losses of, forest ecosystems.

The status of "old-growth" forest, and the effect of harvesting and the habitat fragmentation resulting from harvesting of the old-growth ecosystem, has been the subject of great debate both in Canada and the United States. The most well known case is that of the northern spotted owl in the Pacific Northwest of the United States. Three characteristics of the spotted owl have led to its prominent status: its apparent dependence on an old-growth forest habitat (Carey, 1985), its very large home range requirements (Sisco & Gutierrez, 1984), and its demographic fragility (Dawson, Ligon, Murphy et al., 1987). These characteristics demand that if survival of the species is to be ensured, a very conservative habitat management system involving the preservation of large areas of old-growth forest must be undertaken. An Interagency Science Committee was formed to design a management system for spotted owls, and it recommended the establishment of a network of conservation areas averaging about 25,000 ha and separated by no more than 20 km (Wood, 1991). However, the strategy has come under criticism, and Lande (1988) has postulated that because of its population genetics foundation, the strategy is more likely to lead to the extinction of the owl, because it is unlikely that viable populations will be maintained. Lande argued that the demographics of the species should provide the basis for the strategy, because maintenance of demographic processes is much more critical to the long-term survival of a species than population genetic factors. In this instance, the spotted owl is fulfilling the role of an "indicator species." The owl itself is rather less important than the ecosystem it represents. The marbled murrelet (Brachyramphus marmoratus) appears to perform a similar function as an indicator of old-growth ecosystem conditons in the coastal forest of British Columbia (Rodway, Savard, & Regehr, 1991). Although about 70% of the coastal forest is still classified as old-growth (A. MacKinnon, British Columbia Forest Service, personal communication), very few of the larger watersheds on Vancouver Island remain unaffected by logging (Moore, 1990).

Clearcutting obviously causes enormous changes to the ecosystem, with dramatic consequences to many organisms. For this reason, clearcutting has been the target of environmental groups and the subject of much research. The effects of clearcutting on vertebrates is highly variable. Some species, for example moose, benefit from a mosaic of food and cover habitats, such as might result from a series of relatively small clearcuts (Timmermann & McNichol, 1988; Payne, McNichol, Eason et al., 1988). Of the common fur-bearing mammals in the boreal forest zone of Ontario, only the marten (*Martes americana*) appears to be reliant on old-growth forest (Thompson, 1988, 1991). Most other fur-bearing species reach peak densities in early or mid-successional habitats, mainly in response to the buildup of prey species (Thompson, 1988).

Although studies have demonstrated significant decrease in some species of songbirds following clearcutting, other species are dependent on early successional stages (Welsh, 1988, 1991). The passerines constitute the dominant family of birds in the boreal forest, and although some species are no longer present after clearcutting, others take their place, such that overall species diversity of all bird species is virtually unchanged (Welsh & Fillman, 1980). Many species of songbirds are clearly dependent on specific forest ecosystems, as well as on specific successional stages. A study of species abundance in northern Ontario in relation to the Forest Ecosystem Classification system demonstrated quite narrow ecosystem preferences for many species (Welsh, 1991). Maintaining species diversity for birds and other fauna therefore requires a continuing supply of forest ecosystems in various successional stages.

Obviously, the impact of clearcutting is not a simple reduction in species diversity, but rather a differential effect, with some old-growth dependent species of plants and animals being lost from the system, presumably until suitable environmental conditions are restored. However, the scale and pattern of clearcuts are also vitally important in determining the response of different species to disturbance. The resultant spatial heterogeneity and fragmentation of the landscape have significant impact of species diversity and abundance. Fragments smaller than a species' home range will result in a reduction in abundance of that species, or its complete elimination. The ecosystem itself may then be threatened with destruction. Research on boreal fragmentation in Sweden has demonstrated that 1 ha reserves that are required by law are composed entirely of edge habitat, and therefore do not function as forest ecosystems (Dr. Per-Anders Esseen, personal communication). Because edge effects penetrate some distance into the remaining forest, even a low intensity of clearcutting can cause substantial habitat fragmentation (Hansen, Spies, Swanson et al., 1991). Pielou (1990) has also called attention to the potential loss of genetically distinct demes, and the consequent impact on genetic diversity of many species as a result of clearcutting.

It is important to contrast secondary forests resulting from clear-cut harvesting with those derived from natural disturbances. In the former, as almost all standing biomass is usually removed, the resulting secondary forest is structurally very simple and uniform. In contrast, natural secondary forest usually retains a substantial component of surviving old-growth individuals, standing dead trees, etc., which provide much greater structural diversity and therefore species diversity.

Concern with forest fragmentation and the resultant extensive edge effects, together with a recognition of the importance of structural diversity in forests, has led to the development of a philosophy termed "The New Forestry" in the Pacific Northwest of the United States (Franklin, 1989; Maser, 1990; Hansen et al., 1991). Part of this philosophy calls for large cut blocks, to minimize edge effects. Large trees, snags, and fallen logs are retained on the harvested area to provide habitat for cavity-nesting, amphibians, fungi, and other microorganisms.

In British Columbia, a substantial amount of effort is being devoted to designing management strategies that are effective in maintaining biodiversity. The provincial government has recently developed a management strategy proposal for the coastal region that integrates a system of reserves with appropriate practices on managed land outside reserves. The principles of ecosystem management and a consensus-building approach to decision making are key factors in the strategy (Pojar & Hamilton, 1993). The ecological reserve component of this management strategy seeks to protect a minimum of 10 to 12% of old-growth forest in each forested site series, with rare ecosystems being overrepresented

(British Columbia Ministry of Forests, 1993). Linkages among reserves will be maintained or established and managed specifically for that purpose. Although reserves are permanent, linkages are temporary features, with individual linkages being replaced by others over time. The major objective of forest management outside reserves is to maintain a variety of several stages, with not less than 30% of the total forest area less than 20 years old, and a variable cut block size, with an average of 40 ha. Maintenance of structural diversity through retention of dead wood, snags—at least ten large diameter snags per hectare, green trees, and coarse woody debris are also features of the stand-level component of this management strategy.

Harvesting is not the only forest management operation affecting diversity: several others are reviewed by Hunter (1990). Of particular interest to those involved in tree improvement is regeneration, particularly the use of genetically improved and clonal planting stock. Despite dramatic increases in annual planting programs in recent years, only 20% of land that is harvested or lost to fires and pests each year is currently regenerated by planting (Forestry Canada, 1990). Most of the planting stock still comes from "wild" collections, although seed orchard seed is now approaching or meeting demands is some provinces. The proportion of planting stock that is clonal remains tiny, but more than a hundred hectares of experimental black spruce (*Picea mariana*) stecklings have been panted in northern Ontario, and Quebec is undertaking a similar program.

A typical seed orchard in Canada may consist of several hundred open-pollinated families, resulting in tens of thousands of possible interfamily pollinations. The permutations in clonal orchards may be fewer, but greater fecundity is likely to result in a more even representation of possible crosses. Temporal and spatial variation in fecundity of natural forests reduces the potentially infinite number of crosses to a much smaller effective number, which may be similar to those resulting from seed orchards. (See Boyle, Liengsiri, & Piewluang, 1990, for an example of low heterozygosity in open-pollinated seed from a natural forest.) Planting with seed from seed orchards managed to promote outcrossing is unlikely to reduce genetic diversity of the planted species, but see El-Kassaby (1993) for a discussion of other factors. Except in very simple ecosystems, however, tree species and age structure diversity are likely to be reduced as a result of site preparation and planting.

As discussed previously, however, frequency of abundnce are not the only important aspects of diversity, spatial heterogeneity or pattern may also have important ecological consequences. The spatial arrangement of genetic diversity on planted sites will be completely artificial. This is of particular concern when dealing with clonal planting stock, when choices must be made as to the size and arrangement of monoclonal blocks, or clonal representation in an intimate mixture. Again, although arbitrary guidelines can be established, as has been done in Ontario, the ecological consequences of such choices can be determined only through a comprehensive program of research. The assumption is often made that natural regeneration guarantees maintenance of diversity. This is true only when harvesting operations mimic natural disturbances in a given ecosystem. Thus, the New Forestry style of harvesting practices, with retention of trees, snags, and logs, may adequately mimic a natural disturbance such as fire in the coastal forest. Applying the same design to boreal or Great Lakes–St. Lawrence forests may not be justifiable.

Research on Forest Biodiversity in Canada

The substantial differences in constitutions and land tenure systems between Canada and the United States have resulted in fewer legislative pressures in Canada to incorporate diversity issues into planning and management. As a result, research on biodiversity in Canadian forests has not been as active as in the United States. However, current domestic and international initiatives addressing biodiversity have led to a rapid expansion in research effort devoted to diversity issues. A major impetus to such research has been the federal government's "Green Plan," released in December 1990, through which substantial sums of money have been identified to support activities related to biodiversity. Research related to biodiversity has been supported by the Green Plan under both the Forest Practices and National Forest Genetic Resources Centre (NFGRC) initiatives.

Although research on biodiversity of Canadian forests in now conducted throughout the country, much of it is focused on old-growth forests, especially the old-growth forests of the west coast (see Table 18.2 for a representative listing of current projects). There is a very broad spectrum of research on biodiversity currently being conducted, with the involvement of a diverse array of agencies, including federal and provincial governments, universities, and industry. Despite cooperation among agencies, it is the view of this author that there is still an overall lack of coordination, largely due to the independence of the relevant agencies. The result is a lack of sites with comprehensive estimates of genetic and species diversity, including temporal and spatial variation, and the effects of forestry practices. Without such information, scientifically tenable assessments of the impact of various human or natural events cannot be made.

A more complete understanding of ecosystem functioning and the role of individual species is required in order to assess the consequences human impacts. Namkoong (1992) argued for a process-oriented approach by proposing that the goal for management of natural ecosystems should be the maintenance of ecological integrity, and conservation of biodiversity should be seen as a means of attaining that goal, or an indicator,

Table 18.2. Current biodiversity research programmes in Canada's west coast forests

Subject	Agency	Reference
General		
Predicting biological diversity on forest land in British Columbia.	UBC (Forestry)	Daust and Bunnell, 1992
Occurrence and ecology of various groups of vertebrate in the Nimpkish Valley.	BCE	
Abundance and diversity of various groups of vertebrates in coastal alder and conifer stands.	BC MoF	
Impact of harvesting, site preparation and other practices on wildlife.	BC MoF/BCF	
Influence of vegetation management on the diversity of plants and small mammals.	BC MoF	
Fungi		
Fungi of old-growth forests in British Columbia.	UBC (Soil Science)	Gamiet and Birch, 1992
Diversity of mycorrhizal fungi in old-growth and mature second growth stands of Douglas fir on Vancouver Island.	University of Victoria	Goodman, 1993
Vascular plants		
Multi-locus and multi-genome analysis of genetic diversity in western red cedar and Douglas fir.	UBC (Forestry)	
Genetic diversity in mountain hemlock.	Forestry Canada/ BC MoF/ CP Forest Products	
Sustainable development of natural sources of the emerging anti-cancer agent, taxol: assessing cultivation options.	Forestry Canada	
Changes in plant diversity in Douglas fir stand following the conversion of old-growth to second growth.	Arenaria Research/ Forestry Canada	Fraser and Ryan, 1993
Microorganisms/invertebrates		
Nematode diversity and distribution in different forest soil habitats.	UBC (Soil Science) Forestry Canada	Panesar and Marshall, 1993
Succession of Collembola in stumps of Douglas fir seres representing clear-cuts to old-growth forest.	Forestry Canada	Setala and Marshall, 1993
Taxonomic survey of springtails collected from northern Vancouver Island sites.	UBC (Soil Science)/ Forestry Canada	Baumbrough et al., 1992
Soil fauna communities in cedar-hemlock and hemlock-amabilis fir forest types on northern Vancouver Island.	UBC (Soil Sciece)	Battigelli and Berch, 1992

Variations in carabid community structure associated with forest successional stages.	UBC (Zoology)	Craig, 1993
Community ecology of insect/arthropod fauna from an old-growth forest with special emphasis on canopy fauna.	University of Victoria	Winchester, 1993
Terrestrial molluscs: their role in the ecological dynamics of forest floor communities.	University of Alberta	Boag, 1993
Vertebrates, general		
Comparison of bird, small mammal, and amphibian abundance and diversity in old-growth and second-growth forests of coastal British Columbia.	BC MoF/CWS	Seip and Savard, 1992
Amphibians		
Differential resource use, interspecific interactions, and the effect of logging on the distribution and abudance of terrestrial salamanders on southern Vancouver Island.	University of Victoria	Davis, 1993
Amphibian associations in old-growth and second-growth communities, with specific reference to coarse woody debris.	UBC (Zoology)	F. Bunnell, personal communication
Birds		
Bird communities in old-growth and regenerating forests of western Vancouver Island.	Bryant Services/ Canadian Wildlife Service/MacMillan Bloedel Ltd.	Bryant et al., 1992
Breeding bird communities in managed second-growth stands, riparian areas, and old-growth forests on northeastern Vancouver Island.	CWS/MacMillan Bloedel	Eckert et al., 1992
Utilization of residual patches of old-growth timber by forest birds.	CWS	Millikin, 1993
Genetics, population status and habitats of barn owls.	UBC (Zoology)	
Mammals		
Relations between small mammals and coarse woody debris.	BCE/UBC (Forestry)	F. Bunnell, personal communication
Habitat selection for marten in the Nanaimo River Valley.	BCE/Simon Fraser University	
Impact of forest development on grizzly bears.	BCE/BC MoF	
Den requirements and forest management impacts on black bears.	BCE/Canadian Pacific Forest Products	

rather than as the goal itself. Lande (1988) also stressed the improtance of considering processes, in designing conservation strategies. In reviewing recovery plans for two endangered bird species in America, Lande postulated that strategies based on demographic processes would have a far greater chance of success than the existing strategies based on the population genetics of the two species.

Biodiversity is arranged in distinct spatial patterns at all levels. The distribution of genetic diversity is often found to be non-random (e.g., Linhart, Mitton, Stwrgeon et al., 1981). Species distribution is also usually associated with discrete patches of habitat. Although on a broad scale the distribution of habitat is clearly realted to environmental variables, the significance of small-scale patterns and the nonrandom distribution of genetic diversity is not well understood (Lovejoy, 1988). Human manipulations typically tend to simplify or eliminate spatial patterns. Natural mixed forests are often replaced by plantations of a single, or a few, species. The genetic diversity of these plantations may be random, overdispersed, highly structured or, in the extreme case of monoclonal plantations, absent. The risks associated with a reduction in biodiversity are well known from past experience and ecological theory, and policy guidelines are usually devised to address this issue, but management guidelines governing pattern and distribution are generally more concerned with convenience than with any ecological principles.

Increasingly, theories derived in fields of science such as mechanics and engineering are being modified and adapted for use in ecology. One such theory governing the movement of particles through a matrix, "percolation theory" (Stauffer, 1985), predicts that if there exists a critical set of sites through which energy or material flows (the "backbone"), the loss of other sites will have no significant effect, whereas the loss of a backbone site would have far greater consequences (Gardner & Turner, 1991). This concept, which is of obvious relevance to maintenance of ecosystem processes, infers that, the identification of "keystone habitat patches" may be more significant that the classification of species as being keystone species. Experimental or empirical evidence for the validity of such a concept has not yet been obtained.

Future Initiatives

In the past few years there have been a large number of international policy initiatives, most of them with the involvement of foresters, that have conducted valuable analyses of actions required to manage and conserve biodiversity. Such initiatives include the report of the World Commission on Environment and Development (the "Bruntland Commission"), published in 1987, various products of the 1992 United Nations Conference on the Environment and Development (UNCED), the

"Caracas declaration" resulting from the Fourth World Congress on National Parks and Protected Areas in 1992, and innumerable national reports and strategies. Several initiatives of particular relevance for forestry having global, as well as national and local, applicability are briefly reviewed below.

International

The United Nations Convention on Biodiversity, which was signed by more than 150 nations at the "Earth Summit" in Rio de Janeiro in 1992 and came into force in Decemeber 1993, calls for the development of national strategies for biodiversity. The Convention is legally binding on all parties to it, and the possibility exists for negotiation of protocols dealing with specific issues. Forestry is one issue that may be subject to such a protocol. Canada is actively pursuing this goal through the development of jurisdictional (federal and provincial) and sectoral plans, including forestry. The National Forest Sector Strategy (Canadian Council of Forest Ministers, 1992) concentrates heavily on biodiversity, and will form a basis for the forest component of the national Biodiversity Strategy.

Various other international initiatives, while not legally binding, serve to focus attention on biodiversity and serve as a framework for action. The World Conservation Union (IUCN), in partnership with UNEP and the World Wide Fund for Nature (WWF), has published *Caring for the Earth: A Strategy for Sustainable Living* (World Conservation Union et al., 1991). The document constitutes a development of the *World Conservation Strategy*, published by the same agencies in 1980. Chapter 14 deals with forest lands, and lists 10 "Action Items," which include establishing a comprehensive system of protected natural forests, increasing national capacity to manage forests sustainably; adopting environmental policies that protect ecological services and biodiversity; and expanding efforts to conserve forest genetic resources.

The World Resources Institute (WRI), in partnership with IUCN and UNEP, has also produced the *Global Biodiversity Strategy* (World Resources Institute et al., 1991), following a number of consultative meeting with governments, industrial, and nongovermental organizations around the world. The strategy is obviously closely related to *Caring for the Earth*, with two of the three partners also responsible for the latter document. However, while *Caring for the Earth* deals with all aspects of sustainable living, the *Global Biodiversity Strategy* covers specifically the conservation of biodiversity. The strategy notes that among the fundamental causes of biodiversity loss are the steadily narrowing spectrum of traded products, economic systems and policies that fail to value the enviroment and its resources, deficiencies in knowledge and its application, and legal and institutional systems that promote unsustainable exploitation.

Many of the strategy's 85 actions relate to forests and forestry. Some of the more prominent ones are abandon forestry practices that encourage resource degradation, reduce pressure on fragile ecosystems by using land already under cultivation more efficiently, incorporate biodiveraity conservation into the management of all forests, undertake national biodiversity inventories, and promote basic and applied research on biodiversity conservation.

Americas's Interhemispheric Geo-Biosphere Organization (AMIGO) has recently developed a proposal for a project, involving monitoring and inventorying biodiversity along four east–west transects, two in North America and two in South America. This project should make a major contribution in filling one of the research gaps previously noted, the lack of a comprehensive study for any one forest site in the coastal forests.

Provincial Program

The newly developed coastal forest management plan has already been described. The British Columbia Ministry of Forests is also conducting a number of associated projects, including the development of a genetic conservation strategy for economically important tree species, the compilation of an inventory of biodiversity "hotspots," and the preparation of gap analyses (E. Hamilton, British Columbia Ministry of Forests, personal communication). In addition, the British Columbia Ministry of the Environment is developing a new management direction for its wildlife program: "Managing Wildlife to 2001" (British Columbia Environment, 1991). One of the six "challenges" outlined in this program is managing wildlife for diversity and sustainability. A number of different activities are included under this challenge, including legislation for wildlife, inventory, and evaluation of species and habitats, and stronger control of habitat.

The coastal forests of British Columbia are particularly rich in all components of biodiversity, compared with other regions of Canada. The current status of forest biodiversity in the region is also much better than many other parts of the world. However, there is now a recognition of the major impact of past forest practices on aspects of biodiversity that are not as obviousnor as dramatic as the species extinctions that tend to foster public concern. In recent years there has existed a somewhat confrontational interaction between forest sector agencies and environmental organizations. This is now being replaced by progressive consensus-building decision-making processes, such as those being fostered by the Commission on Resources and the Environment (CORE), under the British Columbia Ombudsman. Such a constructive approach to forest management is the only appropriate way to deal with the highly complex and interactive processes that are responsible for maintenance of biodiversity.

It is clear that conservation of globally significant ecosystems, such as Canada's coastal rainforests, will increasingly come under international scrutiny. Those calling for increased efforts to conserve the remaining unlogged watersheds on the west coast are no longer merely individuals or local environmental groups, but international nongovernmental organizations. This was well illustrated in the resolution passed at the recent General Assembly of the IUCN, calling on the Canadian and British Columbia governments to increase conservation efforts, while taking into account the opinions of environmental groups. The lack of reference to input from aboriginal people led the Canadian delegation to abstain from joining the consensus. However, the message from such a resolution is that the effectiveness of conservation measures are no longer simply subject to provincial, or even national, judgment.

References

Banfield, A.W.F. 1974. The Mammals of Canada. Toronto: University of Toronto Press.

Battigelli, J.P., and Berch, S.M. 1992. Soil fauna communities in cedar-hemlock and hemolck-amabilis fir forest types on northern Vancouver Island. *NW Environ J*, 8, 213–214.

Baumbrough, B., Berch, S.M., and Marshall, V.G. 1992. Taxonomic survey of springtails collected from northern Vancouver Island sites. *NW Environ J*, 8, 215–217.

Boag, D.A. 1993. Terrestrial molluscs: Their role in the ecological dynamics of forest floor communities. In D. Pollard (ed.), *Proceedings of a Workshop on Forest Ecosystem Dynamics, February 1993* (pp 37–40). Victoria: Forestry Canada.

Boyle, T.J.B., Liengsiri, C., and Piewluang, C. 1990. Genetic organization in upland and lowland populations of black spruce. *Heredity*, 65, 393–399.

British Columbia Environment. 1991. Managing Wildlife to 2001: A Discussion Paper. Victoria: British Columbia Wildlife Branch.

British Columbia Ministry of Forests. 1993. Biodiversity guidelines; Coastal stand-level biodiversity and landscape-level biodiversity. *For Plan Can*, 9(1), 33–41.

Bryant, A.A., Savard, J.-P.L., and McLaughlin, R.T. 1992. Bird communities in old-growth and regenerating forests of western Vancouver Island. *NW Environ J*, 8, 224–225.

Bunnell, F.L., and Williams, R.F. 1980. Subspecies and diversity—The spice of life or prophet of doom. In R. Stace-Smith, L. Johns, and P. Joslin (eds.), *Threatened and Endangered Species and Habitats in British Columbia and the Yukon: Proceedings of a Symposium* (pp. 246–259). Victoria: British Columbia Ministry of Environment, Fish and Wildlife Branch.

Calder, J.A., and Taylor, R.L. 1967, *Flora of the Queen Charlotte Islands*. Part I: *Systematics of the Vascular Plants*. Ottawa: Canada Department of Agriculture Research Branch Monograph No. 4.

Canadian Council of Forest Ministers. 1992. *Sustainable Forests: A Canadian Commitment*. Hull, Quebec: Forestry Canada.

Canadian Wildlife Service. 1990. COSEWIC review. *Recovery*, 2(1), 4–5.

Carey, A.B. 1985. A summary of the scientific basis for spotted owl management. In R.J. Gutierez and A.B. Carey (eds.), *Ecology and Management of the Spotted Owl in the Pacific Northwest* (pp. 100–114). U.S.D.A., Forest Service General Technical Report PNW-185.

Christensen, N.L., and Peet, R.K. 1984. Measures of natural diversity. In: J.L. Cooley and J.H. Cooley (eds.), *Natural Diversity in Forest Ecosystems: Proceedings of the Workshop* (pp. 43–56). Athens: Institute of Ecology, University of Georgia.

Cohen, J.I., and Potter, C.S. 1993. Conservation of biodiversity in natural habitats and the concept of genetic potential. In C.S. Potter, J.I. Cohen, and D. Janczewski (eds.), *Perspectives on Biodiversity: Case Studies of Genetic Resource Conservation and Development* (pp. xix–xxiii). Washington, DC: AAAS Press.

Copes, D.L. 1981. Isoenzyme uniformity in western red cedar seedlings from Oregon and Washington. *Can J For Res*, 11, 451–453.

Craig, K. 1993. Variation in carabid community structure associated with forest successional stages. In D. Pollard (ed.), *Proceedings of a Workshop on Forest Ecosystem Dynamics, February 1993* (pp. 22–24). Victoria, BC: Forestry Canada.

Daust, D.K., and Bunnell, F.L. 1992. Predicting biological diversity on forest land in British Columbia. *NW Environ J*, 8, 1990–220.

Davis, T.M. 1993. Differential resource use, interspecific interactions and the effect of logging on the distribution and abundance of terrestrial salamanders on southern Vancouver Island. In D. Pollard (ed.), *Proceedings of a Workshop in Forest Ecosystem Dynamics, February 1993* (pp. 25–27). Victoria BC: Forestry Canada.

Dawson, W.R., Ligon, J.D., Murphy, J.R., Myers, J.P., Simberloff, D., and Verner, J. 1987. Report of the advisory committee on the spotted owl. *Condor*, 89, 205–229.

Demarchi, D.A. 1991. Is it Coke or is it Pepsi? Ecosystem classification in B.C. *Bioline*, 10(1), 2–5.

Demarchi, D.A. 1992. Biophysical habitat classification in British Columbia: A system for mapping mountainous ecosystems. In G.B. Ingram and M.R. Moss (eds.), *Landscape Approaches to Wildlife and Ecosystem Management: Proceedings of the Second Symposium of the Canadian Society for Landscape Ecology and Management* (pp. 39–46). Morin Heights, Canada: Polyscienice Publications.

Eckert, C.D., Savard, J.-P.L., and McLaughlin, R.T. 1992. Breeding bird communities in managed second growth stands, riparian areas, and old-growth forests on northeastern Vancouver Island, British Columbia. *NW Environ J*, 8, 228–229.

El-Kassaby, Y.A. 1990. Genetic variation within and among conifer populations: Review and evaluation of methods. In H.H. Hattemer and S. Fineschi (eds.), *Biochemical Markers in the Population Genetics of Forest Trees* (pp. 59–74). The Hague: SPB Academic Publishing.

El-Kassaby, Y.A. 1993. Domestication and genetic diversity—should we be concerned? *For Chron*, 68, 687–700.

El-Kassaby, Y.A., and Sziklai, O. 1982. Genetic variation of allozyme and quantitative traits in a selected Douglas fir (*Pseudotsuga menziesii* var. *menziesii* [Mirb.] Franco) population. *For Ecol Manag*, 4, 115–126.

Environment Canada. 1991. *The State of Canada's Environment*. Ottawa: Environment Canada.

Findley R. 1990. Will we save our own? *Nat Geographic*, 178(3), 106–136.

Foresry Canada. 1990. *Forestry Facts*. Ottawa: Forestry Canada.
Franklin, J.F. 1988. Structural and functional diversity in temperate forests. In E.O. Wilson (ed.), *Biodiversity* (pp. 166–175). Washington, DC: National Academy Press.
Franklin, J.F. 1989. Toward a New Forestry. *Am Forests*, 95(11/12), 37–44.
Fraser, D.F., and Ryan, M.W. 1993. Changes in plant diversity in Douglas fir stands following conversion from old-growth to second growth. In: D. Pollard (ed.), *Proceedings of a Workshop on Forest Ecosystem Dynamics, February 1993* (pp. 16–18). Victoria, BC: Forestry Canada.
Gamiet S., and Berch, S.M. 1992. Fungi of old-growth forests in British Columbia. *NW Environl J*, 8, 168–170.
Gardner, R.H., and Turner, M.G. 1991. Future directions in quantitative landscape ecology. In M.G. Turner and R.H. Gardner (eds.), *Quantitative Methods in Landscape Ecology* (pp. 519–526). New York: Springer-Verlag.
Goodman, D. 1993. Diversity of mycorrhizal fungi in old-growth and mature second growth stands of Douglas fir on Vancouver Island. In D. Pollard (ed.), *Proceedings of a Workshop on Forest Ecosystem Dynamics, February 1993* (pp. 41–42). Victoria, BC: Forestry Canada.
Goward, T. 1990. Rare and endangered lichens in B.C. *Bioline*, Spring 1990, 13–15.
Goward, T. 1991. Epiphytic lichens: Going down with the trees. In S. Rautio (ed.), *Community Action for Endangered Species: Proceedings of a Public Symposium, Vancouver, BC* (pp. 153–158). Federation of British Columbia Naturalists and Northwest Wildlife Preservation Society.
Hamrick, J.L., Mitton, J.B., and Linhart, Y.B. 1981. Levels of genetic variation in trees: influence of life history characteristics. In M.T. Conkle (ed.), *Proceedings of a Symposium on Isozymes of North American Forests Trees, and Forest Insects* (pp. 35–41). U.S.D.A., Forest Service, General Technical Report PSW 48.
Hansen, A.J., Spies, T.A., Swanson, F.J., and Ohmann, J.L. 1991. Conserving biodiversity in managed forests. *Bioscience*, 41, 382–392.
Harris, L.D. 1984. *The Fragmented Forest*. Chicago: University of Chicago Press.
Heilman, P.E. 1990. Forest management challenged in the Pacific Northwest. *J For*, 88(11), 16–23.
Hopper, S.D., and Burgman, M.A. 1983. Cladistic and phenetic analyses of phylogenetic relationships among populations of Eucalyptus caesia. *Aust J Bot*, 31, 35–49.
Hosie, R.C. 1979. *Native Trees of Canada*. Don Mills, ON: Fitzhenry & Whiteside.
Hunter, M.L. 1990. *Wildlife Forests and Forestry: Principles of Managing Forests for Biodiversity*. Englewood Cliffs, NJ: Prentice-Hall.
Lande, R. 1988. Genetics and demography in biological conservation. *Science*, 241, 1455–1460.
Li, P., and Adams, W.T. 1988. Range-wide patterns of allozyme variation in Douglas fir (*Pseudotsuga menziesii* var. *menziesii* [Mirb.] Franco). *Can J For Res*, 18, 149–161.
Linhart, Y.B., Mitton, J.B., Sturgeon K.B., and Davis, M.L. 1981. An analysis of genetic architecture in populations of ponderosa pine. In M.T. Conkle (ed.), *Proceedings of the Symposium on Isozymes of North American Forest Trees and Forest Insects, Berkeley CA, July 1979* (pp. 53–59). U.S.D.A., Forest Service, General Technical Report PSW-48.
Liu, E.H., and Godt, M.J.W. 1983. The differentiation of populations over short distances. In C.M. Schonewald-Cox, S.M. Chambers, B. MacBryde, and W.L. Thomas (eds.), *Genetics and Conversation* (pp. 78–95). Menlo Park, CA: Benjamin/Cummings.

Lovejoy, T.E. 1988. Diverse considerations. In E.O. Wilson (ed.), *Biodiversity* (pp. 421–427). Washington, DC: National Academy Press.
Lowe, J.J. 1991. Canada's forest inventory: The sustainable commercial timber base and its growth rate. In D.G. Brand (ed.), *Canada's Timber Resources* (pp. 33–36). Forestry Canada Information Report PI-X-101.
MacKinnon, A., Meidinger, D., and Klinka, K. 1992. Use of the biogeoclimatic ecosystem classification system in British Columbia. For Chron, 68, 100–120.
Maser, C. 1990. *The Redesigned Forest*. Toronto: Stoddart.
Milliken, R. 1993. Utilization of residual patches of old-growth timber by forest birds. In D. Pollard (ed.), *Proceedings of a Workshop in Forest Ecosystem Dynamics, February 1993* (p. 43). Victoria, BC: Forestry Canada.
Moore, K. 1990. Where is it and how much is left? The state of the temperate rainforest in British Columbia. *For Plan Can*, 6(4), 15–19.
Moran, G.F., and Hopper, S.D. 1983. Genetic diversity and the insular population structure of the rare granite rock species Eucalyptus caesia Benth. *Aust J Bot*, 31, 161–172.
Munro, W.T., and Low, D.J. 1980. Preliminary plan for the designation of threatened and endangered species in British Columbia. In R. Stace-Smith, L. Johns, and P. Joslin (eds.), *Threatened and Endangered Species and Habitats in British Columbia and the Yukon: Proceedings of Symposium* (pp. 65–78). Victoria: British Columbia Ministry of Environment, Fish and Wildlife Branch.
Musick, H.B., and Grover, H.D. 1991. Image textural measures as indices of landscape pattern. In M.G. Turner and R.H. Gardner (eds.), *Quantitative Methods in Landscape Ecology* (pp. 77–104). New York: Springer-Verlag.
Namkoong, G. 1992. Creating policy on genetic diversity. In R. Szaro (ed.), *Biodiversity in Managed Landscapes: Theory and Practice*. Oxford University Press. NY.
Nevo, E. 1987. Genetic variation in natural populations: Patterns and theory. *Theor Pop Biol*, 13, 121–177.
Orchard, S.A. 1980. The status of reptiles in British Columbia. In R. Stace-Smith, L. Johns, and P. Joslin (eds.), *Threatened and Endangered Species and Habitats in British Columbia and the Yukon: Proceedings of a Symposium* (pp. 152–159). Victoria: British Columbia Ministry of Environment, Fish and Wildlife Brach.
Orchard, S.A. 1992. Amphibian population declines in British Columbia. In C.A. Bishop and K.E. Pettit (eds.), *Declines in Canadian Amphibian Populations: Designing a National Monitoring Strategy* (pp. 10–13). Canadian Wildlife Service, Occasional Paper 76.
Panesar, T.S., and Marshall, V.G. 1993. Nematode diversity and distribution in different forest soil habitats. In D. Pollard (ed.), *Proceedings of a Workshop on Forests Ecosystem Dynamics, February 1993* (pp. 28–31). Victoria: Forestry Canada.
Pielou, E.C. 1990. Depletion of genetic richness is not "harmless" consequence of clearcutting. *For Plan Can*, 6(4), 29.
Pojar, J. 1980. Threatened forest Ecosystems of British Columbia. In R. Stace-Smith, L. Johns, and P. Joslin (eds.), *Threatened and Endangered Species and Habitats in British Columbia and the Yukon: Proceedings of a Symposium* (pp. 28–39). Victoria: British Columbia Ministry of Environment, Fish and Wildlife Branch.
Pojar, J., and Hamilton, E. 1993. Trying to maintain biological diveristy in coastal British Columbia: A case study. Unpublished paper, presented at the *Commonwealth Forestry Conference, Kuala Lumpur*, 1993.
Pojar, J., Klinka, K., and Meidinger, D.V. 1987. Biogeoclimatic ecosystem classification in British Columbia. *For Ecol Manage*, 22, 119–154.

Rodway, M.S., Savard, J.-P.L., and Regehr, H.M. 1991. *Habitat Use of Activity Patterns of Marbled Murrelets at Inland and At-Sea Sites in the Queen Charlotte Islands, British Columbia*. Canadian Wildlife Service, Technical Report Series, 122.
Romme, W.H. 1982. Fire and landscape diversity in subalpine forests of Yellowstone National park. *Ecol Monogr*, 52, 199–221.
Sampson, J.F., Hopper, S.D., and James, S.H. 1988. Genetic diversity and the conservation on Eucalyptus crucis Meiden. *Aust J Bot*, 36, 447–460.
Schofield, W.B. 1989. Phytogeography of the mosses of North America. In R.J. Taylor and A.E. Leviton (eds.), *The Mosses of North America* (pp. 131–170). San Francisco, CA: Pacific Division, American Association for the Advancement of Science.
Schofield, W.B. 1988. Biogeography and the bryophytic characterization of biogeoclimatic zones of British Columbia, Canada. *Can J Bot*, 66, 2673–2686.
Scott, J.M., Csuti, B., Smith K., Estes, J.E., and Caicco, S. 1991. Gap analysis of species richness and vegetation cover: an integrated biodiversity conservation strategy. In K.A. Kohm (ed.), *Balancing on the Brink of Extinction: The Endangered Species Act and Lessons for the Future* (pp. 282–297). Washington, DC: Island Press.
Seip, D.R., and Savard, J.-P.L. 1992. Comparison of bird, small mammal and amphibian abundance and diversity in old-growth and second growth forests of coastal British Columbia. *NW Environ J*, 8, 230–231.
Setala, H., and Marshall, V.G. 1993. Succession of Collembola in stumps of Douglas fir seres representing clearcuts to old-growth forest. In D. Pollard (ed.), *Proceedings of a Workshop on Forest Ecosystem Dynamics, February 1993* (pp. 19–21). Victoria, BC: Forestry Canada.
Shafter, C.L. 1990. *Nature reserves: Island Theory and Conservation Practice*. Washington DC: Smithsonian Institution Press.
Simpson, G.G. 1969. Species density of North American recent mammals. *Syst Zool*, 13, 57–73.
Sisco, C., and Gutierrez, R.J. 1984. Winter ecology of radio tagged spotted owls in Six Rivers National Forests, Humboldt County, CA: U.S.D.A., Forest Service, Six Rivers National Forest Final Report.
Stauffer, D. 1985. *Introduction to Percolation Theory*. London: Taylor & Francis.
Taylor, R.L., and MacBryde, B. 1977. *Vascular Plants of British Columbia: A Description Resource Inventory*. Vancouver, BC: The Botanical Garden Technical Bulletin, No. 4, University of British Columbia Press.
Thompson, I.D. 1988. Habitat needs of furbearer in relation to logging in boreal Ontario. *For Chron*, 64, 251–261.
Thompso, I.D. 1991. Could marten become the spotted owl of eastern Canada? *For Chron*, 67, 136–140.
Timmermann, H.R., and McNichol, J.G. 1988. Moose habitat needs. *For Chron*, 64, 238–245.
Turner, S.J., O'Neill, R.V., Conley, W., Conley, M.R., and Humphries, H.C. 1991. Pattern and scale: Statistics for landscape ecology. In M.G. Turner and R.H. Gardner (eds.), *Quantitative Methods in Landscape Ecology* (pp. 17–50). New York: Spinger-Verlag.
Turner, A.M., Rubec, C.D.A., and Wiken, E.B. 1992. Canadian ecosystems: A System approach to their conservation. In: J.H.M. Willison, S. Bondrup-Nielsen, C. Drysdale, T.B. Herman, N.W.P. Munro, and T.L. Pollock (eds.), *Science and Management of Protected Areas: Proceedings of an International Conference, Wolfville, NS*. Amsterdam: Elsevier.
Watts, R. 1991. Pressure builds to incorporate concept of biodiversity in management of forests. *Victoria Times-Colonial*, 8 April 1991, p. A14.

Welsh, D.A. 1988. Meeting habitat needs of non-game forest wildife. *For Chron*, 64, 262–266.
Welsh, D.A. 1991. Managing forested land for the conservation of biological diversity and the production of timber. International Association of Landscape Ecologists' Conference, Ottawa, ON (unpubl. abstract).
Welsh, D.A., and Fillman, D.R. 1980. The impact of forest cutting on boreal bird populations. *Am Birds*, 34, 84–94.
Winchester, N. 1993. Coastal Sitka spruce canopies: Conservation of biodiversity. Bioline Fall, 1993.
Wood, G.W. 1991. Owl conservation strategy flawed. *J For*, 89(2), 39–42.
World Commission on Environment and Development (WCED). 1987. *Our common future*. Oxford, England: Oxford University Press.
World Conservation Union. 1990. A framework for the classification of terrestrial and marine protected areas. Switzerland: Gland.
World Conservation Union, United Nations Environment Programme, and World-wide fund for Nature. 1991. *Caring for the Earth*. Gland, Switzerland: World Conservation Union.
World Resources Institute, World Conservation Union, and United Nations Environment Programme. 1991. *The Global Biodiversity Strategy*. Washington, DC: World Resource Institute.
Yeh, F.C., and El-Kassaby, Y.A. 1980. Enzyme variation in natural populations of Sitka spruce (*Picea sitchensis*). I. Genetic variation patterns among trees from 10 IUFRO provenances. *Can J For Res*, 10, 415–422.

4. Conclusion

19. Afterword

Richard G. Lawford and Paul B. Alaback

In spite of the broad-scale similarities between the western coasts of North and South America, there are important distinctions arising from the finer-scale variability in the physical and biotic patterns along these coasts. In particular, forcing factors such as coastal ocean currents, strong westerly winds in the upper atmosphere at mid-latitudes, and mountain ranges near the coastlines exert similar influences on the abiotic environmental features of these areas. However, local and regional biotic community composition and structure vary markedly between North and South America. The coastal rainforests of these two continents offer a unique opportunity to study a wide range of regional unique physical and biological interactions that occur in regions with similar macroscale environmental constraints.

There is evidence that the evolutionary history of temperate forests in South America is quite distinct from that of North America, which also helps to explain many of the fundamental differences in community composition between the two areas. Although the current climates in the two regions may be giving rise to convergence in biotic structure and function, these characteristics may diverge in the future because of the differential forcing factors of climate change and land-use patterns between the two regions.

Physical Processes

The interconnections between elements of the physical environment have been clearly documented in the first section of this book. At mid- and high latitudes, the atmospheric circulation serves as a driving force, producing distinct precipitation and temperature patterns along the western coasts of both North and South America and leading to well-defined patterns of runoff and ecosystem productivity as the rain falling on the surface is redistributed by the local topography. These same atmospheric circulation patterns control the coastal patterns of upwelling and, in turn, are influenced by the larger oceanic forcing effects such as El Niño. In spite of the common influence of the large-scale oceanic thermal forcing in each hemisphere, the presence of an ice-covered continent at the South Pole, the smaller land mass in the Southern Hemisphere, and the greater disruption to the flow introduced by the higher Andes mountains combine to produce some significant deviations in the hydroclimatic patterns along the western coasts of the two continents.

Regional-scale effects are important in atmosphere–ocean and atmosphere–hydrosphere linkages. Upwelling and marine biological activity are linked to the atmosphere through the production of dimethylated sulphur and ammonia. These gases are transported by wind and precipitation to forest areas, where they play an important role in supplying chemicals and nutrients to plants and soils. These inputs vary between North and South America because the atmospheric inputs are either supplemented by the geological inputs (common in the glaciated seasonal rainforests of North America) or are modified very little by geological inputs (more common in the unglaciated perhumid rainforest zone of South America). Clearly, forest productivity and type are strongly influenced by both the direct variability of temperature and precipitation patterns and the indirect effects such as geochemical and nutrient cycling.

Biota

The biota associated with the high- and temperate-latitude rainforests of North and South America are widely divergent, reflecting fundamental differences in evolutionary history and biogeography. The long history of separation from other temperate forests has allowed Chilean forests to develop many unique species and families, and its more equitable climate appears to have been a key factor in allowing the development of a high level of plant species richness. The connection of southern temperate rainforests to the tropics in early geological history appears to have key significance in explaining many of today's plant and animal community structures. While the southern rainforest flora are relatively species rich and provide many fascinating parallels with the northern flora, the mammalian fauna is relatively species poor and lacks many of the niche mammals present in North America. Likewise, the forest birds and freshwater biota of Chile are quite distinctive, not only in terms of evolutionary history but also in terms of the niches occupied and overall community structure. Of key importance are the uniqueness and specialization of the rainforest fauna and flora in Chile, as contrasted with the generality and widespread nature of many biota associated with northern rainforests. Tighter interrelationships of plant and animal species, as illustrated by the key role of a few species of birds to seed dispersal and flower pollination, also suggest a greater potential sensitivity of Chilean forests to environmental change.

The unique cold, wet climate of high- and temperate-latitude rainforests in both hemispheres appears to be a key factor leading to convergent tendencies in both structure and function. This unique climate has lead to analogous soils (despite geological differences) and disturbance regimes and to similar environmental stresses for biota. Similarities in geography and offshore processes may also have contributed to convergence in the interrelationships of terrestrial and marine ecosystems in both hemispheres. The key role of climate in driving these convergences underscores the critical importance of examining the sensitivity of parallel environments to climate change.

Human Activities

The management of high- and temperate-latitude rainforests to ensure their long-term sustainability poses a challenge for a number of governments in the Americas. The results reported in this volume demonstrate that many tools and extensive expertise exist to assist in making resource-management decisions. Furthermore, there are examples from Canada and Tasmania where conservation principles and policies have been formulated and plans for their implementation in a coordinated way have been developed. However, the ever increasing demands of industry for access to coastal rainforests for logging purposes and for other development activities are a serious and growing threat to the ecosystem integrity and biodiversity of these systems. To balance the pressures for growth and exploitation, stronger institutions with the requisite mandates, authorities, and commitment are needed to develop more environmentally sound policies and to implement corrective actions. Furthermore, there is a clear need to develop on a priority basis a better understanding of the interactions between human activities and institutional policies and their effects on forests and forest ecosystems in both North and South America.

Comparative Studies

Collectively, the regional comparative studies reported in this volume demonstrate that a great deal of insight regarding the controls and functioning of ecosystems can be gained by studying ecosystems with similar environmental features in different regions. Throughout the chapters the reader has seen how comparisons between areas in North America and similar areas in South America have yielded surprises—in some cases because the histories of the regions were different, in others because the effects of small changes in climatic and geochemical gradients were amplified at the local scale. Natural scientists do not have the luxury of laboratory studies, where every parameter can be controlled and varied in a systematic way. An alternative, scientifically rigorous approach is to take two areas with similar large-scale climatic and geophysical controls and to study their responses to external forcings. By documenting the differences in ecosystem responses between the two areas, it is possible to better understand the significance of variations in geologic history, local climatic variability, and human activity.

Regional comparative studies also provide a basis for generalizing the results of relationships derived from site-specific studies. While it may be possible to demonstrate consistent relationships between biotic community structure and physical forcing factors in regions of similar geologic history, it is a much more rigorous test of these relationships to compare between sites with different geological histories. In some cases, comparative studies provide the only practical means to compare key relationships and ecosystems. For example, to examine the role of geologic history or cultural influences on ecosystem processes, comparative studies are the only means to acquire independent data sets.

Interdisciplinary Research

Contributors to this volume have demonstrated the synergism that takes place between different disciplines when total systems are studied. To advance our understanding of the dependencies among various facets of an ecosystem, it is necessary to shape our scientific approach to the problem so that it transcends disciplinary paradigms and represents process-based interactions that characterize natural functions. Ecosystem research must be interdisciplinary to be successful in providing an understanding of the way in which these processes are integrated. This understanding requires the atmospheric insight of climatologists and atmospheric scientists, the geochemical considerations of geologists and limnologists, the understanding of plant and plant community functions and plant–animal links provided by biologists, and the human dimension aspects of these problems described by social scientists. Furthermore, this new understanding not only depends on the involvement of scientists from different disciplines but requires systems to be examined with shared paradigms and with common approaches that integrate the findings into both quantitative and qualitative frameworks. Scientific studies must become fully integrated if they are going to address fully the complexity of interactions between physical environments and biological communities and to predict the effects of human influences of ecosystems. One of the major challenges in the full integration of disciplines arises from the need to link processes at different scales, because biota respond at local scales, while most physical processes are local manifestations of much larger-scale interactions.

Multiscale Approaches

One theme that frequently arises in this volume relates to the importance of scale in the analysis and understanding of ecosystems. Variability occurs at local scales with different forest types along gradients of moisture and nutrient availability or biotic responses to disturbances at all scales. Variations in forest structure and function lead to different responses of the forest to the climate and, in turn, different feedbacks from the forest to the atmosphere. On the one hand, forest ecosystems are affected by fragmentation introduced through human activities at the landscape level; on the other hand, they are influenced by continental and ocean basin–scale atmospheric controls arising from sea surface temperature anomalies. Large-scale climatic fluctuations also lead to local extremes in runoff, affect the intensity of geochemical cycling, and influence the development and functioning of plant communities. Scale effects need to be accounted for in the collection and evaluation of field data. Data obtained by remote sensing represent areal averages that may or may not be related to the biological characteristics observed on the earth's surface. Furthermore, these data do not include the three dimensional effects of vegetation structure, a key parameter for forest and wildlife relationships.

Variability also affects larger scales. Although logging is site specific, its effects on water chemistry and erosion may occur a number of kilometers away from the place where it is taking place. Local emissions of air pollutants to the atmosphere may be carried by the prevailing winds to sites tens and even hundreds of kilometers away. Logging dispersed over a large enough area will create patches in the forest that lead to local mesoscale circulation patterns that can affect the local distribution of precipitation and at the same time can disrupt the continuity of natural habitat for animals and binds.

The functioning of natural systems involves the ongoing dynamic scaling up of local effects and scaling down of larger-scale atmospheric effects, which combine to produce complex ecosystem behavior. More comprehensive models for examining the combined effect of processes operating at different scales and the movement of variability from one scale to another are technically feasible, but they need to be tested and tailored to systems operating under a number of environmental constraints. Conceptually, these interactions can be addressed by using hierarchical frame-works, while practical applications can rely on existing models and scaling analysis.

Socioeconomic Considerations

During the past centuries, the relative geographical isolation of high- and temperate-latitude rainforests in both hemispheres has been a key factor in their conservation and persistence as intact landscapes. In the future, this geographic isolation is unlikely to be effective in buffering these forests from the impacts of the emerging social and economic pressures that have so overwhelmed warmer and drier ecosystems throughout the globe. Widespread investments by multinational corporations and others during the past few decades have already brought about many fundamental and rapid changes in high-latitude rainforest landscapes. While the relative uniqueness of pristine landscapes and associated wildlife may now provide greater opportunities for alternative economic-development scenarios in temperate rainforest regions, it will be difficult to implement such alternative solutions without developing a solid scientific foundation that is well integrated with social and cultural considerations. The parallel environments in which the contrasting cultures of North and South America occur could provide a unique opportunity to examine how social and cultural factors affect the relationship of natural resources and economic development to ecosystem processes. A better understanding of these basic human and environmental relationships is needed to provide policymakers with a much more solid information base upon which to develop sustainable development strategies for these ecological distinct regions.

Future Research

Asymmetries in economic and cultural factors between Chile and North America and asymmetries in the resilience of ecosystems to human activities provide an ideal opportunity to test the effects of human activities on ecosystem processes. As these two regions change over time, studies need to be carried out at the forest-site scale as well as landscape and regional scales to examine differential global change influences on the interactions among oceanic, atmospheric, and terrestrial exchanges. Furthermore, models for assessing the consequences of these changes for biological community structure and functioning need to be developed and validated. In addition, research programs that incorporate and quantify the effects of human activities on biotic patterns and physical processes should be given priority.

Index

B

Ecological Studies

Volume 68
Stable Isotopes in Ecological Research (1989)
P.W. Rundel, J.R. Ehleringer, and K.A. Nagy (Eds.)

Volume 69
Vertebrates in Complex Tropical Systems (1989)
M.L. Harmelin-Vivien and F. Bourlière (Eds.)

Volume 70
The Northern Forest Border in Canada and Alaska (1989)
J.A. Larsen

Volume 71
Tidal Flat Estuaries: Simulation and Analysis of the Ems Estuary (1988)
J. Baretta and P. Ruardij (Eds.)

Volume 72
Acidic Deposition and Forest Soils (1989)
D. Binkley, C.T. Driscoll, H.L. Allen, P. Schoeneberger, and D. McAvoy

Volume 73
Toxic Organic Chemicals in Porous Media (1989)
Z. Gerstl, Y. Chen, U. Mingelgrin, and B. Yaron (Eds.)

Volume 74
Inorganic Contaminants in the Vadose Zone (1989)
B. Bar-Yosef, N.J. Barrow, and J. Goldshmid (Eds.)

Volume 75
The Grazing Land Ecosystems of the African Sahel (1989)
H.N. Le Houérou

Volume 76
Vascular Plants as Epiphytes: Evolution and Ecophysiology (1989)
U. Lüttge (Ed.)

Volume 77
Air Pollution and Forest Decline: A Study of Spruce (*Picea abies*) on Acid Soils (1989)
E.-D. Schulze, O.L. Lange, and R. Oren (Eds.)

Volume 78
Agroecology: Researching the Ecological Basis for Sustainable Agriculture (1990)
S.R. Gliessman (Ed.)

Volume 79
Remote Sensing of Biosphere Functioning (1990)
R.J. Hobbs and H.A. Mooney (Eds.)

Volume 80
Plant Biology of the Basin and Range (1990)
C.B. Osmond, G.M. Hidy, and L.F. Pitelka (Eds.)

Volume 81
Nitrogen in Terrestrial Ecosystem: Questions of Productivity, Vegetational Changes, and Ecosystem Stability (1990)
C.O. Tamm

Volume 82
Quantitative Methods in Landscape Ecology: The Analysis and Interpretation of Landscape Heterogeneity (1990)
M.G. Turner and R.H. Gardner (Eds.)

Volume 83
The Rivers of Florida (1990)
R.J. Livingston (Ed.)

Volume 84
Fire in the Tropical Biota: Ecosystem Processes and Global Challenges (1990)
J.G. Goldammer (Ed.)

Volume 85
The Mosaic-Cycle Concept of Ecosystems (1991)
H. Remmert (Ed.)

Volume 86
Ecological Heterogeneity (1991)
J. Kolasa and S.T.A. Pickett (Eds.)

Volume 87
Horses and Grasses: The Nutritional Ecology of Equids and Their Impact on the Camargue (1992)
P. Duncan

Volume 88
Pinnipeds and El Niño: Responses to Environmental Stress (1992)
F. Trillmich and K.A. Ono (Eds.)

Ecological Studies

Volume 89
Plantago: A Multidisciplinary Study (1992)
P.J.C. Kuiper and M. Bos (Eds.)

Volume 90
Biogeochemistry of a Subalpine Ecosystem: Loch Vale Watershed (1992)
J. Baron (Ed.)

Volume 91
Atmospheric Deposition and Forest Nutrient Cycling (1992)
D.W. Johnson and S.E. Lindberg (Eds.)

Volume 92
Landscape Boundaries: Consequences for Biotic Diversity and Ecological Flows (1992)
A.J. Hansen and F. di Castri (Eds.)

Volume 93
Fire in South African Mountain Fynbos: Ecosystem, Community, and Species Response at Swartboskloof (1992)
B.W. van Wilgen et al. (Eds.)

Volume 94
The Ecology of Aquatic Hyphomycetes (1992)
F. Bärlocher (Ed.)

Volume 95
Palms in Forest-Ecosystems of Amazonia (1992)
F. Kahn and J.-J. DeGranville

Volume 96
Ecology and Decline of Red Spruce in the Eastern United States (1992)
C. Eagar and M.B. Adams (Eds.)

Volume 97
The Response of Western Forests to Air Pollution (1992)
R.K. Olson, D. Binkley, and M. Böhm (Eds.)

Volume 98
Plankton Regulation Dynamics (1993)
N. Walz (Ed.)

Volume 99
Biodiversity and Ecosystem Function (1993)
E.-D. Schulze and H.A. Mooney (Eds.)

Volume 100
Ecophysiology of Photosynthesis (1994)
E.-D. Schulze and M.M. Caldwell (Eds.)

Volume 101
Effects of Land-Use Change on Atmospheric CO_2 Concentrations: South and South East Asia as a Case Study (1993)
V.H. Dale (Ed.)

Volume 102
Coral Reef Ecology (1993)
Y.I. Sorokin (Ed.)

Volume 103
Rocky Shores: Exploitation in Chile and South Africa (1993)
W.R. Siegfried (Ed.)

Volume 104
Long-Term Experiments With Acid Rain in Norwegian Forest Ecosystems (1993)
G. Abrahamsen et al. (Eds.)

Volume 105
Microbial Ecology of Lake Plußsee (1993)
J. Overbeck and R.J. Chrost (Eds.)

Volume 106
Minimum Animal Populations (1994)
H. Remmert (Ed.)

Volume 107
The Role of Fire in Mediterranean-Type Ecosystems (1994)
J.M. Moreno and W.C. Oechel (Eds.)

Volume 108
Ecology and Biogeography of Mediterranean Ecosystems in Chile, California, and Australia (1994)
M.T.K. Arroyo, P.H. Zedler, and M.D. Fox (Eds.)

Ecological Studies

Volume 109
Mediterranean Type Ecosystems: The Function of Biodiversity (1994)
G.W. Davis and D.M. Richardson (Eds.)

Volume 110
Tropical Montane Cloud Forests (1994)
L.S. Hamilton, J.O. Juvik, and F.N. Scatena (Eds.)

Volume 111
Peatland Forestry (1995)
E. Paavilainen and J. Päivänen

Volume 112
Tropical Forests: Management and Ecology (1995)
A.E. Lugo and C. Lowe (Eds.)

Volume 113
Arctic and Alpine Biodiversity: Patterns, Causes and Ecosystem Consequences (1995)
F.S. Chapin III and C. Körner (Eds.)

Volume 114
Crassulacean Acid Metabolism: Biochemistry, Ecophysiology and Evolution (1995)
K. Winter and J.A.C. Smith (Eds.)

Volume 115
Islands: Biological Diversity and Ecosystem Function (1995)
P.M. Vitousek, H. Andersen, and L. Loope (Eds.)

Volume 116
High-Latitude Rainforests and Associate Ecosystems of the West Coast of the Americas (1995)
R.G. Lawford, P.B. Alaback, and E. Fuentes (Eds.)

Volume 117
Global Change and Mediterranean-Type Ecosystems (1995)
J.M. Moreno and W.C. Oechel (Eds.)

Volume 118
Impact of Air Pollutants on Southern Pine Forests (1995)
S. Fox and R.A. Mickler (Eds.)

Printed in the USA
CPSIA information can be obtained
at www.ICGtesting.com
LVHW010812280224
773030LV00003B/202

9 780387 944876